Digital Communication

DIGITAL COMMUNICATION

By

Abhishek Yadav
B.E. M.E.
Assistant Professor
IIMT Engineering College,
Merrut (U.P.)

Anoop Singh Poonia
Assistant Professor
Head, Department of Electronics and Communication, Swami Keshvanand Institute of Technology and Management
Gramothan, Ramnagaria, Jagatpura
Jaipur (Rajasthan)

Dr. K.C. Roy
Assistant Professor
Department of Electronics and Communication, Maulana Azad College of Engineering and Technology, Neora, Patna (Bihar)

UNIVERSITY SCIENCE PRESS
(An Imprint of Laxmi Publications Pvt. Ltd.)
An ISO 9001:2015 Company

BENGALURU • CHENNAI • GUWAHATI • HYDERABAD
KOCHI • KOLKATA • LUCKNOW • RANCHI
NEW DELHI

Printed and bound in India

First Edition : 2009 ; *Edition* : 2025

ISBN : 978-81-3180-732-3

Branches

✆	Bengaluru	080-26 75 69 30
✆	Chennai	044-24 34 47 26
✆	Guwahati	0361-254 36 69
✆	Hyderabad	040-27 55 53 83
✆	Kochi	0484-405 13 03
✆	Kolkata	033-40 04 77 79
✆	Lucknow	0522-430 36 13
✆	Ranchi	0651-224 24 64

PUBLISHED IN INDIA BY

UNIVERSITY SCIENCE PRESS

(An Imprint of Laxmi Publications Pvt. Ltd.)

An ISO 9001:2015 Company

113, GOLDEN HOUSE, GURUDWARA ROAD, DARYAGANJ,

NEW DELHI - 110002, INDIA

Corporate Office : B-33, Sector-5, Noida,

Gautambuddha Nagar (U.P.) - 201 301, INDIA

Telephone : +91-120-44 46 900

www.laxmipublications.com

info@laxmipublications.com

Contents

PREFACE

'Communication' is the core subject of "*Electronics & Communication*" branch of Engineering. Now-a-days, it is a very demanding field also. The book *Digital Communication* has been designed to make students well familiar with the *Analog Communication* and *Digital Communication*.

From the very beginning we will study the *fundamentals of communication* that cover the basic block diagram of Digital Communication System. We will also study the basics of information theory, which is the root of the Digital Communication.

The book further covers *digital coding techniques* that include PCM, DPCM, ADCM, different types of signalling formats for digital signals that will help to reduce the noise in the signals during transmission through the channels. *ISI (inter symbol interference)* and *optimum filters* that are very important parameters for healthy digital communication system, are also discussed along with detailed analysis of line coding, digital signal detection, matched filters and information theory and coding techniques.

One of the advantages of the digital system is that we can control and sometime remove the errors in the received signal by using appropriate coding techniques. So, in the last part of the book we will study different *error control codings*.

We personally feel as a student that Communication field is very interesting and we hope that this book will help the teachers and students to study this subject with great interest. In the market so many books are available but here we have tried to keep all the facts and fundamentals in a single place. We are not saying that we have discovered any new fact about this area but as an honest teacher, we have tried to put all the noble thoughts in a single book with simple language. We will feel lucky if students and respected teachers are benefited from our effort. We are grateful to a number of individuals who helped us at various stages of writing this book. We are specially thankful to the publishers for their constant encouragement to complete this text. In spite of our sincere efforts there may be some misprints, omissions and errors. Any suggestion towards the improvement of this book will be highly appreciated.

—Authors

SYLLABUS

6EC4 DIGITAL COMMUNICATION

UNIT 1: **PCM & DELTA MODULATION SYSTEMS :** Uniform and Non-uniform quantization. PCM and delta modulation, Signal to quantization noise ratio in PCM and delta modulation. DPCM, ADM, T1 Carrier System, Matched filter detection. Error probability in PCM system.

UNIT 2: **BASE BAND TRANSMISSION:** Line coding (RZ,NRZ): Polar, Bipolar, Manchester, AMI. Inter symbol interference, Pulse shaping, Nyquist criterion, Raised cosine spectrum.

UNIT 3: **DIGITAL MODULATION TECHNIQUES :** Geometric interpretation of signals, Orthogonalization. ASK, BPSK, BFSK, QPSK, MSK modulation techniques and Coherent detection of these techniques. Calculation of error probabilities.

UNIT 4: **INFORMATION THEORY :** Amount of Information, Average Information, Entropy, Information rate, Increase in Average information per bit by coding, Shannon's theorem and Shannon's bound, Capacity of a Gaussian Channel, BW-S/N trade off.

UNIT 5: **CODING:** Coding and decoding of Information, Hamming code, Single Parity-Bit Code, Linear Block code, cyclic code & convolutional code.

1

INTRODUCTION TO DIGITAL COMMUNICATION

1.1 INTRODUCTION

We have studied under analog communication, that traditionally the basic communication systems used were analog. But from the last decade, we see the remarkable growth in the field of **digital communication**. In digital communication, transmission of information takes place in digital form. The major reason of growth of digital communication recently is the availability of wideband channel such as optical fiber, satellite channels etc. But the concept of digital communication is very old. It is as old as Samuel Morse introduced telegraphy in 1837.

The reason of development of digital communication are simply the advantages of digital with respect to analog.

Some *advantages* of digital communication are listed below:

(*i*) Now a common format of digital can be used for different variety of signals. For example we can transmit voice, video, photo by using the binary digits.

(*ii*) By using error correcting and error detecting codes, we can improve the noise performance of system.

(*iii*) Increased immunity to noise and external interference.

(*iv*) An improved security of message through encryption.

(*v*) Regeneration of the signal is easier. Further, multiplexing in time code is possible. In fact, regenerators clean up the signal increasing SNR.

But some *disadvantages* of digital communications are also these, such as:

(*i*) Now we need the wideband channels due to increased bandwidth of signal.

(*ii*) In digital system, the systems are very complex.

(*iii*) Synchronization is required.

So due to some advantages that are of very much importance, we use digital communication on the cost of costly channels and complex systems but analog communication systems are still these and cannot be replaced completely.

First of all clear about signal, analog signal, discrete signal and digital signal.

1.2 SIGNAL

If we plot a dependent variable with respect to an independent variable then it is called signal.

If we have only one independent variable and one dependent variable then it is called one dimensional signal. Generally signal considered by us such as sinusidual signal is the example of one dimensional signal.

If we have more than one dependent and (or) independent variable than it is called multi dimensional signal.

Example of multi dimensional signal is video signal.

1.2.1 Analog Signal

If we have a continuous variation of dependent variable with respect to a continuous variations of independent variable, then it is called Analog signal. It is also called continuous time signal.

In other words Analog signals are defined for every value of time and they take on continuous values in a given time interval.

For example, signals,

$$S_1(t) = 50 \sin 100\pi t$$

is *analog* or *continuous time signal.*

1.2.2 Discrete Signal

If we have a continuous variation of dependent variable with respect to a discrete value of independent variable, then it is called *discrete signal.*

Thus, now discrete signals are defined for discrete value of time so they can be referred as discrete time signal. It doesn't mean that dependent variable don't have any value at the values other than defined by discrete value of undependent variable. It may have ('or' not) values at those instant also; but we define them for some discrete values. For example,

$$S(nt) = 50 \sin 100\pi nt$$

where $n = 0, 1, 2, 3, 4, \ldots$

So now this signal is defined for $n = 0, 1, 2, 3, 4,$ instants and we don't better about the values of signal $s(nt)$ at $n = 0.5, 1.5, \ldots$ etc.

1.2.3 Digital Signal

Now take a step forward. Take a finite set of possible values of dependent variable and pick up only some of them; definite these pick up value will be defined for some possible values of continuous variable; then signal is called discrete time; discrete valued signal. In general language it is called Digital signal. So we can conclude :

(*i*) Analog signal -	Continuous dependent variable	Continuous independent variable.
(*ii*) Discrete signal-	Continuous dependent variable	Discrete independent variable.
(*iii*) Digital signal -	Discrete dependent variable	Discrete independent variable.

Now let us take the example of practical life.

(*a*) If we are looking the live cricket match at the television continuously; then simple we have ball-by-ball information and it is the example of Analog signal.

(*b*) If we are engaged in the preparation of examination and can't see the match continuously. Now we see the match of one over after very five overs. Thus we have information of 5th, 10th, 15th... Over (not ball by ball) then it is called discrete signal. It should be remember that match is going on ball by ball. Thus match (that is dependent variable of time) is continuous but we are taking its discrete information (at discrete time: after every 5 over).

(*c*) Now take the example of your 'tape recorder' you listen a song for 5 minutes and after that stop the tape; after one hour you again come and play the tape, you will listen the song from just these where you stopped that before one hour.

So tape song (that is dependent variable of time) is discrete and you are also taking its discrete information (at discrete time; after every one hour). So it is the example of digital signal.

If the practical life we may have digital signals-such as computer data is digital in nature; but mostly analog signal such as audio and video signals.

If we want to use a digital system, first of all we should know that how can a analog signal can be converted into digital signal so that we may use digital system.

1.3 ANALOG TO DIGITAL CONVERSION

To convert a analog data into a digital data; we have following three stages : [Fig 1.1]

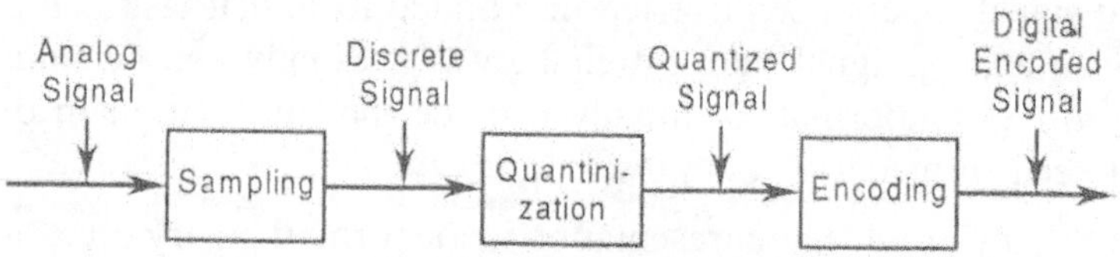

Fig. 1.1 *Analog to digital (A/D) converter*

(*i*) **Sampling :** Sampling is a process that converts a analog signal into a discrete signal, or in other words we can say that-

"It converts a continuous signal into a discrete time signal by taking samples of the continuous time signals at discrete-time instant." It should be cleared that the discrete form of the signal must be in such a way that the basic information of continuous signal is preserved by its discrete version. So a continuous signal $S(t)$ is converted into a discrete signal $S(nT_S)$ in such a way that $S(nT_S)$ preserves completely the information of $S(t)$.

(*ii*) **Quantinization :** Quantinization is the process that converts a continuous valued discrete time signal into a discrete valued discrete time signal. In other words, it converts the discrete signal into digital signal.

(*iii*) **Encoding :** It encodes the digital output of quantinizer to a binary sequence of certain bits.

So we have converted a analog signal into a digital signal and that can easily used by digital system.

1.4 ELEMENTS OF DIGITAL COMMUNICATION

Fig. 1.2 shows the basic Block diagram of a Digital Communication System.

The **source of information,** in most of the practical systems, are analog in nature such as voice signal, video signal. These signal are also non-electrical quantity so cannot be processed directly.

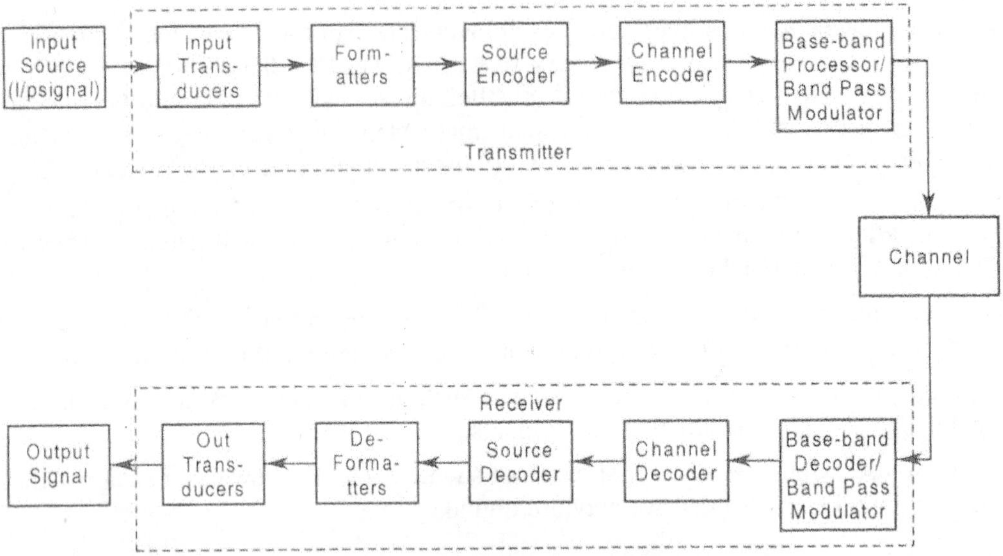

Fig. 1.2. *Element of digital communication system*

Input transducers converts the applied non-electrical quantity into electrical quantity.

Now the received signal should be converted into digital form that task is initiated by formatter; **formatter** converts an analog signal into a digital signal. (Simply can be said a analog to digital converter). If the source of information already provide the information in digital form such as computer' we don't need formatter.

Source encoder: Simply provide a representation in the form of binary digits to the applied digital signal.

Channel encoder plays a very important role. According to the channel properties, channel encoder provides some reduency bits to the message so that at the receiver side, a error free signal can be reproduced.

For very low-speed wireless transmission directly base-band signal can be transmitted. Otherwise, we have to employ some features such as pulse shaping, line coding so that data can easily travel in a noisy atmosphere this task is performed by **Baseband Processor.**

Now ever for high speed transmission as in case of analog transmission, digital data is modulated with the help of high frequency carrier. This task is performed by the **Baseband Modulator.**

The **Communication Channel** is the medio via which the signal is transmitted from the transmitting end to the receiving end. Communication channel is the part of model at which the maximum noise is added to the signal.

At the receiver side we just reciprocal sequence of transmitting end. We first off all have a **Baseband Decoder/Baseband Demodulator.** By Baseband Decoder by pulse shaped 'or' line coded data, original data is recovered back. If at the transmitter side we have used Baseband Modulator, at the receiver side; by demodulator, the data is demodulated to get original signal.

At the receiver side, by the **Channel Decoder,** the reduency bits are removed back to the signal.

By using **Source Decoder,** we are able to decode the digital digits so that the original data is now available in digital form.

Now **Deformatter** converts the digital data into analog and that data is applied to a appropriate output device.

1.5 HISTORY OF COMMUNICATIONS SYSTEMS

Everyday, in our work and in our leisure time, we come in contact with and use a variety of modern communication systems and communication media, the most common being the telephone, radio, television, and the Internet. Through these media we are able to communicate (nearly) instantaneously with people on different continents, transact our daily business, and receive information about various developments and events of note that occur all around the world. Electronic mail and facsimile transmission have made it possible to rapidly communicate written messages across great distances.

Most of these modern-day communication systems were invented and developed during the past century. Here, we present a brief historical review of major developments within the last two hundred years that have had a major role in the development of modern communication systems.

1.5.1 Historical Review

Telegraphy and telephony. One of the earliest inventions of major significance to communications was the invention of the electric battery by Alessandro Volta in 1799. This invention made it possible for Samuel Morse to develop the electric telegraph, which he demonstrated in 1837. The first telegraph line linked Washington with Baltimore and became operational in May 1844. Morse devised the variable-length binary code given in Table 1.1, in which letters of the English alphabet were represented by a sequence of dots and dashes (code words). In this code, more frequently occurring letters are represented by short code words, while letters occurring less frequently are represented by longer code words.

Table 1.1 Morse code

A	....	*N*	....		
B	....	O	....		
C	....	P	....		
D	....	Q	....	1	....
E	....	R	....	2	....
F	....	S	....	3	....
G	....	T	....	4	....
H	....	U	....	5	....
I	....	V	....	6	....
J	....	W	....	7	....
K	....	X	....	8	....
L	....	Y	....	9	....
M	....	Z	....	0	....

(*a*) Letters (*b*) Numbers

Period (.)	....	Wait sign (AS)	....
Comma (,)	....	Double dash (break)	....
Interrogation(?)	....	Error sign	....
Quotation Mark(")	....	Fraction bar (*f*)	....
Colon (:)	....	End of message (AR)	....
Semicolon(;)	....	End of transmission (SK)	....
Parenthesis( )	....		

(*c*) Punctuation and special characters

The *Morse code* was the precursor to the variable-length source-coding methods that are described in Chapter 6. It is remarkable that the earliest form of electrical communications that was developed by Morse, namely *telegraphy*, was a binary digital communication system in which the letters of the English alphabet were efficiently encoded into corresponding **variable-length code** words having binary elements.

Nearly forty years later, in 1875, Emile **Baudot** developed a code for telegraphy in which each letter was encoded into fixed-length binary code words of length 5. In the *Baudot code* the binary code elements have equal length and are designated as mark and space.

An important milestone in telegraphy was the installation of the first transatlantic cable in 1858 that linked the United States and Europe. This cable failed after about four weeks of operation. A second cable was laid a few years later and became operational in July 1866.

Telephony came into being with the invention of the telephone in the 1870s. Alexander Graham Bell patented his invention of the telephone in 1876, and in 1877 established the Bell Telephone Company. Early versions of telephone communication systems were relatively simple and provided service over several hundred miles. Significant advances in the quality and range of service during the first two decades of the twentieth century resulted from the invention of the carbon microphone and the induction coil.

The invention of the triode amplifier by Lee De Forest in 1906 made it possible to introduce signal amplification in telephone communication systems and, thus, to allow for telephone signal transmission over great distances. For example, transcontinental telephone transmission became operational in 1915.

Two world wars and the Great Depression during the 1930s must have been a deterrent to the establishment of transatlantic telephone service. It was not until 1953, when the first transatlantic cable was laid, that telephone service became available between the United States and Europe.

Automatic switching was another important advance in the development of telephony. The first automatic switch, developed by Strowger in 1897, was an electromechanical step-by-step switch. This type of switch was used for several decades. With the invention of the transistor, electronic (digital) switching became economically feasible. After several years of development at the Bell Telephone Laboratories, a digital switch was placed in service in Illinois in June 1960.

During the past thirty years there have been numerous significant advances in telephone communications. Fiber optic cables are rapidly replacing copper wire in the telephone plant and electronic switches have replaced the old electromechanical systems.

1.5.2 Wireless Communications

The development of wireless communications stems from the works of Oersted, Faraday, Gauss, Maxwell, and Hertz. In 1820, Oersted demonstrated that an electric current produces a magnetic field. On August 29, 1831, Michael Faraday showed that moving a magnet in the vicinity of a conductor produces an induced current. Thus, he demonstrated that a changing magnetic field produces an electric field. With this early work as background, James C. Maxwell in 1864 predicted the existence of electromagnetic radiation and formulated the basic theory that has been in use for over a century. Hertz verified Maxwell's theory experimentally in 1887.

In 1894, a sensitive device that could detect radio signals, called the *coherer*, was used by its inventor Oliver Lodge to demonstrate wireless communication over a distance of 150 yards at Oxford, England. Guglielmo Marconi is credited with the development of *wireless telegraphy*. Marconi demonstrated the transmission of radio signals at a distance of approximately 2 kilometers in 1895. Two years later, in 1897, he patented a radiotelegraph system and established the Wireless Telegraph and Signal Company. On December 12, 1901, Marconi received a radio signal at Signal Hill in Newfoundland, which was transmitted from Cornwall, England, a distance of about 1700 miles.

The invention of the vacuum tube was especially instrumental in the development of radio communication systems. Fleming invented the vacuum diode in 1904 and De Forest invented the vacuum triode amplifier in 1906, as previously indicated. The invention of the triode made radio broadcast possible in the early part of the twentieth century. *Amplitude Modulation* (AM) broadcast was initiated in 1920 when radio station KDKA, Pittsburgh, went on the air. From that date, AM radio broadcasting grew rapidly across the country and around the world. Edwin Armstrong, as we know it today, invented the super heterodyne AM radio receiver, during World War I. Another significant development in radio communications was the invention of *Frequency Modulation* (FM), also by Armstrong. In 1933, Armstrong built and demonstrated the first FM communication system. However, the use of FM was slow to develop compared with AM broadcast. It was not until the end of World War II that FM broadcast gained in popularity and developed commercially.

The first television system was built in the United States by V.K. Zworykin and demonstrated in 1929. Commercial television broadcasting began in London in 1936 by the British Broadcasting Corporation (BBC). Five years later the Federal Communications Commission (FCC) authorized television broadcasting in the United States.

1.5.3 The Past Fifty Years

The growth in communications services over the past fifty years has been phenomenal. The invention of the transistor in 1947 by Walter Brattain, John Bardeen, and William Shockley; the integrated circuit in 1958 by Jack Kilby and Robert Noyce; and the laser by Townes and Schawlow in 1958, have made possible the development of small-size, low-power, low-weight, and high-speed electronic circuits which are used in the construction of satellite communication systems, wideband microwave radio systems, and light wave communication systems using fiber optic cables. A satellite named Telstar I was launched in 1962 and used to relay TV signals between Europe and the United States. Commercial satellite communication services began in 1965 with the launching of the Early Bird satellite.

Currently, most of the wire line communication systems are being replaced by fiber optic cables which provide extremely high bandwidth and make possible the transmission of a wide variety of information sources, including voice, data, and video. Cellular radio has been developed to provide telephone service to people in automobiles, buses, and trains. High-speed communication networks link computers and a variety of peripheral devices literally around the world.

Today we are witnessing a significant growth in the introduction and use of personal communications services, including voice, data and video transmission. Satellite and fiber optic networks provide high-speed communication services around the world. Indeed, this is the dawn of the modern telecommunications era.

REVIEW QUESTIONS

1. What do you understand by signal?
2. Differentiate between analog and digital signals with examples.
3. Draw the Block diagram explaining the process of analog to digital conversion.
4. Draw the basic Block diagram of a digital communication system.
5. What do you understand by element of electrical communication system?
6. State the advantage of digital communication.
7. What are the Limitations of digital communications systems?
8. Give the history of electrical communication.

2

Sampling of Base Band Signal

2.1 INTRODUCTION

As we have discussed in Chapter 1, the first step of a digital system is the sampling of analog signal.

The output of sampling system is discrete time signal. It means that instead of taking complete signal, we take the samples of signal at different time interval. (Generally this time interval is uniform for a system).

But samples are taken in such a way that sampled signal can be used instead of signal to reproduce the original signal.

Thus the stead of sampling is very important because it is the first step of digital communication. This is also useful because in Pulse Modulation system (study in next chapter), discrete signals are also used for transmission.

The fundamentals of sampling starts with Nyquist theorem (also known as Nyquist-Shannon sampling theorem).

2.2 NYQUIST-SHANNON SAMPLING THEOREM

2.2.1 Introduction to Sampling Theorem

For a band limited signal, the sampling theorem can be explained as follows.

(*a*) A band limited signal, (that has no frequency component outside the bandwidth 'or' say that no frequency component higher than f_H,) can be completely described by its sample values at uniform intervals less than 'or' equal to $1/2f_H$ second apart. This is *time statement* of sampling theorem.

(*b*) A band limited signal that has no frequency components higher than f_H Hz, may be completely recovered from the knowledge of its samples taken at the rate of $2f_H$ samples per second.

This is *frequency statement* of sampling theorem and also explains the concept of reconstruction.

2.2.2 Mathematical Basis of the Theorem

The mathematical model of the sampling process is to multiply $x(t)$ by the impulse train (also called Dirac comb function), which is an infinite sequence of impulses occurring at integer.

$$T_S = \frac{1}{f_s} \tag{2.1}$$

Multiples of sampling interval. So it can be denoted by

$$\sum_{n=-\infty}^{\infty} \delta(t - nT_s) \tag{2.2}$$

So the sampled output can be represented by

$$g(t) = x(t) \sum_{n=-\infty}^{\infty} \delta(t - nT_s) = \sum_{n=-\infty}^{\infty} x(nT_s)\,\delta(t - nT_s) \tag{2.3}$$

We have assumed that $x(t)$ is a band limited signal and its Fourier transform is $x(f)$

$$x(t) \leftrightarrow x(f) \tag{2.4}$$

The Fourier transform of impulse train is another impulse train

$$\sum_{n=-\infty}^{\infty} \delta(t - nT) \leftrightarrow \frac{1}{T_S} \sum_{K=-\infty}^{\infty} \delta\left(f - \frac{K}{T_S}\right)$$

$$= \frac{1}{T_S} \sum_{K=-\infty}^{\infty} \delta(f - Kf_s) \tag{2.5}$$

Now in time domain Multiplication is equivalent to convolution in frequency domain

So $$g(t) = x(t) \sum_{n=-\infty}^{\infty} \delta(t - nt) \tag{2.6}$$

$$g(f) = x(f) * \frac{1}{T_s} \sum_{K=-\infty}^{\infty} \delta(f - kf_s)$$

$$= \frac{1}{T_s} \sum_{K=-\infty}^{\infty} X(f - kf_s) \tag{2.7}$$

By the shifting property of impulse function

Sampled output

$$g(f) = \frac{1}{T_s} \sum_{K=-\infty}^{\infty} X(t - kf_s) \tag{2.8}$$

Now we can draw the spectrum of the received expression.

From Fig. 2.1, it is clear that if we want to recover $x(f)$ from $g(f)$, then each sample of $g(f)$ must be seperable 'or' in other words must not be overlapped. From Fig. 2.1 (g), it is clear that it is possible when

$$\boxed{f_s \geq 2f_H} \tag{2.9}$$

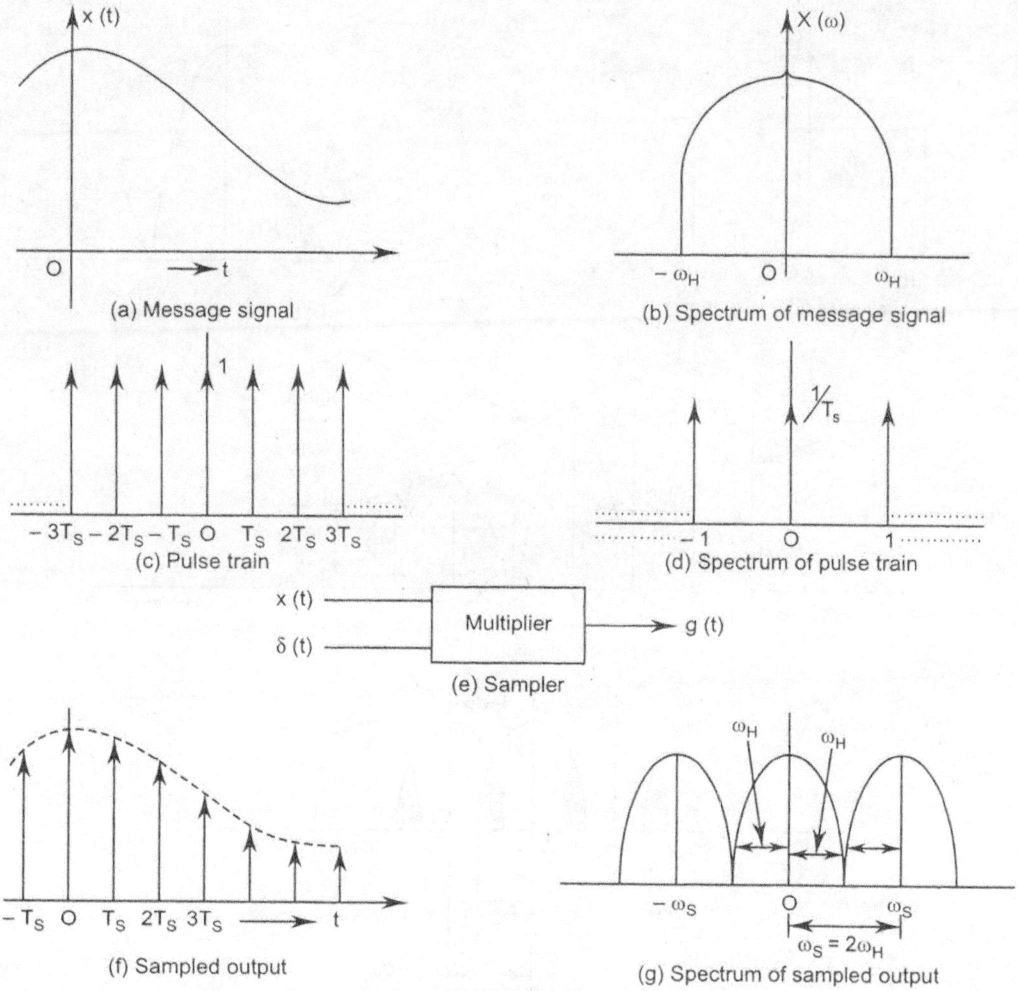

Fig. 2.1 *Sampling of signal*

2.2.3 Critical Frequency

The critical frequency is defined as twice the bandwidth of the signal. If the sampling frequency is at the critical frequency, exactly twice the highest frequency of the input signal, then phase mismatches between the sampler and the signal will distort the signal. For example, sampling cos (πt) at $t = 0, 1, 2, \ldots$ will give you the discrete signal cos (πn), as desired. However, sampling the same signal at $t = 0.5, 1.5, 2.5, \ldots$ will give you a constant zero signal. These two sets of samples, which differ only in phase and not frequency, give dramatically different results because they sample at the critical frequency, instead of strictly above the critical frequency. This result provides a rationale for the strict inequality of the sampling condition, and why the sampling rate must exceed the critical frequency.

2.2.4 Aliasing

If the sampling condition is not satisfied, then frequencies will overlap to each other; as shown in Fig. 2.2 (*d*) if

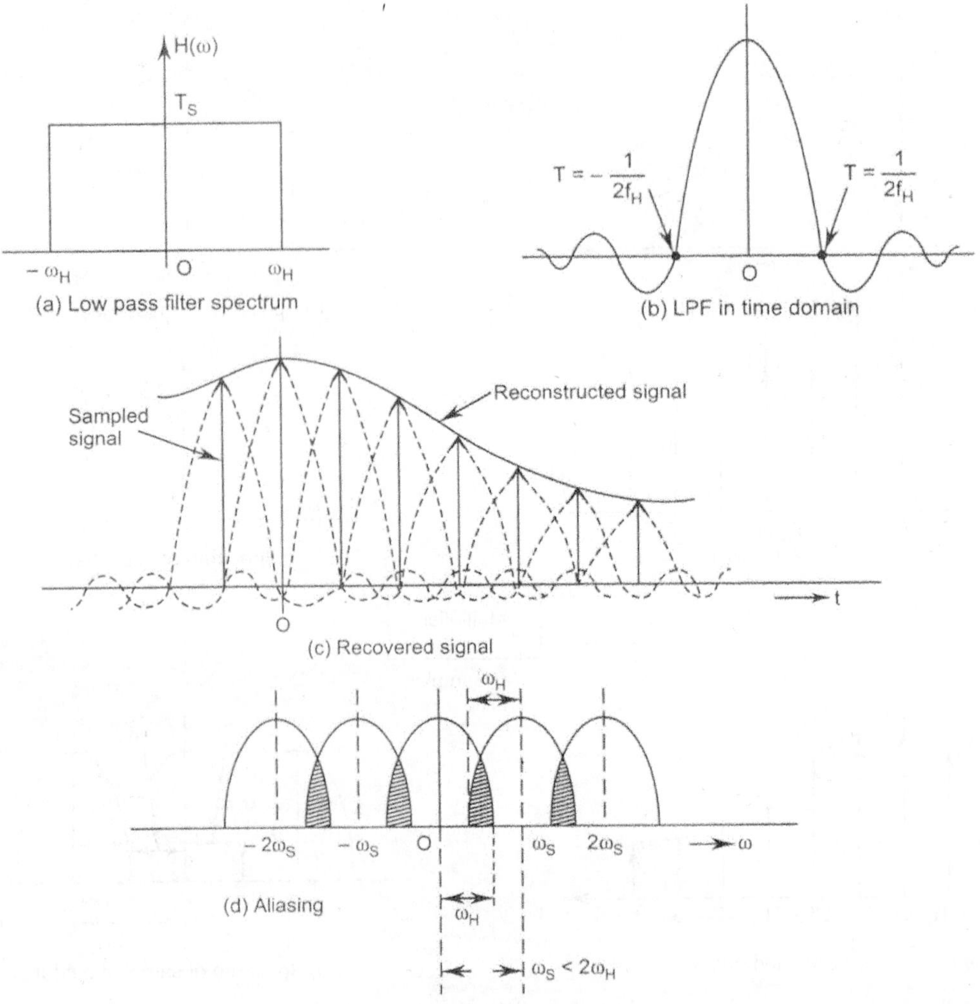

Fig. 2.2 *Reconstruction*

$$f_s \geq 2f_H \qquad \text{(Eq.2.9)}$$

The high frequency components of signal take the identity of low frequency components of signal. This overlap is called **aliasing**.

To prevent the aliasing, two things can readily be done

(*i*) Increase the sampling rate. (*ii*) Introduce an antialiasing filter.

The antialising filter is to restrict the bandwidth of signal to satisfy the sampling condition. This holds in theory, But practically it is not possible. It is not satisfiable in reality because a signal will have some energy outside of Bandwidth. However this energy can be small enough that the aliasing effects are negligible.

So for general operation; due to aliasing it will be difficult to reconstruct to signal from its samples.

These are also situations where overlap is allowed to occur, and due to the particular shape of X(f), the samples coincide with Null region of X(f). Thus signal's spectrum is not effected by the aliasing. This process is called under sampling.

2.2.5 Reconstruction of Sampled Signal

For the reconstruction of sampled signal, we use a low pass filter. This is also called interpolation filter.

The frequency response of a low pass filter has been shown in Fig. 2.2(*a*). The time response of frequency response of low pass filter has been also drawn.

In domain the transfer function frequency = Rectangular function

$$H(f) = T_s \times \text{rect}\left(\frac{\omega}{2\pi f_H}\right) \tag{2.10}$$

So in time domain

$$h(t) = F^{-1}[H(\omega)].$$

$$h(t) = 2 f_H T_s \sin C(2\pi f_H t) \tag{2.11}$$

But $$T_s = \frac{1}{2 f_H}.$$

So $$2 f_H . T_s = 1 \tag{2.12}$$

$$\boxed{h(t) = \sin C(2\pi f_H t)} \tag{2.13}$$

From Fig. 2.2, it may be observed that $h(t) = 0$ at all Nyquist sampling instant $t = \pm n/2 f_H$ except at $t = 0$

Now apply $g(t)$ to this low pass filter having transfer function $h(t)$. Then we recover $x(t)$

$$x(t) = g(t) \times h(t) \tag{2.14}$$

$$= \sum_{n=-\infty}^{\infty} \delta(t - nT_s) \times h(t)\, x(nT_s)$$

$$= \sum_{n=-\infty}^{\infty} h(t - nT_s)\, x(nT_s)$$

$$= \sum_{n=-\infty}^{\infty} \sin C\left[2\pi f_H (t - nT_s)\right] x(nT_s)$$

$$= \sum_{n=-\infty}^{\infty} \sin C(2\pi f_H t - n\pi)\, x(nT_s)$$

$$= \sum_{n=-\infty}^{\infty} x(nT_s) . \sin C(2\pi f_H t - n\pi) \tag{2.15}$$

This is called **interpolation formula** of reconstruction.

Example 1. *Find the Nyquist rate and Nyquist interval for the signal.*

$$x(t) = \frac{1}{2\pi} \cos(8000\,\pi t). \cos(2000\,\pi t)$$

Solution. We have signal

$$x(t) = \frac{1}{2\pi} \cos(8000\,\pi t). \cos(2000\,\pi t)$$

$$= \frac{1}{2\pi} \times \frac{1}{2}\left[2.\cos(8000\,\pi t).\cos(2000\,\pi t)\right]$$

$$[\because \quad 2\cos A\cos B = \cos(A+B) + \cos(A-B)]$$

$$= \frac{1}{4\pi}\left[\cos(8000\,\pi t + 2000\,\pi t) + \cos(8000\,\pi t - 2000\,\pi t)\right]$$

$$= \frac{1}{4\pi}\left[\cos(10000\,\pi t) + \cos(6000\,\pi t)\right]$$

Where $\omega_1 t = 10000\,\pi t$ and $\omega_2 t = 6000\,\pi t$

$$2\pi f_1 = 10000\,\pi \qquad 2\pi f_2 = 6000\,\pi$$

$$f_1 = 5000\text{ Hz} \qquad f_2 = 3000\text{ Hz}$$

According to Nyquist theorem

$$f_s = 2 \times \text{Maximum frequency component}$$

$$= 2 \times 5000\text{ Hz}$$

$$\boxed{f_s = 10{,}000\text{ Hz}}$$

Nyquist interval $T_s = \frac{1}{f_s} = \frac{1}{10000\text{ Hz}}$

$$T_s = 10^{-4}\text{ sec}$$

$$\boxed{T_s = 0.1\ m\text{ sec}}$$

Example 2. *A continuous signal is given by x (t) = 16 cos 200 πt. Determine*

(a) Maximum Sampling Rate, Nyquist rate required to avoid aliasing.

(b) If the sampling frequency

$$f_s = 800\text{ Hz}.$$

What is discrete-time signal x [n] after sampling.

Solution. (*a*) for signal

$$x(t) = 16\cos 200\,\pi t$$

Here $\omega = 200\,\pi$

$$2\pi f = 200\,\pi$$

$$f = 100\text{ Hz}$$

So sampling rate $> 2f$

$$> 200\text{ Hz}$$

$$\text{Nyquist rate} = 200\text{ Hz}$$

(*b*) The frequency of discrete signal

$$f = \frac{\text{Frequency of continuous signal}}{\text{Sampling frequency}}$$

$$f = \frac{100}{800}$$

$$= \frac{1}{8}$$

So the discrete signal is given by

$$x[n] = 16 \cos 2\pi fn$$

$$= 16 \cos 2\pi \times \frac{1}{8n}$$

$$\boxed{x[n] = 16 \cos \frac{\pi}{4\omega}}$$

Example 3. *Determine the Nyquist rate for following signal.*

$$x(t) = 10 \cos 60\pi t + 5 \cos 15\pi t - 2 \sin 25\pi t$$

Solution. By signal

$$\omega_1 = 60\pi$$
$$2\pi f_1 = 60\pi$$
$$f_1 = 30 \text{ Hz}$$

Again

$$\omega_2 = 15\pi$$
$$2\pi f_2 = 15\pi$$
$$f_2 = 7.5 \text{ Hz}$$

Again

$$\omega_3 = 25\pi$$
$$2\pi f_3 = 25\pi$$
$$f_3 = 12.5 \text{ Hz}$$

Thus

$$\text{Nyquist Rate} = 2f_H$$
$$= 2 \times 30$$
$$= 60 \text{ Hz}$$

Example 4. *Consider the analog signal*

$$V = 2 \cos (200\pi t)$$

Determine the minimum required sampling rate to avoid aliasing. Determine the discrete-time signal obtained by sampling the signal at a rate of 100 Hz. Determine the sampled signal when sampling rate is 45 Hz.

Solution. (*a*) For the given signal

$$\omega = 200\pi$$

So $f_H = 100.$

So

$$\text{sampling rate} = 2 f_H$$
$$= 2 \times 100 = 200 \text{ Hz}$$

(*b*) Discrete signal frequency for 100 Hz

$$f = \frac{100}{200} = \frac{1}{2} \text{ Hz}$$

So

$$\text{discrete signal} = 2.\cos 2\pi fn$$
$$= 2.\cos(\pi n)$$

(*c*) Discrete signal frequency for 45 Hz

$$f = \frac{45}{200} = \frac{9}{40}$$

So $\qquad$ discrete signal $= 2 \,.\, \cos 2\pi f n$

$$= 2 \,.\, \cos \frac{9\pi}{20} n$$

2.3 SAMPLING TECHNIQUES AND PRACTICAL ASPECTS OF SAMPLING

We have discussed the Basics of sampling. Now there are many ways of sampling including basic three types.

(*i*) Instantaneous, Ideal 'or' impulse sampling

(*ii*) Natural sampling and

(*iii*) Flat top sampling.

Now we will discuss their practical circuit of generation and detection.

2.3.1 Instantaneous Sampling

The sampling discussed till now by using a impulse train to sample a continuous signal is the example of instantaneous sampling. This is also called *ideal sampling*.

In practice sampling of an analog signal is accomplished by means of high speed switching transistor circuit (Fig. 2.3(*d*)).

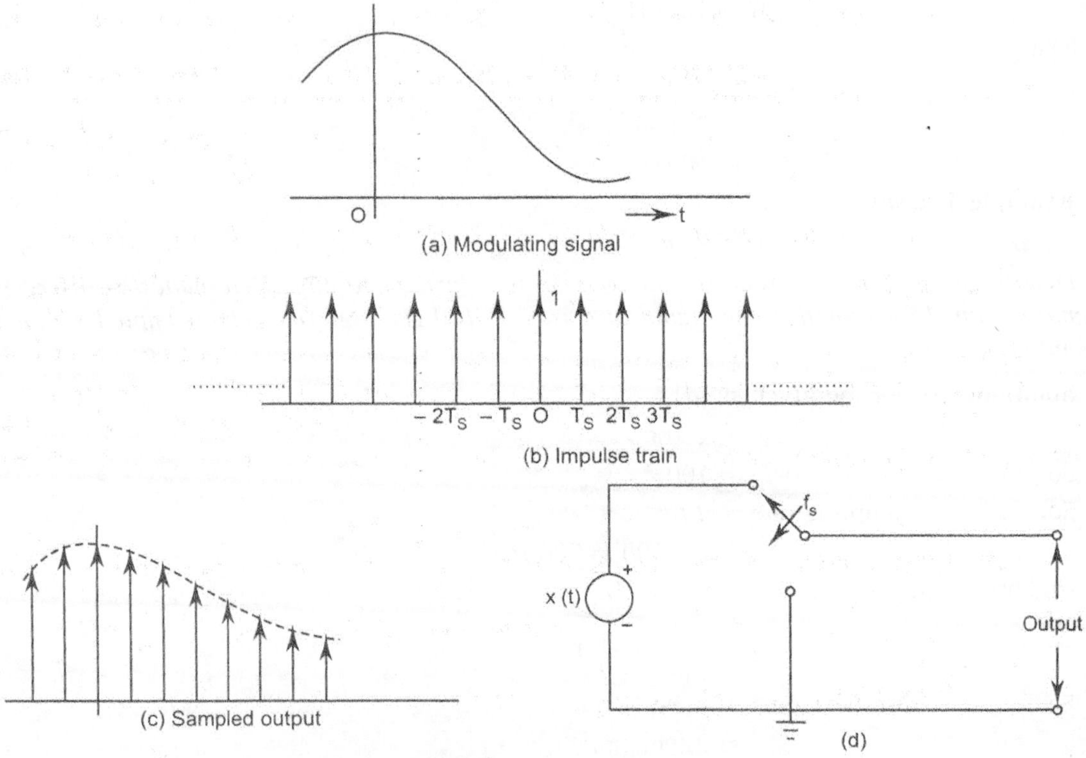

Fig. 2.3 *Instantaneous sampling*

To realize an ideal impulse function; the speed of operation must be very fast; in other words the switch shown in Fig. 2.3(*d*) should be closed for $t = 0$. Thus the output of circuit (also called switching sampler) is shown by equation.

$$g(f) = \frac{1}{T_s} \sum_{K=-\infty}^{\infty} X(f - kf_s) \quad (2.16)$$

$$g(f) = f_s \sum_{K=-\infty}^{\infty} X(f - kf_s) \quad (2.17)$$

As discussed till now.

But practically we will find some deviation from the ideal sampling in the practical case; as practical switching circuit will take some finite duration in switching operation. In other word the ideal impulse train cannot be generated. So we will now discuss the practical sampling methods named by natural sampling and flat top sampling.

In the ideal sampling we have one more disadvantage that the impulse has width τ approaching to zero. Due to this the power content of impulse is negligible and they cannot be use for transmission purpose.

2.3.2 Natural Sampling

Now in practical case we have pulse trains of some finite width (duration) τ.

An analog signal $x(t)$ is applied to a switching circuit controlled by a sampling signal $c(t)$; that consists of an infinite succession of pulses of width τ and amplitude A. The output of sampler is $g(t)$;

Now as in case of ideal sampler; switching operation extracts the analog signal $x(t)$; at per determined duration T, Where T is the sampling period

$$T = \frac{1}{f_s}; \text{ Mathematically}$$

$$g(t) = x(t).\, c(t) \quad (2.18)$$

$x(t)$ is the band limited signal

$$x(t) \leftrightarrow x(\omega)$$

$c(t)$ is a pulse train and can be represented by complex Fourier series in time domain

$$c(t) = f_s \tau A \sum_{n=-\infty}^{\infty} \sin c\,(nf_s \tau).\, e^{j2\pi f_s . nt} \quad (2.19)$$

So; Sampled output

$$g(t) = x(t) \times c(t)$$

$$= x(t) \times (f_s \tau A) \sum_{n=-\infty}^{\infty} \sin c\,(nf_s \tau).\, e^{j2\pi f_s . nt} \quad (2.20)$$

Now in frequency domain

$$g(f) = x(\omega) * F\left[f_s\,(\tau A) \sum_{n=-\infty}^{\infty} \sin c\,(nf_s \tau)\,.\, e^{j2\pi f_s . nt} \right]$$

$$\boxed{g(f) = f_s \tau A \sum_{n=-\infty}^{\infty} \sin c\,(nf_s \tau)\, X(f - nf_s)} \quad (2.21)$$

Now consider a case that $A\tau \to 1$ and

$$\tau \to 0$$

Then
$$g(f) = f_s \sum_{n=-\infty}^{\infty} . X(f - nf_s) \tag{2.22}$$

That is identical to the ideal sampling now we have spectrum identical to that of ideal sampling, but now with amplitude $\tau A \sin C(nfs\tau)$. In the case of ideal sampling the basic shape of spectrum remains constant over whole range of spectrum (Fig. 2.4).

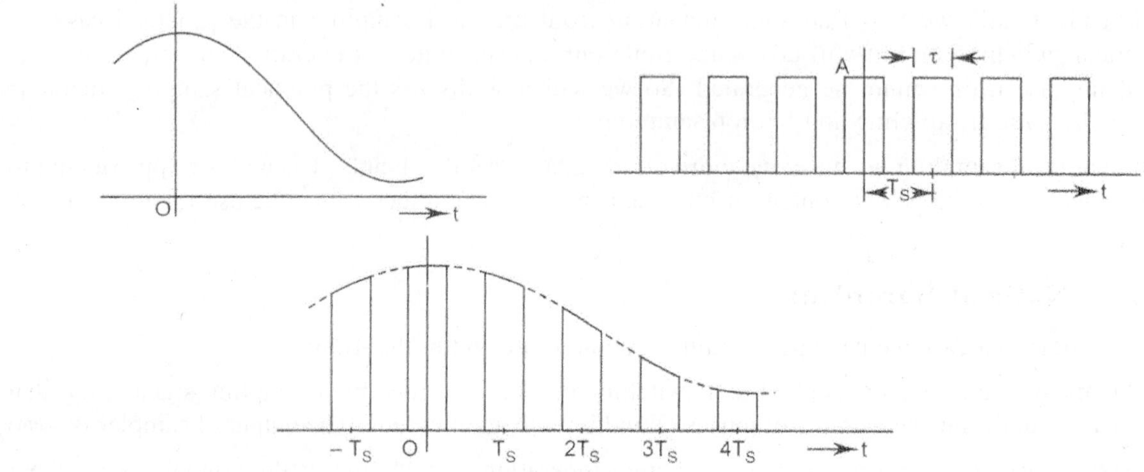

Fig. 2.4 *Natural sampled*

2.3.3 Flat Top Sampling

Generally in practice, the flat top sampling in mostly used due to easy in generation the practical used circuit is given below. This circuit generates flat top sampling by sample and hold operation.

The S/N circuit consists of two FET. (Acts as a switch) and a capacitor. The sampling switch is closed for a short duration by a pulse train applied at gate Q_1 of transistor. The capacitor is quickly charged upto a voltage equal to the instantaneous sample value of the incoming signal. It holds the charge until the discharging pulse is applied at gate Q_2. Because the capacitor holds the charge so the capacitor voltage doesn't follow the shape of applied signal. Thus at output we have flat-top samples for an applied input (Fig. 2.5).

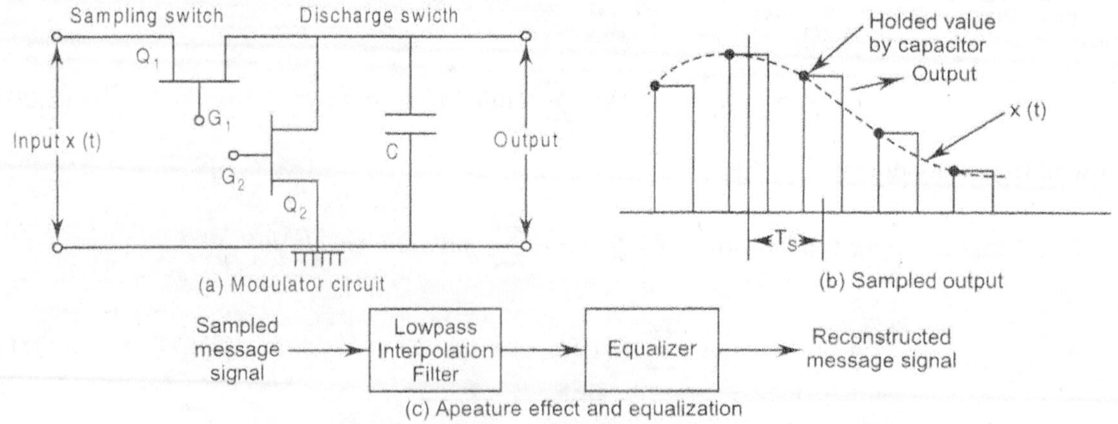

Fig. 2.5 *Flat top sampling*

Thus as discussed till now; flat top sampling can be assumed that a analog signal $x(t)$ is sampled instantaneously at the rate $T_s = \frac{1}{f_s}$ and the duration of each sample is lengthened to T (the time for which it holds charge).

Mathematically $s(t)$ to denote the sequence of flat top pulses generated in this way, we may write

$$\boxed{s(t) = \sum_{n=-\infty}^{\infty} x(nT_s)\, h(t - nT_s)} \tag{2.23}$$

The $h(t)$ is a rectangular pulse of unit amplitude and duration T, defined as follows-

$$h(t) = \begin{Bmatrix} 1 & 0 < t < T \\ 0 & t < 0 \text{ and } t > T \end{Bmatrix} \tag{2.24}$$

$$h(t) = \text{rect}\left(\frac{t}{T} - \frac{1}{2}\right) \tag{2.25}$$

The output of instantaneous sampling for signal $x(t)$ is given by.

$$s_I(t) = [s(t)]_{\text{instantaneous}} = \sum_{n=-\infty}^{\infty} x(t)\,.\,\delta(t - nT_s) \tag{2.26}$$

At any time $t = nT_s$.

$$= \sum_{n=-\infty}^{\infty} x(nT_s)\,.\,\delta(t - nT_s) \tag{2.27}$$

Now can value the $s_I(t)$ with pulse $h(t)$ is given by

$$s_I(t) * h(t) = \int_{-\infty}^{\infty} s_I(t)\,.\,h(t - \tau)\,.\,d\tau \tag{2.28}$$

$$= \int_{-\infty}^{\infty} \sum_{n=-\infty}^{\infty} x(nT_s)\,.\,\delta(t - nT_s)\, h(t - \tau)\,.\,d\tau$$

$$= \sum_{n=-\infty}^{\infty} x(nT_s) \int_{-\infty}^{\infty} \delta(t - nT_s).\, h(t - \tau).\, d\tau$$

By the shifting property of Delta function

$$\boxed{s_I(t) * h(t) = \sum_{n=-\infty}^{\infty} x(nT_s)\, h(t - nT_s)} \tag{2.29}$$

It is clear that

$$\boxed{s(t) = s_I(t) * h(t)} \tag{2.30}$$

Thus we can say, "flat top sampling mathematically is the convolution of instantaneously sampled output with a rectangular pulse".

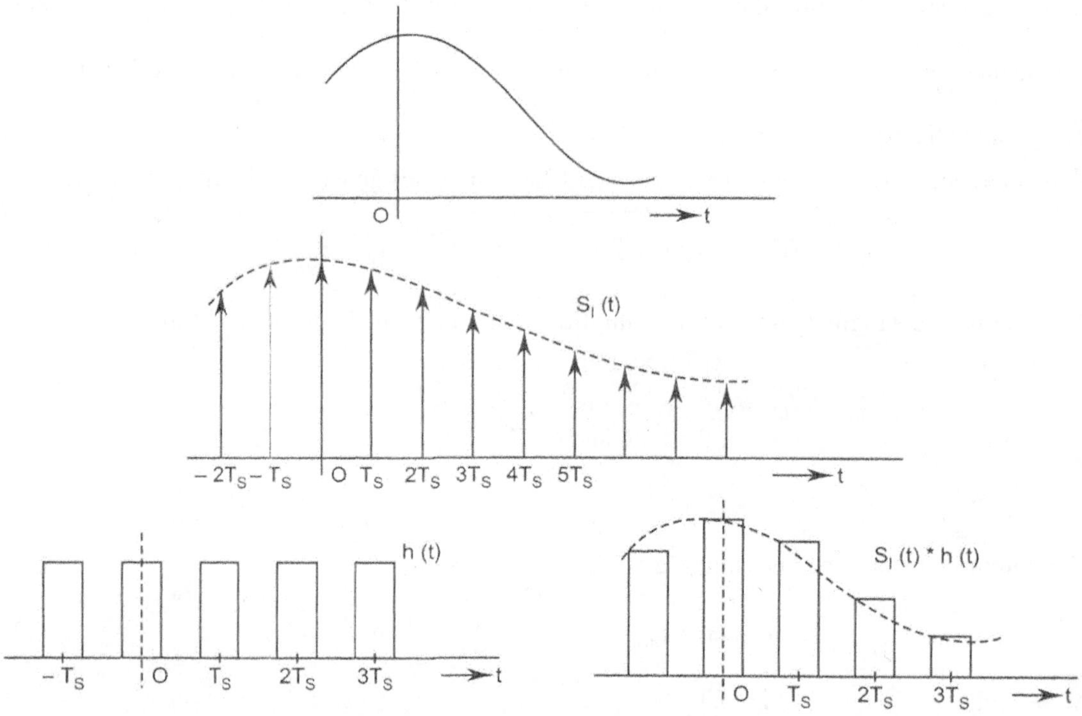

Fig. 2.6 *Flat top sampling process*

Now take the Fourier transform of both sides

$$s(f) = s_I(f)\, H(f) \tag{2.31}$$

Where $s(t) \leftrightarrow s(f)$

$$s_I(t) \leftrightarrow s_I(f) = f_s \sum_{n=-\infty}^{\infty} X(f - nf_s)$$

$$H(f) \leftrightarrow h(t)$$

So

$$\boxed{s(f) = f_s \sum_{n=-\infty}^{\infty} X(f - nf_s)\, H(f)} \tag{2.32}$$

2.3.4 Aperture Effect and Equalization

Now consider the equation

$$s(f) = s_I(f)\, H(f) \qquad \text{(Eq.2.31)}$$

Where $H(f) \leftrightarrow h(t)$

and

$$h(t) = \text{rect}\left(\frac{t}{T} - \frac{1}{2}\right)$$

So

$$\boxed{H(f) = T \,.\, \text{sin c}\,(fT) \,.\, e^{-j\pi fT}} \tag{2.33}$$

Thus we can say that now at the signal; we have applied amplitude distortion as well as a delay of T/2.

So we apply a equalizer circuit at output to recover the signal; the equalizer provides compensation for aperture effect; the Amplitude response of equalizer can be given.

$$\left|H_{eq}(f)\right| = \frac{1}{|H(f)|} \tag{2.34}$$

$$= \frac{1}{T} \cdot \frac{\pi f T}{\sin \pi f T}$$

$$\left|H_{eq}(f)\right| = \frac{1}{T \cdot \sin c\,(fT)} \tag{2.35}$$

An equalizer can be realized with the help of a passive Network where transfer function has a frequency dependence of the form that is inverse of sin c function (Fig. 2.7).

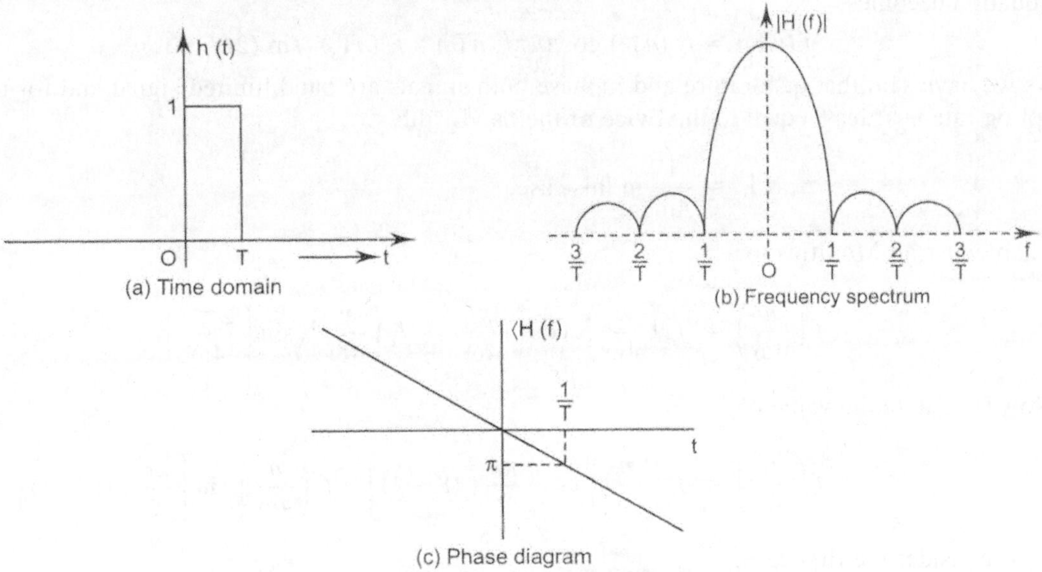

Fig. 2.7 *Equalizer*

2.4 SAMPLING THEOREM FOR BAND PASS SIGNAL

Up to this point of discussion of sampling, we have assumed that all signals under consideration are band limited (also called low pass signal) signals. But in communication, we sometime come across the band pass signals.

By Band Pass signals we means a continuous signal with frequency content concentrated in a narrow band of frequencies called Bandwidth around some frequency f_c, called **center frequency**.

We have till now discussed that for proper reconstruction of signal from its samples, sampling must be done at the rate of twice of its Maximum frequency component (that was really the Bandwidth of signal). We will see here; this criterion equally works good for Band Pass signals also.

2.4.1 Mathematical Approach for Band Pass Signal Sampling

To study the sampling of a band pass signal; we have assumed a band pass signal as shown in Fig. 2.8(*a*). We know that a Band pass signal can be represented in terms of in phase and quadrature components, we may write here.

$$f(t) = f_c(t).\cos 2\pi f_c t - f_s(t).\sin 2\pi f_c t \tag{2.36}$$

where $f_c(t)$ is the inphase component and $f_s(t)$ is the quadrature component both are band limited signals.

Let us assume that the highest frequency component in the band pass signal $f(t)$. In our case if we assume the Bandwidth of signal is ω; Then $f_c + \omega$ is an integer multiple of the bandwidth 2ω; Mathematically we write

$$f_c + \omega = K.2\omega. \tag{2.37}$$

where K = positive integer.

That is

$$f_c = (2K - 1)\,\omega \tag{2.38}$$

Now we know that for a sampling period T_s; $t = nT_s$ shows the total number of samples in time t. So equation becomes,

$$f(nT_s) = f_c(nT_s)\cos(2\pi f_c.nT_s) - f_s(nT_s).\sin(2\pi f_c.nT_s) \tag{2.39}$$

As we have said that quadrature and inphase both signals are band limited signal and for that the sampling rate is at least equal to the twice of the bandwidth.

$$T_s = \frac{1}{4\omega} \text{ in this case.} \tag{2.40}$$

Then equation Modifies to.

$$f\left(\frac{n}{4\omega}\right) = f_c\left(\frac{n}{4\omega}\right).\text{ccs}\left(\frac{2\pi f_c}{4\omega}\right) - f_s\left(\frac{n}{4\omega}\right).\sin\left(\frac{2\pi f_c}{4\omega}\right). \tag{2.41}$$

Now substitute the value of f_c;

$$f\left(\frac{n}{4\omega}\right) = f_c\left(\frac{n}{4\omega}\right).\cos\left[\frac{\pi n}{2}(2K-1)\right] - f_s\left(\frac{n}{4\omega}\right).\sin\left[\frac{\pi n}{2}(2K-1)\right] \tag{2.42}$$

Now consider the first case;

(*a*) For even value of n; we may put

$$n = 2\,m \;(m \text{ is another integer})$$

We can easily say that the quadrature component will be zero as it carries sine term and sine term is a odd term that will zero for even values of n; so we have only cosine terms;

It can be easily verified that for all values of K.

$$\cos\left[\frac{\pi n}{2}.(2K-1)\right] = \cos\left[\pi m(2K-1)\right]$$

$$= (-1)^m \tag{2.43}$$

Use this Relation we may have for even values of ω;

$$f\left(\frac{n}{4\omega}\right) = (-1)^m.f_c\left(\frac{m}{2\omega}\right) \tag{2.44}$$

(*b*) For odd values of ω put

$$(n = 2\,m - 1)$$

Again for odd values of m; the cosine terms (containing in phase components) will be zero; and only sine terms will be available. So it can be easily verified that for all values of K;

$$\sin\left[\frac{\pi n}{2}(2K-1)\right] = \sin\left[\frac{\pi}{2}(2m-1)(2K-1)\right]$$
$$= (-1)^{m+K} \quad (2.45)$$

So using these relations, the equation for odd values of ω is given by

$$f\left(\frac{2m-1}{4\omega}\right) = (-1)^{m+K+1} f_s\left(\frac{2m-1}{4\omega}\right) \quad (2.46)$$

So we can say that inphase and quadrature phase both components can be easily recovered; if samples are taken at the rate of twice of its Bandwidth; then; we have reproduction interpolation formulas for bath.

(*i*) Interpolation formula for the reconstruction of inphase component

$$f_c(t) = \sum_{n=-\infty}^{\infty} f_c\left(\frac{m}{2\omega}\right).\sin c\,(2\omega t - m) \quad (2.47)$$

(*ii*) Interpolation formula for the reconstruction of quadrature component.

$$f_s(t) = \sum_{n=-\infty}^{\infty} f_c\left(\frac{2m-1}{4\omega}\right)\sin\left(2\omega t - m + \frac{1}{2}\right) \quad (2.48)$$

So to obtained complete signal just add in phase and quadrature phase component.

$$f(t) = \sum_{n=-\infty}^{\infty} f_c\left(\frac{m}{2\omega}\right)\sin c\,(2\omega t - m)\cos 2\pi fct$$
$$+ \sum_{n=-\infty}^{\infty} f_c\left(\frac{2m-1}{2\omega}\right)\sin c\left(2\omega t - m + \frac{1}{2}\right).\sin 2\pi fct \quad (2.49)$$

Or use relations and writing again

$$f(t) = \sum_{n=-\infty}^{\infty} (-1)^m f\left(\frac{m}{2\omega}\right)\sin c\,(2\omega t - m).\cos 2\pi fct$$
$$+ \sum_{n=-\infty}^{\infty} (-1)^{m+K} f\left(\frac{2m-1}{4\omega}\right).\sin c\left(2\omega t - m + \frac{1}{2}\right).\sin 2\pi fct \quad (2.50)$$

Now again by the use of relations.

$$T_s = \frac{1}{4\omega} \text{ and}$$
$$f_c = (2K-1)\omega$$

$$f(t) = \sum_{n=-\infty}^{\infty} f\left(\frac{m}{2\omega}\right).\sin c\,(2\omega t - m).\cos(2\pi fc\,(t - 2mt_s))$$

$$+\sum_{n=-\infty}^{\infty} f\left(\frac{2m-1}{4\omega}\right).\operatorname{sinc}\left(2\omega t - m + \frac{1}{2}\right).\cos\left(2\pi f_c\left(t-(2m-1)T_s\right)\right)$$

$$\boxed{f(t) = \sum_{n=-\infty}^{\infty} f(nT_s).\operatorname{sin c}\left(2\omega t - \frac{n}{2}\right).\cos\left(2\pi f_c\left(t - nt_s\right)\right)} \tag{2.51}$$

That proves that if a band Pass signal having the highest frequency, a complete integer multiple of bandwidth; It can be completely recovered if the samples are taken at a rate equal to the twice of the bandwidth of signal $f(t)$.

Now if the upper cut off frequency is not the integer multiple of Bandwidth so the Ratio (Fig. 2.8(*b*)).

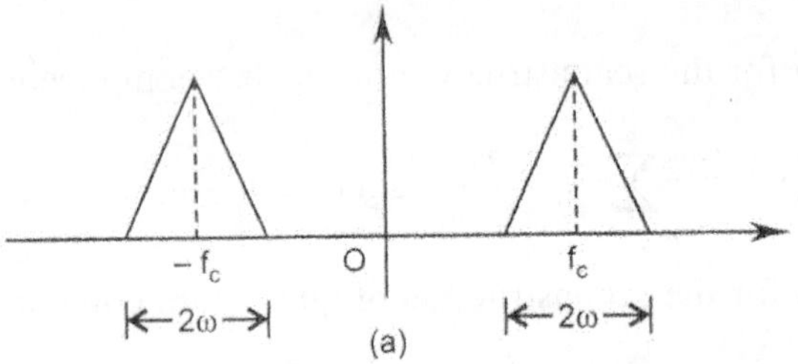

Fig. 2.8 *(a) Spectrum of a band pass signal*

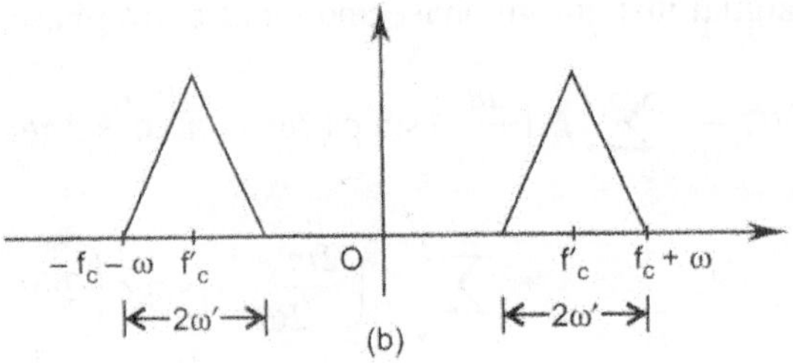

Fig. 2.8 *(b) Shifting of bandwidth 2ω and center frequency so that* $\frac{f_c+\omega}{2\omega}$*, is a complete integer*

$$\frac{f_c+\omega}{2\omega} = r' \text{ (Integer Part of Ratio)} \tag{2.52}$$

Now take the value of $(f_c + \omega)$, constant and increase the value of 2ω such that it becomes a complete integer.

$$\frac{f_c+\omega}{2\omega'} = r \tag{2.53}$$

Simply it can be said that r is nothing but the integer part of (r'). If suppose ratio $r' = 13.11$. Then, we will proceed to take it at $r = 13$. By increasing the value of 2ω; so the

(*a*) New value of center frequency

$$f_c' = f_c + \omega - \omega'$$

$$\boxed{f_c' = \left(1 - \frac{1}{2r}\right)(f_c + \omega)} \tag{2.54}$$

(b) $$\text{Sampling Rate} = \frac{1}{4\omega'} \tag{2.55}$$

$$\text{Sampling frequency} = 4\omega'$$

But $$f_c + \omega = 2\omega r' = 2\omega' r$$

'or' $$\omega' = \omega\left(\frac{r'}{r}\right)$$

So sampling frequency

$$\boxed{f_s = 4\omega\left(\frac{r'}{r}\right)} \tag{2.56}$$

SOLVED EXAMPLES

Example 1. *Find the Nyquist rate and the Nyquist interval for each of the following signals :*

(i) $m(t) = 5 \cos 1000\pi t \cos 4000\pi t$ *(ii)* $m(t) = \dfrac{\sin 200\pi t}{\pi t}$

(iii) $m(t) = \left(\dfrac{\sin 200\pi t}{\pi t}\right)^2$.

Solution. *(i)* Given: $m(t) = 5 \cos 1000\, \pi t \cos 4000\, \pi t$

$= 2.5(\cos 3000\pi t + \cos 5000\pi t)[\because\ 2 \cos A \cos B = \cos(A + B) + \cos(A\,.\,B)]$

Thus, $m(t)$ is band-limited with $f_M = 2500$ Hz. Hence, the Nyquist rate is 5000 Hz, and the Nyquist interval is $\frac{1}{5000}$ sec $= 0.2$ *ms*.

(ii) We have $\dfrac{\sin at}{\pi t} \leftrightarrow p_a(\omega) = \begin{cases} 1 & |\omega| < a \\ 0 & |\omega| > a \end{cases}$

Thus we see that $m(t)$ is a band-limited signal with $f_M = 100$ Hz. Hence, the Nyquist rate is 200 Hz, and the Nyquist interval is $\frac{1}{200}$ sec .

(iii) From the frequency convolution theorem, we find that the signal $m(t)$ is also band-limited and its bandwidth is twice that of the signal of part *(ii)*, that is, 200 Hz. Thus, the Nyquist rate is 400 Hz, and the Nyqist interval is $\frac{1}{400}$ sec .

Example 2. *Given the signal*

$$m(t) = 10 \cos 2000\pi t \cos 8000\pi t$$

(i) What is the minimum sampling rate based on the low-pass uniform sampling theorem ?

(ii) Repeat (i) based on the bandpass sampling theorem.

Solution. (*i*) $m(t) = 10 \cos 2000\pi t \cos 8000\pi t = 5 \cos 6000\pi t + 5 \cos 10000\pi t$

$$f_M = 5000 \text{ Hz} = 5 \text{ kHz} , \text{Thus } f_s = 2f_M = 10 \text{ kHz}.$$

(*ii*) $f_u = f_M = 5 \text{ kHz}$ and $f_B = (5 - 3) = 2\text{kHz}$.

$$\frac{f_u}{f_B} = \frac{5}{2} = 2.5 \rightarrow k = 2$$

Based on the bandpass sampling theorem, $f_s = \dfrac{2f_u}{k} = 5\text{kHz}$

Example 3. *An analog signal is sampled at the Nyquist rate f_s and quantized into L levels. Find the time duration τ of 1 bit of the binary encoded signal.*

Solution. Let n be the number of bits per sample. The by $L = 2^n$ or $n = \log_2 L$

$n = [\log_2 L]$

where $[\log_2 L]$ indicates the next higher integer to be taken if $\log_2 L$ is not an integer value; nf_s binary pulses must be transmitted per second. Thus

$$\tau = \frac{1}{nf_s} = \frac{T_s}{n} = \frac{T_s}{[\log_2 L]}$$

where T_s is the Nyquist interval.

Example 4. *A signal $m_1(t)$ is band-limited to 3.6 kHz, and three other signals $m_2(t)$, $m_3(t)$, and $m_4(t)$ are band-limited to 1.2 kHz each. These signals are to be transmitted by means of time-division multiplexing.*

(*i*) *Set up a scheme for accomplishing this multiplexing requirement, with each signal sampled at its Nyquist rate.*

(*ii*) *What must be the speed of the commutator (in samples per second) ?*

(*iii*) *If the commutator output is quantized with L = 1024 and the result is binary-coded, what is the output bit rate ?*

(*iv*) *Determine the minimum transmission bandwidth of the channel.*

Solution. (*i*)

Message	*Bandwidth*	*Nyquist Rate*
$m_1(t)$	3.6 kHz	7.2 kHz
$m_2(t)$	1.2 kHz	2.4 kHz
$m_3(t)$	1.2 kHz	2.4 kHz
$m_4(t)$	1.2 kHz	2.4 kHz

If the sampling commutator rotates at the rate of 2.4×10^3 rotations per second, then in one rotation we obtain one sample from each of $m_2(t)$, $m_3(t)$ and $m_4(t)$ and three samples from $m_1(t)$. This means that the commulator must have at least six poles connected to the signals, as shown in Fig. 2.9.

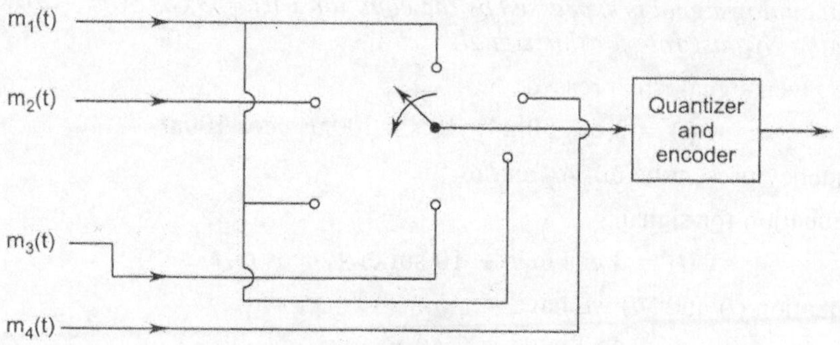

Fig. 2.9

(ii) $m_1(t)$ has 7200 sample/s. And $m_2(t)$, $m_3(t)$, and $m_4(t)$ each have 2400 samples/s. Hence, there are a total of 14400 samples/s.

(iii) $L = 1024 = 2^{10} = 2^n$

Thus, the output bit rate is 10(14400) = 144 kb/s.

(iv) The minimum channel bandwidth is

$$f_B = \frac{1}{2}(7.2 + 2.4 + 2.4 + 2.4) = 7.2 \text{ kHz}$$

Example 5. *Calculate the capacity of a standard telephone channel with a 32 dB signal to noise ratio.*

Solution. Standard telephone channels occupy the frequency range of **300 to 3400 Hz**. Also the actual signal to noise ratio is antilog $\left(\frac{32}{10}\right)$ = antilog (3.2) = 1585

Bandwidth $\Delta f = 3400 - 300 = 3100$ Hz

$$C = \Delta f \log_2\left(1 + \frac{S}{N}\right) = 3100 \times \log_2(1 + 1585) = 3100 \times \log_2(1585)$$

$$= 3100 \times 10.63 = 32.953 \text{ bits/sec}$$

Example 6. *A signal $\cos^2 200\pi t$ is multiplied by the signal $\sin 200\pi t$. The frequency which will be generated in this process include.*

Solution. Signal $\cos^2 2000\pi t$ is multiplied by $\sin 200\pi t$

The new signal is

$$S(t) = \sin 200\pi t \cos^2 200\pi t \sin 200\pi t \left(\frac{\cos 400\pi t + 1}{2}\right) \quad \left[\because \cos^2 A = \frac{1 + \cos 2A}{2}\right]$$

$$= \frac{1}{2}\left[\frac{1}{2}\{\sin(200\pi t - 400\pi t) + \sin(200\pi t + 400\pi t) + \sin 200\pi t\}\right]$$

$$= \frac{1}{2}\left[-\frac{1}{2}\sin 200\pi t + \frac{1}{2}\sin 600\pi t + \sin 200\pi t\right] = \frac{1}{4}[\sin 200\pi t + \sin 600\pi t]$$

Thus the frequency contained in the compounded signal are 100 Hz, 200 Hz and 300 Hz.

Example 7. *An analog signal is expressed by the equation* $x(t) = 3 \cos 50\pi t + 10 \sin 300\pi t - \cos 100\pi t$. *Calculate the Nyquist rate for this signal.*

Solution. The given signal is expressed

$$x(t) = 3 \cos 50\pi t + 10 \sin 300\pi t - \cos 100\pi t \qquad \ldots(i)$$

Let three frequency present be ω_1, ω_2 and ω_3
so that the new equation for signal.

$$x(t) = 3 \cos \omega_1 t + 10 \sin \omega_2 t - \cos \omega_3 t \qquad \ldots(ii)$$

Comparing equation (*i*) and (*ii*) we have

$$\omega_1 t = 50\,\pi t \text{ or } \omega_1 = 50\,\pi$$
$$2\pi f_1 = 50\,\pi \text{ or } f_1 = 25 \text{ Hz}$$

Similarly, for second factor

$$\omega_2 t = 300\,\pi t \text{ or } \omega_2 = 300\,\pi$$
$$2\pi f_2 t = 300\,\pi t \text{ or } 2\pi f_2 = 300\,\pi$$
$$f_2 = 150 \text{ Hz}$$

Again, for third factor

$$\omega_3 t = 100\,\pi t \text{ or } 2\pi f_3 t = 100\,\pi t$$
$$2\pi f_3 = 100\,\pi \text{ or } f_3 = 50 \text{ Hz}$$

Therefore, the maximum frequency present in $x(t)$ is $f_2 = 150$ Hz

Nyquist rate is given as $f_s = 2 f_m$

where f_m = maximum frequency present in the signal

Here $f_m = f_2 = 150$ Hz

Therefore, Nyquist rate $f_s = 2 f_2 = 2 \times 150 = 300$ Hz.

Example 8. *Find the Nyquist rate and the Nyquist interval for the signal*

$$x(t) = \frac{1}{2\pi}\,[\cos(4000\pi t)\cos(1000\pi t)].$$

Solution. Given signal is

$$x(t) = \frac{1}{2\pi}\left[\cos(4000\pi t)\cos(1000\pi t)\right]$$

$$x(t) = \frac{1}{4\pi}\left[2\cos(4000\pi t)\cos(1000\pi t)\right]$$

$$x(t) = \frac{1}{4\pi}\left[2\cos(4000\pi t + 1000\pi t) + \cos(4000\pi t - 1000\pi t)\right]$$

[$\because$ $2\cos A \cos B = \cos(A + B) + \cos(A - B)$]

or

$$x(t) = \frac{1}{4\pi}\left[\cos 5000\pi t + \cos 3000\pi t\right] \qquad \ldots(i)$$

Let the three frequency present in the signal be ω_1 and ω_2 so that the new equation for the signal will be

$$x(t) = \frac{1}{4\pi}\left[\cos \omega_1 t + \cos \omega_2 t\right] \quad \ldots(ii)$$

Comparing equation (*i*) and (*ii*), we have

$$\omega_1 t = 5000\ \pi t \text{ or } 2\pi f_1 t = 5000\pi t$$

$$2 f_1 = 5000 \text{ or } f_1 = 2500 \text{ Hz}$$

Similarly, for second factor

$$\omega_2 t = 3000\ \pi t \text{ or } 2\pi f_1 t = 3000\pi t$$

$$2 f_2 = 3000 \text{ or } f_2 = 1500 \text{ Hz}$$

Therefore, the maximum frequency present on $x(t)$ is $f_1 = 2500$ Hz

Nyquist rate is gives as $f_s = 2 f_m$

Where f_m maximum frequency presents in the signal here $f_m = f_1 = 2500$ Hz

Therefore **Nyquist rate**

$$f_s = 2 f_m = 2 \times 2500 = 5 \text{ KHz}$$

$$f_s = 5000 \text{ Hz} = 5 \text{ KHz}$$

Nyquist interval is given as

$$T_s = \frac{1}{2 f_m} \text{ or } T_s = \frac{1}{2 \times 2500}$$

$$T_s = \frac{1}{5000} \text{ or } T_s = 0.2 \times 10^{-3} \text{ seconds}$$

$$T_s = 0.2 \text{ m/seconds}$$

Example 9. *A leased telephone of bandwidth 3 KHz is used to transmit binary data. Calculate the data ratio (in bits per second) that can be transmitted if we use*

(*a*) *Polar signal with rectangular half-width pulses*

(*b*) *Polar signal with rectangular full width pulses*

(*c*) *Polar signal using Nyquist criterion pulses of $r = 0.25$*

(*d*) *Bipolar signal with rectangular half width pulses.*

Solution. (*a*) $B_T = 2R_b,\ R_b = \frac{B_T}{2} = \frac{3}{2} = 1.5$ kbits/sec.

(*b*) $B_T = R_b \Rightarrow R_b = 3$k bits/sec.

(*c*) $B_T = \frac{1 + r}{2} R_b$

Hence $3000 = \frac{1.25}{2} R_b \Rightarrow R_b = 4.48$ kbits/sec.

(*d*) $B_T = R_b \Rightarrow R_b = 3$k bits/sec

$$P_e = 1 - (1 - 10^{-4})^{1,00,000}$$

$$= 0.999.$$

Example 10. *An analog signal of bandwidth 10 KHz is sampled at a rate of 24 KHz, quantized into 256 levels, and coded using M–ary multiamplitude pulses satisfying the Nyquist criterion with a roll–off factor r = 0.2.A 30–KHz bandwidth is available to transmit the data. Determine the smallest acceptable value of M.*

Solution. Each sample requires 8 bits to be represented ($256 = 2^8$)

So data transmission rate is $24{,}000 \times 8 = 192{,}000$ bits/sec

Given that $B_T = 30$ KHz

Since $R = \dfrac{2}{1+r} B_T = \dfrac{2}{1.2} \times 30{,}000 = 50{,}000$ M–ary pulses/sec

We have 50,000 M–ary pulses/second to transmit 192,000 bits/sec. Hence each pulse must transmit at least

$$\frac{192{,}000}{50{,}000} = 3.84 \text{ bits}$$

So $N = 4$ bits

Therefore $M = 2^N = 2^4 = 16$ is the minimum acceptable value.

Example 11. *An audio signal of bandwidth 5 KHz is sampled at a rate 25% above the Nyquist rate and quantized. The qnatization error is not to exceed 0.2% of the signal peak amplitude. The resulting quantized samples are now coded and transmitted by 4–ary pulses.*

(a) Determine the minimum number of 4–ary pulses required to encode each sample.

Solution. Given that the sampling rate is 25% above the Nyquist rate

Sampling rate $= 2 \times 5000 \times 1.25 = 12500$ Hz

Qualification error $= \dfrac{M_p}{L} = 0.002\, M_p$

$\therefore$ $L = 500$

Because L should be a power of 2, we select $L = 500 \approx 2^9$, $n = 9$ bits/sample as Each M–ary pulse conveys $\log_2 4 = 2$ bits of information

Hence we need $\dfrac{9}{2} = 4.5$ 4–ary pulses/sample.

Example 12. *Consider a PCM system is which 24 signals are to be time division multiplexed. Each signal has a 3–KHz bandwidth. The sampling rate is 33.3 percent higher than the theoretical maximum, and 8 bits are used for each sample.*

(a) Determine the required bit rate.

(b) Find the minimum required transmission bandwidth.

Solution. Signal bandwidth = 3 KHz.

Since sampling rate is 33.3% higher than theoritical maximum rate (*i.e.*, twice of signal bandwidth)

$\therefore$ **Sampling rate** $= 2 \times 3 \times 10^3 (1.33) = 8 \times 10^3$ samples/sec.

Since each samples require 8 bits to be represented

$\therefore$ **Data rate** $= 8 \times 8 \times 10^3$ bps $=$ 64 Kbps.

Since there are 24 signals to be multiplexed.

$\therefore$ **Transmission bandwidth required** $= 24 \times 64$ KHz $=$ 1.536 MHz.

(Here we have assumed that 1 Hz is required to transmit 1 bit)

Example 13. *Twenty-four voice signals are to be multiplexed and transmitted over twisted pair. What is the bandwidth required for FDM ? Assuming a bandwidth efficiency of 1bps/Hz, what is the bandwidth required for TDM using PCM ?*

Solution. Bandwidth of a voice signal is 4 KHz

$\therefore$ Bandwidth requirement for 24 voice signals using FDM will be $= 24 \times 4 = 96$ KHz.

With PCM, each voice signal requires a data rate of 64 Kbps.

$\therefore$ Data rate for 24 voice signals is $= 24 \times 64 = 1.536$ Mbps for bandwidth efficiency of 1bps/Hz

the required bandwidth will be $= \dfrac{1.536}{1} = 1.536$ MHz.

Example 14. *Ten analog signals that are bandlimited to frequencies below 16 kHz are sampled at the nyquist rate, the digitizing error is to be held below 0.2%. The signals are to travel on a synchronous TDM channel. What is the data rate required for the channel ?*

Solution. Bandwidth of the signals limited up to $=$ 16 kHz

Sample at as per Nquist rate $= 2 \times 16$ kHz $=$ 32 kHz

For ten analog signals $= 32 \times 10 = 320$ kHz.

The error needs to be kept below 0.2% therefore

$$\Rightarrow \quad \frac{\Delta V}{2m_p} = \frac{0.2}{100}$$

$$\Rightarrow \text{ Number of levels} \quad L = \frac{2m_p}{\Delta V}$$

$$\text{So} \quad L = \frac{100}{0.2} = 500$$

$$\text{Now} \quad L = 2^n = 500$$

$$\Rightarrow \quad n \log 2 = \log 500$$

$$\text{So} \quad n = 8.9 \simeq 9$$

$$\text{So} \quad \text{data rate} = 320 \times 9 = 2880 \text{ kbps.}$$

REVIEW QUESTIONS

1. What do you understand by sampling? Explain Nyquist theorem of sampling for a band limited signal.
2. What is the aliasing? How do we avoid this and why?
3. State and Proof sampling theorem.
4. Explain the different type of sampling theorem.

5. What is aperture effect? Why is it avoided and how?
6. State and proof the sampling theorem for Band Pass signal.
7. Calculate the
   (*a*) Nyquist rate
   (*b*) Signaling Rate for the signal

$$V = 2 \cos 1000\pi t + 2 \sin 1000\pi t - 4 \cos 500\pi t$$

   Determine the sampled signal if sampling frequency is 145 Hz.
8. Distinguish between instantaneous, nature and flat topped simply techniques.
9. State and prove uniform sampling theorem for low pass signal.
10. State and explain sampling theorem for band pass signal.

3

DIGITAL CODING TECHNIQUES

3.1 INTRODUCTION

Having sampled an analog signal (message signal), the next step in its digital transmission is the generation of a digital representation.

The output of sampling process is a discrete signal now we have to change it in digital form. Now a code is provided to each digital level.

Pulse Code Modulation (PCM) provides one method for digital coded representation of an analog signal.

3.1.1 Advantages of PCM

Despite the inherent complexity of mechanisation, here are a number of marked advantages in the use of PCM due to which it is in modern communication systems.

(*i*) By virtue of the regeneration capacity PCM is distinctly advantageous for systems having many repeater stations. Thus with respect to long distance telephone, this has been the greatest asset.

(*ii*) PCM permits repeating or amplifying the encoded signal without significant distortion being introduced, again owing primarily to the nature of the coded signal.

(*iii*) A PCM system designed for analog message transmission is readily adapted to other input signals, particularly digited data, thereby promoting flexibility and increasing system utilisation.

(*iv*) Compared to other digital communication systems like DM, ADM etc., increase of S/N is more with bit rate is PCM.

(*v*) PCM affords a considerable increase in the output signal-to-noise ratio at the receiver. Thus it is rugged to transmission **noise** and **interference**.

(*vi*) As already stated that the output signal-to-noise ratio increases exponentially with bandwidth, this is much greater improvement than observed for other pulse systems.

(*vii*) The code group is usually so synchronised that the time of occurrence of each pulse is known at the receiver. The only decision which needs to be made by the receiver is whether a pulse is present or not; nothing need to be known concerning its amplitude width, etc. This is an advantage which other pulse systems do not possess.

(*viii*) It increases the reliability of the system on the expansion of bandwidth.

(*ix*) Efficient generation of coded signal along the transmission path.

(*x*) It has both amplitude and time in discrete form, so computer communication is better via PCM.

(*xi*) The uniformity of format for different message signals.

Thus PCM should be given due consideration for applications involving TDM, minimum power, a diversity of message types, (*i.e.*, analog and digital), or any repeater stations. Because most of these factors are present in log–haul telephone transmission.

3.2 PULSE CODE MODULATION

Fig. 3.1. Pulse code Modulation in actual practice is very complex work; in which analog signal (message signal) is passed through several stages to get a coded digital output.

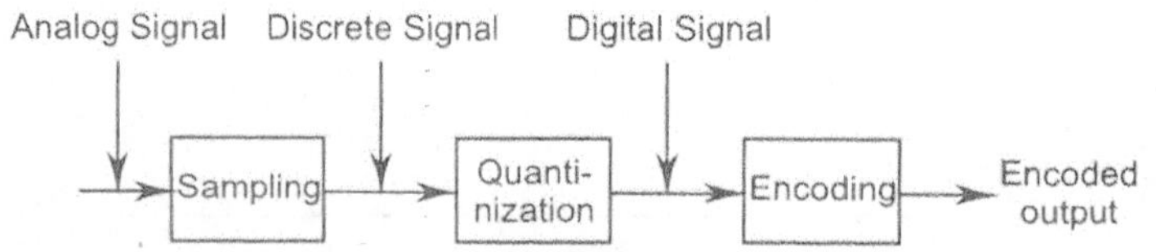

Fig. 3.1 *PCM generation*

But conceptually, the essential operations in the transmitter of a PCM system are sampling Quantizing and encoding as shown in figure. We should note that combinedly is called Analog to Digital (A/D) converter.

One thing should be cleared here that we have used term '**Modulation**'. During the study of Analog communication, we have studied modulation. Modulation means "The change in some parameter of carrier according to the instantaneous value of message signal." Here if we see that the part of sampling; one type of modulation has been taken place.

Now we will study the each block of PCM system one by one.

3.2.1 Sampling

The first step of analog to digital conversion 'or' PCM is the sampling. The incoming message signal (analog signal) is sampled with the use of narrow rectangular pulses (as studied under flat top sampling).

According to Nyquist criterion, to ensure the perfect reconstruction of the message at the receiver, the sampling rate must be greater than twice of the highest frequency component of the message signal. In practice, a low-pass prealias low-pass filter is used before the sampling.

Thus as discussed in sampling, a continuously rising message signal (Analog signal) is converted in to a limited number of discrete values per second (called discrete signal).

3.2.2 Quantinization

An analog signal, such as voice, has a continuous range of amplitudes and therefore its samples covers a continuous amplitude range. In other words, with in a finite amplitude range of signal we find an infinite number of amplitude levels.

However, it is not necessary in fact to transmit the exact amplitude of the samples. Any human ear 'or' eye can detect only finite intensity differences. This means that the original analog signal may be approximated by a signal constructed of discrete amplitudes. The existence of a finite number of discrete amplitude levels in a basic condition of PCM. Clearly, if we assign the discrete amplitude levels with sufficiently close spacing, we may make the approximated signal practically indistinguishable from the original analog signal.

"The conversion of a discrete signal (continuously sampled analog signal) into a digital form is called Quantinization"

3.2.2.1 Uniform Quantinization

Refer to Fig. 3.2. The operation of linear Quantinization can be described as below:

(*i*) Let us consider a signal *m* (*t*), which has peak value V_H and low value V_L (means signal is confined in the range from V_L to V_H). We divide this total range $V_H - V_L$ into M equal intervals each of size Δ.

Δ is called the step size and is given by

$$\Delta = \frac{V_H - V_L}{M} \tag{3.1}$$

In Fig. 3.2 a specific example has been given with M = 8 levels. In center of each of these steps, we locate quantization level denoted by $m_o, m_1, m_2, \ldots, m_7$.

(*ii*) The quantinized signal m_q (*t*) is generated simply by adopting the procedure as we will discuss now. Simply if we see the figure; the message signal within the range of M_o is represented by a single level m_o; for the message signal in the range of M_1, can be represented by only a single level m_1 and so on.

Thus we can easily conclude that at every instant of time message signal *m* (*t*). Does not change at all the time but it only jumps from one level to another level.

For example, till the value of message signal is in the range M_o; the output of Quantinized level is only M_o; and when message signal changes from M_o to M_1 the output of quantinization will jump to output m_1.

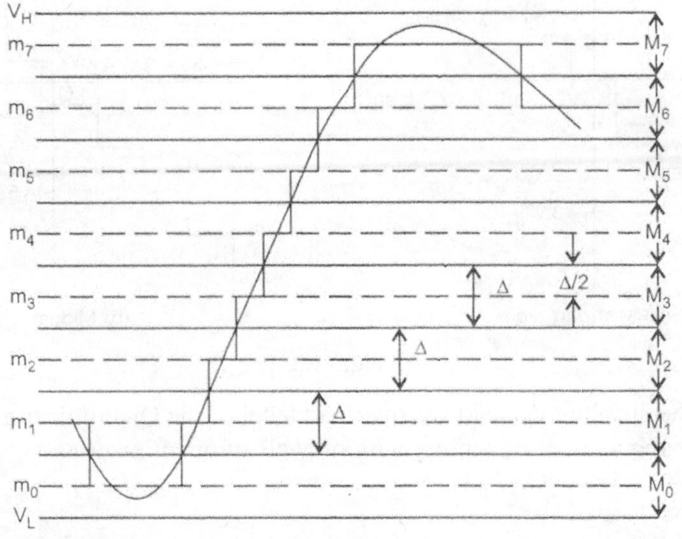

Fig. 3.2 *Uniform quantinization*

(*iii*) Graphically, the quantizing process means that a straight line representing the relation between input and output of a linear analog system is replaced by a transfer characteristic that is stair case like in appearance.

It is called linear Quantinization because the separation between the decision thresholds and the separation between the representation level of the quantinizer have a common value called step size.

According to the staircase-like characteristics of Fig. 3.3(*a*), the decision thresholds of quantinizer are located at $\pm \frac{\Delta}{2}, \pm \frac{3\Delta}{2}, \pm \frac{5\Delta}{2}, \ldots$ and the representation levels are located at $0; \pm \Delta, \pm 2\Delta \ldots$ where Δ is the step size.

A uniform Quantinizer characterized in this value is referred to as a symmetric Quantinizer of midtread type Quantinizer, because the origins lies in the middle of a tread of the staircase Quantinizer.

(*iv*) Fig. 3.3(*b*) shows another staircase-like transfer characteristics, in which the decision thresholds of the Quantinizer are located at $0, \pm \Delta, \pm 2\Delta, \ldots$ and the representation levels are located at $\pm \frac{\Delta}{2}, \pm \frac{3\Delta}{2}, \pm \frac{5\Delta}{2}, \ldots$ where Δ is again step size. A Quantinizer having second characteristics is referred to as a symmetric Quantinizer of the midriser type Quantinizer, because in this case the origin lies in the middle of the riser of the staircase.

This types of quantinizer referred as '**memoryless**'.

"This midtread 'or' midriser type Quantinizer are called memoryless Quantinizer But the Quantinizer output is determined only by the value of a corresponding input sample. This value of output is totally independent of easier (or later) analog samples applied to the input."

(*v*) Here we have include a overload level in the characteristics of linear quantinization characteristics.

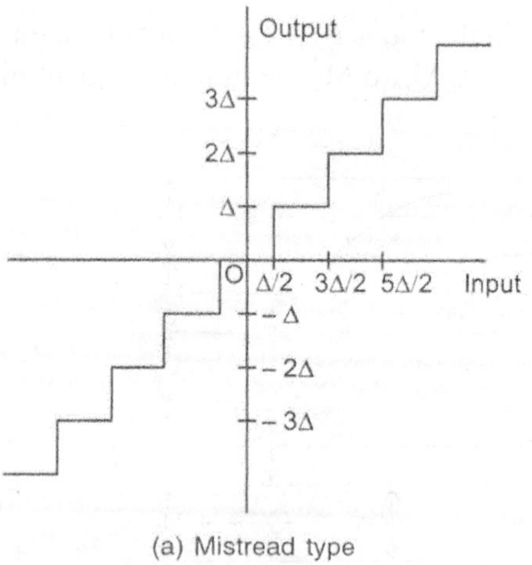

(a) Mistread type

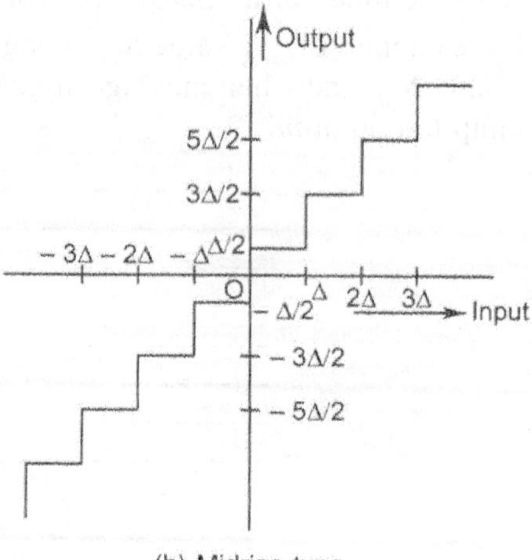

(b) Midrise type

Fig. 3.3

Let $x_{\max}$ denotes the absolute value of the overload level of the Quantinizer and Δ denotes the step size; we may express the number of representation levels of quantinizer as

$$L = 1 + \frac{2x_{max}}{\Delta} \tag{3.2}$$

(It should be noted that number of representation level is add for midtread Quantinizer.)

The amplitude of the quantinizer output in any case must not exceed the overload level, otherwise overload distortion will occur. (Discussed later).

3.2.2.2 Non-uniform Quantinization

Practically a non-linear Quantinization may be viewed as passing the baseband signal through a compressor and then subjecting the compressed signal to uniform quantinizing.

Thus "Non-uniform quantinization is basically a compressor followed by a uniform quantinization."

The input-output characteristics of non-uniform quantinizer has been shown in Fig. 3.4. Note that at low amplitudes the slope is larger than at large amplitudes. A signal transmitted through such a network will have compressed output. Now small amplitude signal will range through more quantinization region than would be the case in the absence of compression of course, the compression produces signal distortion. To the signal distortion at the receiver side we pass the signal through an expander. The **expander** network has an input-output characteristics, which is the inverse of the characteristics of **compressor**.

Thus the process of compressing and expending combindily called companding (Fig. 3.4)

Compressing + Expanding = Companding

A number of companding laws has been proposed. Among these choices two compression laws called **μ-law** and **A-law** has been accepted as desirable standards by CCITT.

The μ-law is used in North America and Japan. The A-law is used in Europe and rest of the world.

(1) μ-law of companding

For a given input V_1; where the expression for compressed output V_2 is given by

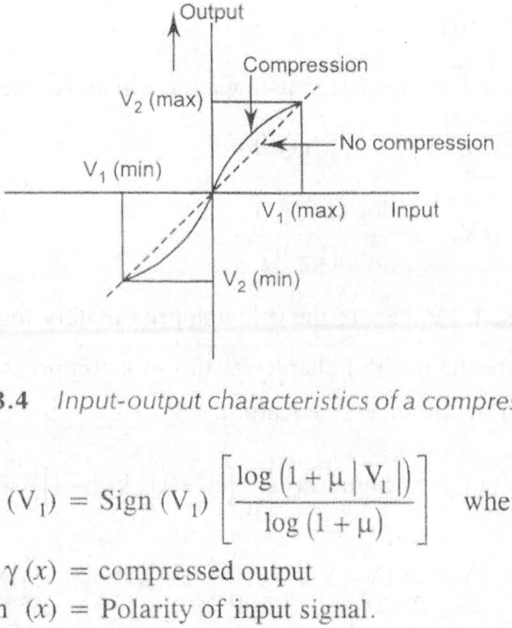

Fig 3.4 *Input-output characteristics of a compressor*

$$V_2(V_1) = \text{Sign}(V_1)\left[\frac{\log\left(1+\mu\left|V_1\right|\right)}{\log(1+\mu)}\right] \quad \text{where } -1 \le V_1 \le 1 \qquad (3.3)$$

where

$\gamma(x)$ = compressed output

Sign (x) = Polarity of input signal.

μ = positive constant which determines the degree of compression.

$\mu = 255$ (8 Bits) in the North America and Japanese standards.

$$|V_2| = |V_2(V_1)| = \frac{\log(1+\mu|V_1|)}{\log(1+\mu)} \quad \text{where} -1 \le V_1 \le 1 \tag{3.4}$$

The plot of normalized output $|V_2|$ with normalized input $|V_1|$ for various values of μ is shown in Fig. 3.5(*a*). At $\mu = 0$ simply it will provide *no* compression so output is a uniform quantinization.

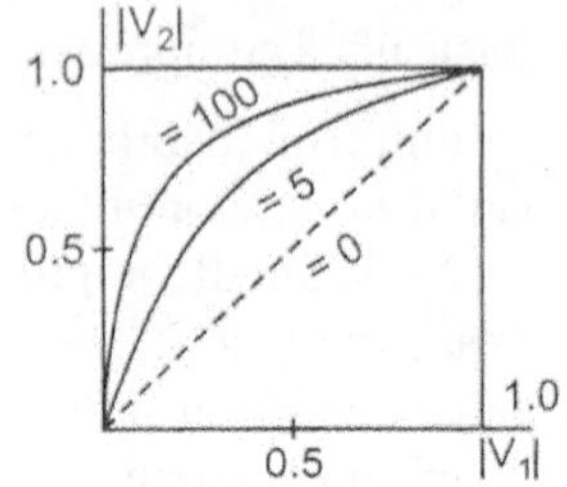

Fig. 3.5 *(a) μ-law*

The quantum step is defined as the reciprocal of the slope of the compression curve given by.

$$\frac{d|V_1|}{d|V_2|} = \frac{\log(1+\mu)}{\mu}\cdot(1+\mu|V_1|) \tag{3.5}$$

Now note the two points :

(*a*) For low input level, $\mu|V_1| << 1$; therefore,

$$|V_2| \simeq \frac{\mu|V_1|}{\log(1+\mu)} \tag{3.6}$$

Thus at low input levels, $\mu|V_1| << 1$; the μ-law approximately linear.

(*b*) For high input level, $\mu|V_1| >> 1$; therefore;

$$|V_2| = \frac{\log(\mu|V_1|)}{\log(1+\mu)} \tag{3.7}$$

Thus at high input levels, $\mu|V_1| >> 1$; the μ-law approximately logarithmic.,

Both points (*a*) and (*b*) are the desired characteristics of a compression law.

μ-law expansion is given by the inverse equation:

$$V_1(V_2) = \text{Sign}(V_2)\left\{\frac{1}{\mu}\left[(1+\mu)|V_2| - 1\right]\right\} \quad \text{where} -1 \le V_2 \le 1$$

$$|V_1| = |V_1(V_2)| = \frac{1}{\mu}\left[(1+\mu)|V_2| - 1\right] \quad \text{where} -1 \le V_2 \le 1 \tag{3.8}$$

(2) A-law of companding

For a given input (V_1) the expression for output (V_2) is given by

$$V_2(V_1) = \text{Sign}(V_1)\begin{cases} \dfrac{A|V_1|}{1+\log A} & 0 \le |V_1| < \dfrac{1}{A} \\ \dfrac{1+\log A|V_1|}{1+\log A} & \dfrac{1}{A} \le |V_1| \le 1 \end{cases} \quad (3.9)$$

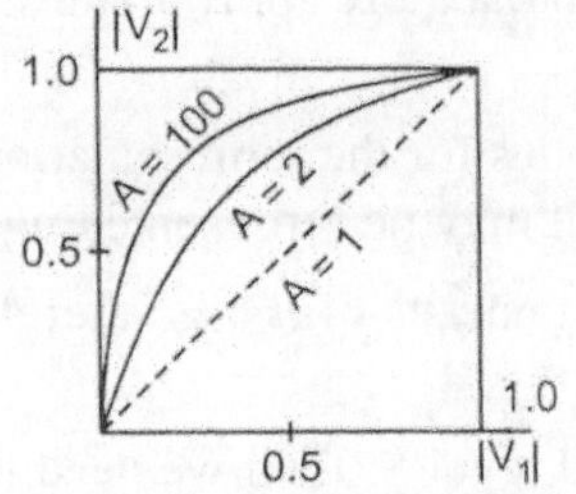

Fig. 3.5 *(b) A-law*

where A is the compression parameter; in Europe; A = 87.7; the value 87.6 is also used.

$$|V_2| = |V_2(V_1)| = \begin{cases} \dfrac{A|V_1|}{1+\log A} & 0 \le |V_1| < \dfrac{1}{A} \\ \dfrac{1+\log A|V_1|}{1+\log A} & \dfrac{1}{A} \le |V_1| \le 1 \end{cases} \quad (3.10)$$

The plot of normalized output $|V_2|$ with normalized input $|V_1|$ has been platted here for different values of A. It should be noted that for A= 1, It behaves just like a *uniform quantinizer.*

Now the value of step is given by

$$\frac{d|V_1|}{d|V_2|} = \begin{cases} \dfrac{1+\log A}{A} & 0 \le V_1 < \dfrac{1}{A} \\ (1+\log A)|V_1| & \dfrac{1}{A} \le |V_1| < 0 \end{cases} \quad (3.11)$$

The A-law expansion formula is given by the following inverse formula.

$$V_2(V_1) = \text{Sign}(V_2)\begin{cases} \dfrac{|V_2|(1+\log(A))}{A} & 0 \le |V_2| < \dfrac{1}{1+\log A} \\ \dfrac{\exp(|V_2|(1+\log A)-1)}{A} + \dfrac{1}{1+\log A} \le |V_2| < 1 \end{cases} \quad (3.12)$$

or

$$|V_2(V_1)| = \begin{cases} \dfrac{|V_2|(1+\log(A))}{A} & 0 \le |V_2| < \dfrac{1}{1+\log A} \\ \dfrac{\exp(|V_2|(1+\log A)-1)}{A} & \dfrac{1}{1+\log A} \le |V_2| 1 \end{cases} \quad (3.13)$$

Both A-Law and μ-Law require non-Linear signal processing and presents considerable implementation problems.

3.2.3 Encoding

Now finally we have generated a analog message signal into digital signal. But now to use this signal practically digital signal needs coding of the generated signal.

Actually generated quantinized signals are not suited to transmission over line 'or' radio path. To take the full advantages of generated digital data, we require the use of an encoding process to translate the generated digital signal into a more appropriate signal format. Now we need some codes that assigned to each discrete values.

There are a number of ways and codes for the representation of digital data but all of them employ binary. In a binary code, each symbol may be represented with the help of 1 and 0.

In a binary code, each code word consists of n bits; therefore using such a code it is possible to represent a total of 2^n distinct values.

For example; if we have quantized level 8; then we need totally 3 bits to represent a code. Now there is a simple technique to represent them has been shown in Table (3.1).

Table 3.1 Encoding of 8 levels

Level	*Encoding*		
m_0	0	0	0
m_1	0	0	1
m_2	0	1	0
m_3	0	1	1
m_4	1	0	0
m_5	1	0	1
m_6	1	1	0
m_7	1	1	1

NOTE Complete explanation will be studied in later.

3.2.4 Transmission Bandwidth of PCM System

Let us assume that Quantinizer use n number of binary digits to represent each level. Then the number of levels that may be presented by n digits, will be

$$L = 2^n \tag{3.14}$$

Thus number of bits per samples $= n$

Number of samples per second $= f_s$

So the number of bits per second

$$= \frac{\text{Bits}}{\text{Sample}} \times \frac{\text{Sample}}{\text{Second}}$$

$$= n \times f_s \tag{3.15}$$

That is signaling rate

$$r = nf_s. \tag{3.16}$$

But we know that

$$f_s \geq 2f_H \text{ By sampling rate}$$

It is seen that

$$\text{PCM Bandwidth} \geq \frac{\text{Signaling Rate}}{2} \geq \frac{r}{2}$$

$$\geq \frac{1}{2} \cdot nf_s \geq \frac{1}{2} \cdot n \cdot (2f_H)$$

$$\boxed{BW \geq nf_H.} \tag{3.17}$$

3.2.5 Signal to Quantinization Noise Ratio for a PCM System (Linear Quantinization)

Suppose we are using Sinusidual Modulating signal swings between $+ V_m$ and $- V_m$.

The number of levels used = L

So the Quantinization step size can be obtained

$$\Delta = \frac{V_m - (-V_m)}{L}$$

$$\Delta = \frac{2V_m}{L} \tag{3.18}$$

The average power for Quantinization can be determined by

$$P_q = \frac{1}{\Delta} \int_{-\Delta/2}^{\Delta/2} q_e^{\,2} \, dq_e \tag{3.19}$$

$$= \frac{1}{\Delta} \left(\frac{q_e^{\,3}}{3} \right)_{-\frac{\Delta}{2}}^{\frac{\Delta}{2}}$$

where q_e = Quantinization error uniformaly distributed uniformly to $-\frac{\Delta}{2}$ to $\frac{\Delta}{2}$

$$P_q = \frac{\Delta^2}{12} \tag{3.20}$$

Substituting the value of Δ.

$$P_q = \left(\frac{2V_m}{L} \right)^2 \frac{1}{12}$$

$$P_q = \frac{V_m^{\,2}}{3L^2} \tag{3.21}$$

So signal to Quantinization error Ratio

$$SNR = \frac{\frac{V_m^{\,2}}{2}}{\frac{V_m^{\,2}}{3L^2}}$$

$$\left(\text{As the signal} = V_m \sin \omega t \text{ has average power} = \frac{V_m^{\,2}}{2} \right)$$

$$\boxed{\text{SNR} = \frac{3L^2}{2}} \tag{3.22}$$

SNR can be represented in decibels.

$$(\text{SNR})_{\text{dB}} = 10 \log_{10}\left(\frac{3}{2}\right) + 20 \log_{10} L$$

$$\boxed{(\text{SNR})_{\text{dB}} = 1.8 + 20 \log_{10} L} \tag{3.23}$$

Example 1. *Consider a audio signal*

$$V(t) = 2 \cos (1000\pi t)$$

(*a*) *Find the signal to Quantinization noise ratio when signal is Quantinized using 8 bit PCM.*

(*b*) *How many bits are required to achieve a signal to Quantinization ratio of at least 30 dB.*

Solution. (*a*) We know that

$$\text{SNR} = 1.8 + 20 \log_{10} L \text{ dB}$$

But $L = 2^n$

So $L = 2^8$

$$\text{SNR} = 1.8 + 20 \log_{10} 2^8$$
$$= 1.8 + 48.16$$
$$= 49.96 \text{ dB}$$

(*b*) Again.

$$\text{SNR} = 1.8 + 20 \log_{10} L$$
$$30 = 1.8 + 20 \log_{10} L$$
$$\log_{10} L = \frac{30 - 1.8}{20}$$
$$\log_{10} L = 1.41$$
$$L = 25.70$$

'or' $2^n = 25.70$

$$n = 4.68$$
$$\boxed{n \simeq 5}$$

Example 2. *A audio signal has a bandwidth of 5.8 MHz encoded by PCM. Given that the total number of bits to represent a level are 10 bits. Determine*

(*a*) *Total number of levels*

(*b*) *Signaling rate/Bit Rate*

(*c*) *Bandwidth requirement*

(*d*) *Output signal to Quantinization Noise Ratio.*

Solution. (*a*) Total number of levels

$$L = 2^n$$

But $n = \text{number of Bits} = 10$

So $L = 2^{10}$

$$\boxed{L = 1024}$$

(*b*) Signaling rate $= nf_s \neq n.2f_H$

$$= 10 \times 5.8 \times 10^6 \times 2$$

$$\boxed{r = 116 \text{ Bits/Sec}}$$

(*c*) $$\text{Bandwidth} \geq \frac{1}{2} r \geq \frac{1}{2} \times 116$$

$$\boxed{\text{BW} \geq 58 \text{ Bits/Second}}$$

(*d*) $$\text{SNR} = 1.8 + 20 \log_{10} \text{L}$$

$$= 1.8 + 20 \log_{10} 2^{10}$$

$$\boxed{\text{SNR} = 62 \; dB}$$

Example 3. *The Bandwidth of an input signal to the PCM is restricted to 4 KHz. The input signal varies in amplitude – 3.8 V to 3.8 V and has the average power of 30 mW. The required SNR is given as 20 dB. Assuming linear quantinization, calculate*

(*i*) *Number of Bits required per sample*

(*ii*) *Output of such 30 PCM coder has been time Multiplexed, calculate total transmission Bandwidth.*

Solution. (*i*) We know that

$$\text{SNR} = \frac{3}{2} \text{L}^2$$

We know that $$\frac{\text{S}}{\text{N}} = 20 \text{ dB}$$

$$= 100$$

So $$100 = \frac{3}{2} \text{L}^2$$

$$\text{L} = 8.16$$

But $$\text{L} = 2^n$$

$$2^n = 8.16$$

So $$n = 3.03$$

$$n \simeq 4$$

(*ii*) Now Signaling Rate $= nf_s$.

$$r = n \times 2f_H$$

$$= 4 \times 2 \times 4$$

$$= 32 \text{ KHz}$$

So $$\text{Bandwidth} \geq \frac{1}{2} r \geq 16 \text{ KHz}$$

But we are using 30 signals.

So total $$\text{B.W.} \geq 16 \times 30$$

$$\boxed{\text{B.W.} \geq 480 \text{ KHz}}$$

Example 4. *The information in an analog waveform with maximum frequency 3 KHz. Is to be transmitted over an M-level PCM system where the number of pulse level is M = 16. Now calculate.*

(*i*) *What would be the max number of bits per sample that should be used in this PCM system?*

(*ii*) *What is the minimum sampling rate and what is resulting bit transmission rate?*

Solution. (*i*) We know total number of levels = 16

But $\quad L = 2^n$

So $\quad 2^n = 16$

$$\boxed{n = 4 \text{ Bits}}$$

(*ii*)
$$\text{Signaling rate} = nf_s$$
$$= n \times 2f_H$$
$$= 4 \times 2 \times 3 \text{ KHz}$$
$$\boxed{r = 24 \text{ K bits/second.}}$$
$$\text{Sampling Rate} = 2f_H$$
$$= 2 \times 3 \text{ KHz}$$
$$\boxed{f_s = 6 \text{ KHz}}$$

Example 5. *Eight message signal having 2KHz bandwidth each are time division multiplexed using. Binary PCM. The error in sampling amplitude cannot be greater than 1% of the peak amplitude. Determine the minimum transmission bandwidth required if raised Cosine pulses with roll off factor $\alpha = 0.2$ are used. The sampling rate must be at least 25% above the Nyquest rate.*

Solution. For
$$\text{PCM} = \frac{\Delta}{2} \le q_e \le \frac{\Delta}{2}$$

$$(q_e)_{\max} = \frac{\Delta}{2} = \frac{m_p}{L} \ge 0.01\, m_p \quad \text{where } (-m_p, m_p) = \text{amplitude}$$

$$L = \text{number of zones each of step size } \Delta$$

$$\Delta = \frac{2m_p}{L}; \text{ Hence } L \ge 100 \quad \text{Choose } L = 128 = 2^7$$

Thus number of bits per sample required is 7.

Nyquist sampling rate $\quad 2f_m = 2 \times 2000 = 4000$ samples/sec

Sampling rate for each signal $f_s = 1.25 \times 4000 = 5000$ samples/sec

There are eight TDM signal, requiring a total $= 8 \times 5000 = 40{,}000$ samples/sec

Each sample is encoded by 7 bits

Resultant bit rate
$$= \frac{1}{T_b} = 7 \times 40{,}000 = 280 \text{ Kb/sec}$$

Now
$$F_B = \frac{1+\alpha}{2T} \text{ Hz} = \frac{1+0.2}{2} \times 280 = \mathbf{168\ KHz}$$

Example 6. *A voice signal with bandwidth of 4 KHz is to be transmitted over 25 KHz base band channel such that $(S/N)_D \ge 18 \times 10^3\, x^2$. Design a PCM system to accomplish this.*

Solution. Here the channel bandwidth $B_T = 25$ KHz

Signal bandwidth $\omega = 4$ KHz

for PCM
$$B_T \ge v\,\omega; \; v \le \frac{25}{4} = 6.25$$

Also $\left\{\frac{S}{N}\right\}_D = 3q^2x^2 \geq 18{,}000x^2$

$$q \geq \sqrt{6{,}000} = 77.5$$

Now we have to find the most appropriate value of μ to satisfy to achieve $v \leq 6.25$ and $q \geq 77.5$.

Now $q = \mu^v$

(i) When $\mu = 2$ and $v = 6$, $q = 2^6 = 64$ which does not satisfy $q \geq 77.5$

(ii) When $\mu = 3$, $v = 6$, $q = 3^6 = 729$, which is too large considering $q \geq 77.5$

(iii) When $m = 3$, $v = 4$, $q = 3^4 = 81$ which properly satisfies $q \geq 77.5$

Now $$f_S \leq \frac{B_T}{v} = \frac{2 \times 25}{4}$$

Thus we have a PCM system with $\mu = 3$, $v = q = 81$ and $f_S = 10$ KHz

Example 7. *A sinusoidal signal is transmitted with PCM, an output (S/N) ratio is required to be 55.8dB. Find the number of representation levels required to achieve this performance.*

Solution. We have $\frac{S}{N}$ in dB as;

$$\left(\frac{S}{N}\right)_{dB} = 6N + 1.76 \text{ dB } i.e., 55.8 = 6N + 1.76$$

$$55.8 = 6N + 1.8 \quad \therefore \quad N = 9 \text{ bits}$$

$$\text{No. of levels} = 2^n = 29 = 512$$

Example 8. *Television signal with a bandwidth of 4.2 MHz is transmitted using binary PCM. The number of quantization levels is 512*

Calculate, the following :

(a) Code word length *(b) Transmission bandwidth*

(c) Final bit rate *(d) Output signal to quantization noise ratio.*

Solution. Given that the bandwidth is 4.2 MHz, that is, highest frequency component will have frequency of 4.2 MHz, therefore, $\omega = 4.2$ MHz

Quantization levels $q = 512$

(a) Number of bits and quantization levels are releated in binary PCM as, $q = 2^V$

$$i.e., 512 = 2^v,\ \log 512 = v \log 2 \text{ or } v = \frac{\log 512}{\log 2} = 9 \text{ bits}$$

Thus the code word length is 9 bits, is the value.

(b) The transmission channel bandwidth is given as,

$$B_T \geq v\omega \geq 9 \times 4.2 \times 10^6 \text{ Hz}$$

$$B_T \geq 37.8 \text{ MHz}$$

(c) The final bit rate will equal to signaling rate is the result. Signaling rate is given as, $r = vf_s$

The sampling frequency $f_s > 2\omega$ by sampling theorem.

$$\therefore \quad f_s \geq 2 \times 4.2 \text{ MHz} \quad \text{since } \omega = 4.2 \text{ MHz}$$

$$\therefore \quad f_s \geq 8.4 \text{ MHz}$$

Substituting this value of 'f_s' in equation for signaling rate, $r = 9 \times 8.4 \times 10^6$

$= 75.6 \times 10^6$ bit/sec is the result

Transmission bandwidth can be obtained as,

$$BT \geq \frac{1}{2} r \geq \frac{1}{2} \times 75.6 \times 10^6 \text{ bits/sec}$$

or $B_T \geq 37.8$ MHz (which is same as the value obtained earlier)

(*d*) The signal to noise ratio is given by;

$$\left(\frac{S}{N}\right)_{dB} \leq 4.8 + 6v \text{ dB}$$

$$\leq 4.8 + 6 \times 9 \geq 58.8 \; dB \text{ is the result}$$

Example 9. *A signal of $x(t) = 3 \sin (500\, t)$ is sampled and quantised using 10 bit PCM*

(*a*) *Find the (S/N) ratio and step size and number of levels also.*

(*b*) *If the above signal is quantised using PCM, find the number of bits required to achieve a (S/N) ratio of at least 60 dB. Find the number of levels and step-size.* [Raj. Univ. 1994, 1998]

Solution. We have; $\frac{S}{N}$ in dB as;

(*a*) $$\left(\frac{S}{N}\right)_{dB} = 6N + 1.76 = 6(10 + 1.76) = 61.76 \; dB$$

$$2^N = 10^{10} = 1024; \; 3(-3) = 6(V_{p-p})$$

And step size $$= \frac{6}{1024} = 5.8 \times 10^{-3} \text{ V} = 5 \times 10^{-3}\text{V}$$

(*b*) $$60 = 6N + 1.76$$

$$N = 9.706 \cong 10$$

Number of levels $$= 2^{10} = 1024$$

The step-size $$= \frac{6}{1024} = 5 \times 10^{-3} \text{ V}$$

Example 10. *The information in an analog waveform with maximum frequency $f_m = 3$ kHz is to be transmitted over an M-level PCM system where the number of pulse levels is M = 16. The quantization distrotion is specified not to exceed 1% of peak to peak analog signal.*

(*a*) *What is the maximum number of bits per sample that should be used in this PCM system ?*

(*b*) *What is the minimum sampling rate and what is the resulting bit transission rate ?*

Solution. (*a*) The number of levels given here are M = 16, therefore, q = M = 16

Bits and levels in binary PCM are related as below :

$$q = 2^v \qquad [v = \text{bits}]$$

$$16 = 2^v \text{ or } v = 4, \text{ is the value}$$

(*b*) We have $f_m = \omega = 3$ kHz

$$56 \times 10^3 = 7.f_s$$

∴ Sampling frequency, $f_s = 8 \times 10^3$Hz

using sampling theorem, $f_s \geq 2\omega$; we get

Maximum frequency that can be handled will be :

$$\omega \le \frac{f_s}{2} \le \frac{8000}{2} \quad \therefore \quad \omega \le 4000 \text{ Hz is the result.}$$

Example 11. *A PCM system uses a uniform quantizer followed by a 7-bit binary encoder. The bit rate of the system is equal to 50×10^6 bits/sec.*

(*a*) *What is the maximum message bandwidth for which the system operates satisfactorily ?*

(*b*) *Determine the output signal to quantization noise ratio when a full load sinusoidal modulating wave of frequency 1 MHz is applied to the input.* [Raj. Univ. 1997, 2002]

Solution. (*a*) Assume that the message bandwidth be WHz, hence sampling frequency should be, $f_s \ge 2\omega$

Number of bits, $v = 7$ bits

The signaling rate is given by $r = v f_s \quad \therefore \quad r = 2 \times 2\omega$

$\therefore$ $50 \times 10^6 \ge 14\omega$ (substituting value of r)

$\therefore$ $\omega \le 3.37$ MHz is the result.

Thus the maximum message bandwidth is 3.57 MHz.

(*b*) The modulating wave is sinusoidal. For such signal, the signal to quantization noise ratio is given by

$$\left(\frac{S}{N}\right)_{dB} = 1.8 + 6v$$

Substituting for $v = 7$, we get

$$= 1.8 + 6 \times 7 = 43.8 \text{ dB is the result.}$$

Example 12. *Consider an audio-signal with spectral components limited to the frequency band of 300 to 3300 Hz. A PCM signal is generated with a sampling rate of 8000 samples. The required output signal-to-quantizing noise ratio is 30 dB.*

(*a*) *What is the minimum number of uniform quantizing levels needed, and what is the minimum number of bits per sample needed ?*

(*b*) *Calculate the minimum system bandwidth required.* [Raj. Univ. 1994, 2003]

Solution. (*a*) We have the $\frac{S}{N}$ as $\left(\frac{S}{N_q}\right)_{0dB} = 1.76 + 20\log L \ge 30$

$$L \ge \frac{1}{20}(30 - 1.76) = 1.412 \rightarrow L \ge 25.82$$

So the minimum number of uniform quantizing levels requird is 26.

$n = [\log_2 L] = [\log_2 26] = [4.7] = 5$ per sample. The minimum number of bits per sample is 5.

(*b*) The minimum required system bandwidth is given by;

$$f_{PCM} = \frac{n}{2} f_s = \frac{5}{2}(8000) = 20000 \text{ Hz} = 20 \text{ kHz}$$

$$\left(\frac{S}{N_q}\right)_{0dB} = 20\log L - 10.1 \geq 30;\ \log L \geq \frac{1}{20}(30 - 10.1) = 2.005 \rightarrow L \geq 101.2$$

$$\left(\frac{S}{N_q}\right)_{0\,dB} = 20\ \log\ L - 10.1 \geq 30$$

$$\log\ L \geq \frac{1}{20}(30 - 10.1) = 2.005 \rightarrow L \geq 101.2$$

Thus the minimum no. of quantizing levels need is 102.

$$n = [\log_2 L] = [6.67] = 7$$

The minimum no. of bits per samples is 7. The minimum bandwidth required for this case is

$$f_{PCM} = \frac{n}{2} f_s = \frac{7}{2}(8000) = 28000\text{ Hz} = 28\text{ kHz}$$

Example 13. *A computer puts out binary data at the rate of 56 Kb/sec. The computer output is transmitted using a baseband binary PAM system that is designed to have a raised cosine spectrum. Determine the transmission bandwidth required for each of the following roll off factors.*

(i) $\alpha = 0.25$ *(ii)* $\alpha = 0.5$ *(iii)* $\alpha = 0.75$ *(iv)* $\alpha = 1.0.$

[Raj. Univ. 1997, 2003]

Solution. Given data is r_b = 56 kb/sec = 56,000 b/sec

Now for filter with raised cosine spectrum we have the BW;

BW = Nyquist BW × (1 + α). Where α is the roll off factor

Now the Bandwidth B will be : $\frac{r_b}{2}(1+\alpha)$

(*i*) For α = 0.25,

$$B_{0.25} = \frac{56,000}{2}(1+0.25) = 35,000 = 35\text{ kHz}$$

(*ii*) For α = 0.5,

$$B_{0.5} = \frac{56,000}{2}(1+0.5) = 42,000 = 42\text{ kHz}$$

(*iii*) For α = 0.75,

$$B_{0.75} = \frac{56,000}{2}(1+0.75) = 49,000 = 49\text{ kHz}$$

(*iv*) For α = 1.0,

$$B_{1.0} = \frac{56,000}{2}(1+1) = 56,000 = 56\text{ kHz}$$

Example 14. *Bandwidth of the input to a impulse code modulator is restricted to 4 MHz. The signal varies form – 3.8 to 3.8V and has the average power of 30 mW. The required signal to quantisation noise power ratio is 20 dB. The modulator produces binary output. Assume uniform quantisation.*

(a) *Calculate the number of bits required per sample.*

(b) *Outputs of 30 such PCM orders are time multiplexed. What is the minimum required transmission bandwidth for the multiplexed signal ?* [Raj. Univ. 1993, 1997, 2003]

Solution. Total voltage range = 3.8 – (– 3.8) = 7.6V

Average signal power = 30 mW

$$20\log_{10}\left(\frac{S}{N}\right) = 20 \text{ db, or } \frac{S}{N} = 100 \text{ , Quantisation noise } = \frac{\delta^2}{12}$$

$$\text{Step size} = \frac{7.6\text{V}}{2^n} = \delta = \frac{30\,mW}{\frac{(7.6/2^n)^2}{12}} = \frac{30 \times 10^{-3} \times 12}{7.6/2^n} = 100$$

$$n = 8 \text{ bits.}$$

(*a*) $n = 8$ bits

(*b*) Bandwidth of each channel is 2 × 8 × 4 MHz = 64 MHz.

Bandwidth of 30 such channel is 30 × 2 × 8 × 4 MHz = 1920 MHz.

Each 4 MHz signal is sampled at 2 × 4 MHz and each sample has 8 bits.

Example 15. *Bandwidth in a PCM system for 128 quantization levels, and 8 KHz sampling frequency when 24 telephone channels, each band limited to 3.4 KHz are to be time division multiplexed.*

Solution.

$$n = 24,\ m = 128$$

$$2f_m = 8\text{KHz}$$

$$2^N = 128;\ N = \log_2 128 = 7$$

$$\therefore \quad \text{B.W.} = 2[n\,(N + 1) + 1]\, f_m$$

$$= [24(7 + 1) + 1]\ 8000 \text{ Hz} = 1.54 \text{ MHz}$$

Example 16. *Consider an audio signal comprised of the sinusoidal term given as $x(t) = 3\cos(500\,\pi t)$*

(*i*) *Determine the signal to quantization noise ratio when this is quantized using 10 bit PCM*

(*ii*) *How many bits of quantization are needed to achieve a signal to quantization noise ratio of at least 40 dB ?*

Solution. Here given that $x(t) = 3\cos(500\,\pi t)$

This is sinusoidal signal applied to the quantizes

(*i*) Let us assure that peak value of cosine wave defined by $x(t)$ covers the complete range of quantizer.

$$\text{AM} = 3\text{V covers complete range } \left(\frac{S}{N}\right)_{dB} = 1.8 + 6v$$

Since here 10 bit PCM is used *i.e.*, $v = 10$

$$\left(\frac{S}{N}\right)_{dB} = 1.8 + 6 \times 10 = 61.8 \text{ dB}$$

(*ii*) For sinusoidal signal again we will use the same relation

$$\left(\frac{S}{N}\right)_{dB} = 1.8 + 6v \text{ dB}$$

To get signal to noise ratio of at least 40 dB we can write above equation as

$$1.8 + 6v \geq 40 \text{ dB}$$

Solving this, we get

$$v > 6.36 \text{ bits} = 7 \text{ bits}.$$

Hence at least 7 bits are required to get signal to noise ratio of 40 dB.

Example 17. *A compact disc (CD) records audio signals digitally by using PCM. Assume the audio signal bandwidth to be 15 KHz.*

(a) What is the Nyquist rate?

(b) If the Nyquist samples are quantized into L = 65,536 levels and then binary coded, determine the number of binary digits required to encode a sample.

(c) Determine the number of binary digits per second (bits/sec) required to encode the audio signal.

Solution. (*a*) The bandwidth is 15KHz. The Nyquist rate is 30 KHz.

(*b*) 65,536 = 2^{16}, so that 16 binary digits are needed to encoded each sample.

(*c*) 30,000 × 16 = 480000 bits/sec.

Example 18. *A TV signal (video and audio) has a bandwidth of 4.5 MHz. This signal is sampled quantized and binary coded to obtain a PCM signal.*

(a) Determine the sampling rate if the signal is to be sampled at a rate 20% above the Nyquist rate.

(b) If the samples are quantized into 1024 level determine the number of binary pulsed required to encode each sample.

(c) Determine the binary pulse rate (bits per sec) of the binary coded signal and the minimum bandwidth required to transmit this signal.

Solution. (*a*) The Nyquist rate is $2 \times 4.5 \times 10^6 = 9$ MHz. The actual sampling rate $1.2 \times 9 = 10.8$ MHz

(*b*) $1024 = 2^{10}$ so that 10 bits or binary pulses are need to encode each sample

(*c*) $10.8 \times 10^6 \times 10 = 108 \times 10^6$ or 108 M bits/sec.

Example 19. *Bandwidth of the input to pulse code modulator is restricted to 4 KHz. The input varies from = 3.8V and has the average power of 20 mW. The required signal to quantization noise power ratio is 20 dB. The modulator produces binary output. Assume uniform quantization.*

(i) Calculate the number of bit required per sample

(ii) Output of 30 such PCM codes are time multiplexed. What is the minimum required transmission bandwidth for the multiplexed signal.

Solution. Bandwidth = 4kHz input–input = – 3.8 to 3.8

Average power = 30 m watts

(*i*) $$\frac{S_0}{N_0} = 3\left(2^{2n}\right)\left(\frac{\overline{m^2}}{m_1^2}\right)$$

$$100 = 3 \times 2^{2n} \times \frac{\left(30 \times 10^{-3}\right)^2}{(3.8)^2} \text{ or } 2^n = \frac{100\,(3.8)^2}{3 \times \left(3 \times 10^{-2}\right)^2} = 126.7$$

Number of bits $n = 7$

(*ii*) $n = 7$ for time division multiplexing

Given 30 PCM signal minimum required bandwidth $= 4 \times 30$ KHz $= 120$ MHz.

Example 20. *In a binary PCM system, the output signal-to-quantizing-noise ratio is to be held to a minimum of 40 dB. Determine the number of required levels, and find the corresponding output signal-to-quantizing-noise ratio.*

Solution. In a binary PCM system $L = 2^n$, where n is the number of binary digits.

Then $$\left(\frac{S}{N_q}\right)_{0\,dB} = 10 \log \left(\frac{S}{N_q}\right)_0 = 1.76 + 20 \log L \text{ becomes}$$

$$\left(\frac{S}{N_q}\right)_{0\,dB} = 1.76 + 20 \log 2^n = 1.76 + 6.02n \text{ dB} \qquad \ldots(i)$$

Now $$\left(\frac{S}{N_q}\right)_{0\,dB} = 40 \text{ dB} \rightarrow \left(\frac{S}{N_q}\right)_0 = 10000$$

Thus, from equation$(SNR)_0 = \left(\frac{S}{N_q}\right)_0 = \frac{3}{2} L^2$

$$L = \sqrt{\frac{2}{3}\left(\frac{S}{N_q}\right)_0} = \sqrt{\frac{2}{3}(10000)} = [81.6] = 82$$

and the number of binary digits n is $n = [\log_2 82] = [6.36] = 7$

Then the number of levels required is $L = 2^7 = 128$, and the corresponding output signal-to-quantizing-noise ratio is

$$\left(\frac{S}{N_q}\right)_{0\,dB} = 1.76 + 6.02 \times 7 = 43.9 \text{ dB} \qquad \ldots(ii)$$

NOTE Equation (*ii*) indicates that each bit in the node word of a binary PCM system contributes 3 dB to the output signal-to-quantizing-noise ratio. This is called the 6 dB rule.

Example 21. *A compact disk (CD) recording system samples each of two stereo signals with a 16-bit analog-to-digial coverter (ADC) at 44.1 kb/s.*

(*i*) *Determine the output signal-to-quantizing-noise ratio for a full-scale sinusoid.*

(*ii*) *The bit stream of digitized data is augmented by the addition of error-correction bits, clock extraction bits, and display and control bit fields. These additional bits represent 100 percent overhead. Determine the output bit rate of the CD recording system.*

(*iii*) *The CD can record an hour's worth of music. Determine the number of bits recorded on a CD.*

(*iv*) *For a comparison, a high-grade collegiate dictionary may contain 1500 pages, 2 columns per page, 100 lines per column, 8 words per line, 6 letters per words, and 7b per letter on average. Determine the number of bits required to describe the dictionary, and estimate the number of comparable books that can be stored on a CD.* [Raj. Univ. 1990, 1993, 1999, 2003]

Solution. (*i*) From equation $\left(\frac{S}{N_q}\right)_{0\,dB} = 10 \log\left(\frac{S}{N_q}\right)_0 = 1.76 + 20 \log L$

$$\left(\frac{S}{N_q}\right)_{0\,dB} = 1.76 + 6.02 \times 16 = 98.08\ dB$$

The very high SNR of the disk has the effect of increasing the dynamic range of recording, resulting in the excellent clarity of sound from a CD.

(*ii*) The input bit rate is $2.(44.1)\,(10^3)\,(16) = 1.411(10^6)$ b/s = 1.411 Mb/s

Including the additional 100 percent overhead, the output bit rate is

$$2(1.411)\,(10^6)\text{ b/s} = 2.822\text{ Mb/s}$$

(*iii*) The number of bits recorded on a CD is

$$2.822(10^6)\,(3600) = 10.16(10^9)\,b = 10.16\text{ gigabits (GB)}$$

(*iv*) The number of bits required to describe the disctionary is

$$1500(2)(100)(8)(6)(7) = 100.8(10^6)\,b = 100.8\text{ Mb}$$

Including the additional 100 percent overhead, then, $\dfrac{10.16\,(10^9)}{2\,(100.8)\,(10^6)} = 50.4$

Thus a disk contains the equivalent of about 50 comparable-books storage capacity.

Example 22. *A PCM encoder accepts a signal with a full scale voltage of 10 volts and generates 8-bit codes using uniform quantization. The maximum normalized quantized voltage is* $1 - 2^{-8}$.

Determine :

(*i*) *Normal step size* (*ii*) *Actual step size in volts*

(*iii*) *Actual maximum quantized levels in volts.* (*iv*) *Actual resolution.* [Raj. Univ. 2003]

Solution. Given in PCM encoder. Full scale voltage = 10 volt and 8-bit coder using uniform quantization.

(*i*) *Normal step size*

A total of 2^8 quantization level are possible, so the normalized step size is $2^{-8} = 0.00390625$

(*ii*) *Actual step size in volts*

$$= 10\text{ volts} \times 2^{-8} = 0.0390625$$

(*iii*) *Actual maximum quantized level in volts*

Given to us maximum normalized quantized voltage is $1 - 2^{-8} = 0.9961$. Thus the actual maximum quantized voltage is

$$0.9961 \times 10V = 9.961\text{ Volts.}$$

(*iv*) *Actual resolution*

The normalized step size is 2^{-8}. The maximum error that can occur is one half the step size. Therefore, the normalized resolution is

$$\pm\frac{1}{2}\times 2^{-8} = \pm 0.01953$$

So actual resolution is

$$\pm 0.01953 \times 10 = \pm 0.1953.$$

Example 23. *Consider an audio signal with spectral components in the range 300 to 3000 Hz. Assume that a sampling rate of 7000 samples per second will be used to generate PCM signals :*

(*i*) *For SNR = 30 dB, what is the number of quantization levels needed ?*

(*ii*) *What data rate is required ?* [Raj. Univ. 2002]

Solution. Given here

$$\text{SNR} = 30\text{ dB and sample rate} = 7\text{KHz}$$

we know that $$\left(\frac{S}{N}\right)_{dB} = 1.76 + 6.02\, n$$

$$30 = 1.76 + 6.02n, \quad n = 4.69$$

as n can only be an integer we will take upper value for n $\therefore$ $n = 5$

(*i*) So that number of quantization levels is $2^5 = 32$

(*ii*) Since each sample is represented by 5 bits

$\therefore$ 7000 samples be represented by 7000×5 bits/sec $= 35 \times 10^3 = 35$ kbps.

3.3 DIFFERENTIAL PULSE CODE MODULATION (DPCM)

We have considered the standard PCM, each sample of base band signal is encoded independently of all other.

Now consider the use of PCM for the digitalization of a voice 'or' video signal, the signal is sampled at a rate slightly higher than Nyquist rate. The resulting sampled signal is then found to exhibit a high correlation between adjecent samples. Due to high co-relation between the samples, the signal doesn't change rapidly from one sample to the next with the result that the difference between adjecent samples has a variance that is smaller than the variance of the signal itself. Thus we can say that if we encode these signal output directly, the resulting encoded signal contains redundant information (the repeated information about the signal in the successive digital output). So these symbols (codes) are not required to transmit. So we can remove these redundant bits before encoding and may get a appropriate coding technique.

Now if we think about some practical solution of this problem, a very easy solution of this problem is now to encode the difference between successive samples rather than samples itself. Due to this process we will find the following solution.

(*a*) As now we are encoding the difference between two successive samples so the redundant information is automatically reduced; the samples having same information will produce zero output. If two samples have (averagely) same information nearly. Now instead of two samples only one sample have to be transmitted.

(*b*) Now if we encode the difference of two successive level; then the total number of levels is just reduced very much; so now the required bits to encode them will be reduced.

So "such type of coding; in which instead of samples; the difference of two successive samples are coded in PCM is called DPCM."

3.3.1 Generation of DPCM Making Comparison at Input of Sample and Hold

The Basic Block diagram (Fig. 3.6(*a*)) has been drawn here. Here we use to generate the differential signal at the sample and hold stage. To generate the difference signal between the present value $m(t)$ and its previous value $\hat{m}(t)$. For that purpose we have used a accumulator at the feedback circuit that will simply add up the previous values so that now we have the value at the output of accumulator that is directly equal to the value till now coded that is $\hat{m}(t)$ for present input $m(t)$. Thus the work of accumulator is to add up all previous sampled values to generate a signal that represent the value of complete signal sampled till now.

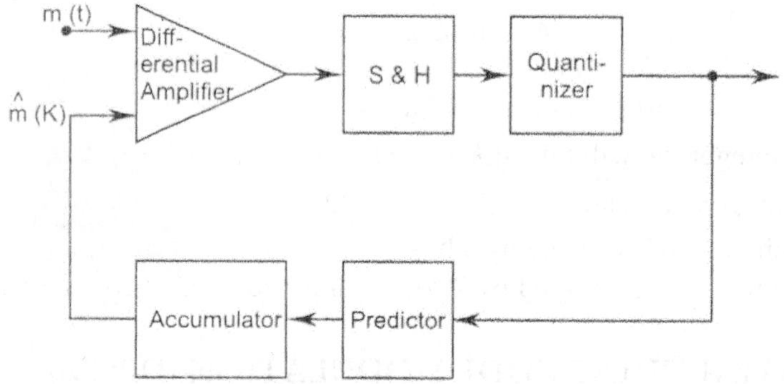

Fig 3.6 *(a) At S & H stage*

Now the value of present signal and the previous signal (output of accumulator) are fed to differential Amplifier. Thus it should be noted that "we are not sending the change in value of $m(t)$ with respect to its previous value, but we sending the difference between $m(t)$ and $\hat{m}(t)$.

Till now we have not discussed the use of predictor. Predictor is used to improve the performance of system; in predictor we have used some algorithm that will provide us. To Predict the next required increment on the basis of all previous stored sample.

Thus by increasing the sampling rate higher than Nyquist rate and a predict filter, DPCM can produce practically one half of the Bit Rate of PCM. That will help use to save spectral space.

At the receiver side; we simply have a predictor and an accumulator (that were used in feedback network to convert a digital output again in analog form). But the analog output of accumulator is not so smooth so we pass it through a low pass filter (Fig. 3.6(*b*)).

Fig 3.6 *(b) DPCM demodulation*

3.3.2 Generation of DPCM by Making Comparison at Input of Quantinizer

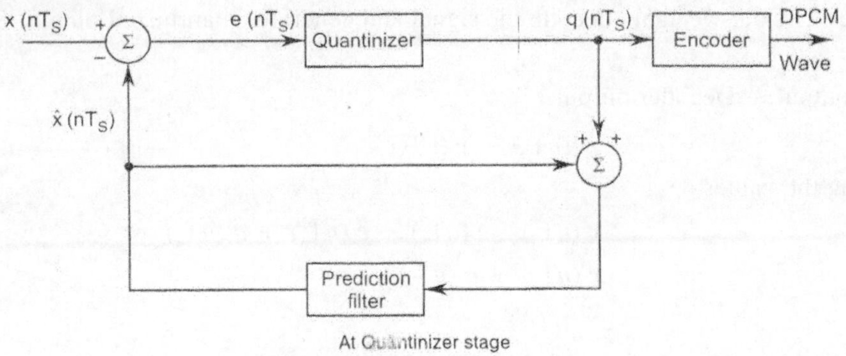

Fig 3.6 *(c) DPCM generation*

Now we have drawn here a circuit (Fig. 3.6(*c*)) that performs the generation of difference signal at Quantinization level. It is simple to see that at first method; we have adopted analog technique to generate a difference signal; but now difference signal is generated at digital (discrete) state.

Is this scheme the input of Quantinizer

$$e\,(nT_s) = x\,(nT_s) - \hat{x}\,(nT_s) \tag{3.24}$$

Thus this is the difference between the present value $x\,(nT_s)$ and its previous value $\hat{x}\,(nT_s)$ predicted by predict filter so it difference also denotes the amount by which predictor fails to predict the input exactly.

The Quantinizer output is denoted by

$$v\,(nT_s) = e\,(nT_s) + q\,(nT_s) \tag{3.25}$$

$$q\,(nT_s) = \text{Quantinization error.}$$

The input value of predict filter

$$u\,(nT_s) = \hat{x}\,(nT_s) + v\,(nT_s)$$

$$= \hat{x}\,(nT_s) + e\,(nT_s) + q\,(nT_s)$$

$$\boxed{u\,(nT_s) = x\,(nT_s) + q\,(nT_s)} \tag{3.26}$$

Thus the input of predict filter is always a Quantinized version of signal $x\,(t)$. So we can say that predictor is a device that predict about the signal (denoted by $\hat{x}\,(nT_s)$ one the basis of present signal $x\,(nT_s)$ and quantinization error and if we again consider the equation.

$$\text{Input of predict filter} = x\,(nT_s) + q\,(nT_s)$$

$$\text{Output of predict filter} = x\,(nT_s) - e\,(nT_s)$$

Thus the error signal generated [$e\,(nT_s)$] is simply showing the variation of difference between signal and its quantinized value called **Quantinization error [$q\,(nT_s)$]**. So if we follow the standards and produces a good quantinized output with very low **quantinization error**, the signal generated [$e\,(nT_s)$] will be very small and we get the most exact and appropriate coding because now the $\hat{m}\,(t)$

and m (t) are very much correlate to each other. Now the Quantinized output is encoded and transmitted.

At the receiver side; we again decode the signal and generate Quantinized output. Simply see that the output we get

Predictor output + Decoder output

$$\hat{x}(nT_s) + v(nT_s) \tag{3.27}$$

Substituting the values-

$$x(nT_s) - e(nT_s) + e(nT_s) + q(nT_s)$$

$$\Rightarrow \quad x(nT_s) + q(nT_s)$$

$$= u(nT_s) \tag{3.28}$$

Thus we have generated a quantinized output of signal that is [x (nT_s)]; it again be very much correct if the value of Quantinization error is very small.

Quantinized output [x (nT_s)] is nothing but an approximated signal (but not smooth) to get better output this signal can be fasted through low pass filter; but in general; we use oversampling (sampling higher than Nyquist rate) so the quantinization error is very less so output signal is very much real to the signal itself so we don't need any low pass filter (Fig. 3.6(d)).

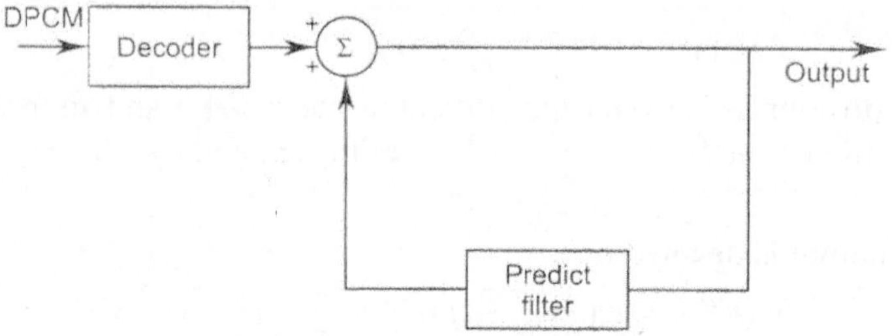

Fig 3.6 *(d) DPCM demodulator*

3.3.3 Summary of DPCM

(i) Sampling rate is always preferred greater than Nyquist rate (called oversampling).

(ii) We used to transmit the (encoded version) difference between the present level of signal to its previous value of signal, we are not sending the variation of signal m (t) only.

(iii) Due to oversampling, it is possible to predict the next value of signal because due to oversampling the correlation between the successive samples of signal is very high and good. So to see the present value of signal we can predict the next value of signal.

This prediction will be good if the present input of signal is pure and good (without Quantinizing error) and if prediction is good the coding will be more accurate.

3.3.4 Advantages of DPCM

(i) Now very less number of levels are required to encode: 8 level PCM is converted into 4 level DPCM; so instead of 3 Bits now we require only 2 Bits for coding. So the required bandwidth is reduced.

(*ii*) Now we see that designing of system is very easy. At the output we don't need any digital to analog conversion as in the case of PCM.

(*iii*) Now the co-relation between successive are very good; so the output samples is far better than that of PCM.

3.3.5 Disadvantages of DPCM

We directly cannot say it the disadvantage of DPCM. It only show the scope of further improvement.

Here we have to generate a positive step size 'or' Negative step size for difference signal. Now it is also possible that now instead of sending the difference value; we will send only the variation of input signal mean to say that now we can send a pulse for any variation and number pulse for number variation of levels of signal because our basic aim is to encode the message signal in digital pulses (o or' 1) form and at receiver side to sense them then we may skip the need of predict filter that is not easy to design and also a cause of error in system. This problem is eliminated in the case of delta modulation.

3.4 DELTA MODULATION

Delta modulation is a special case of DPCM in which the difference signal is encoded into just a single bit.

So we can explain some points on delta modulation;

(*i*) Now we sample the signal at the rate higher than Nyquist rate. Due to this over sampling; now the samples will be more per second. Due to a large number of samples per second; we get better correlation between successive samples.

(*ii*) But now as the number of samples are increased again adopt differential coding technique as discussed in DPCM to reduce the redundant information.

(*iii*) But now represent each samples by a single bit.

In its basic form DM provides a staircase approximation to the over sampled version of an input baseband signal as shown in figure.

The difference between the input and staircase approximation is quantinized in two only levels $\pm\,\delta$, corresponding to positive and negative differences respectively. Thus if the approximation falls below the signal at any sampling epoch, it is increased by δ if, on the other hand, the approximation lies above the signal it is diminished by δ . Provided that the signal does not change too rapidly from sample to sample; we find that the staircase approximation remained with in $\pm\,\delta$ of the input signal.

Note that Δ denotes the absolute value of the two representation levels of the one bit quantinizer used in DM. These two levels are indicated in the transfer characteristics of figure. The step size of the quantinizer is therefore related to Δ by;

$$\Delta = 2\delta \tag{3.29}$$

The basic technique discussed till now is also called **linear Delta Modulation**. [Fig. 3.7(*a*).]

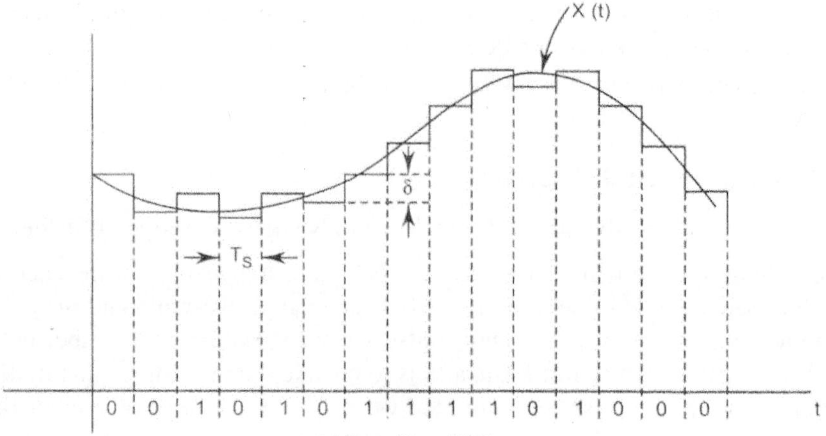

(a) Encoding of DM

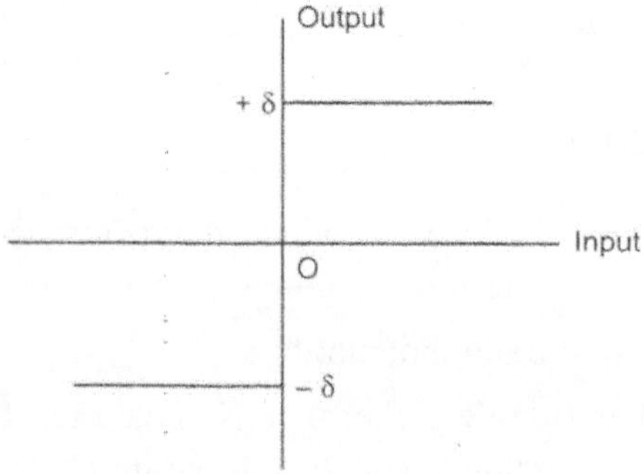

(b) Two level quantinizer

Fig. 3.7 *Delta modulation*

3.4.1 Generation and Detection of Linear Delta Modulation

(*a*) By the comparison at sampling stage

One possible method of generation of delta modulation has been shown these now comparator compare the present signal with the quantinized output (Fig. 3.8(*a*)).

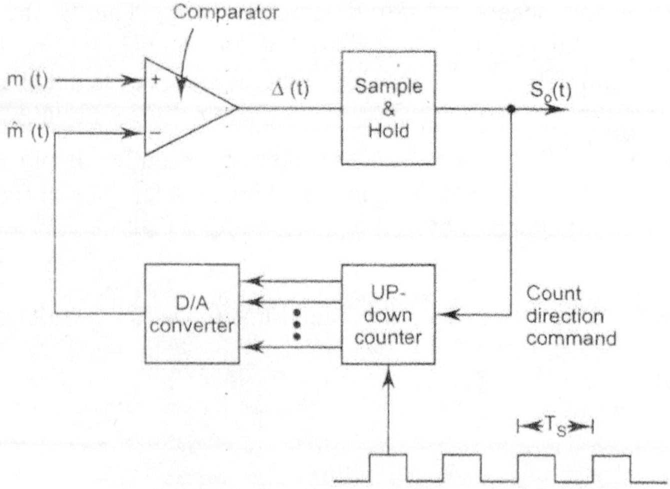

Fig. 3.8 *(a) At S & H input*

As the Quantinized output is in digital form, so it should be converted into analog signal to compare with the message signal.

Now comparator as the name suggests, compare the two applied signal; if the value of quantinized signal is less than the present value of signal; the output of comparator is – δ.

If the value of message signal is higher than the quantinized output we get at output + δ.

Here we don't require any value to be predict as in the case of DPCM, we require only to know whether the next signal level is larger 'or' smaller than the quantinized output.

Here we see that we have introduced **up-down counter;** the frequency of up-down counter should be synchronized with the sampling frequency. The count direction means (increment in the count 'or' decrement of count) is determined by the voltage levels at the "count direction command" input of counter. When the output is at high level (increased by δ), the counter counts up; when the output is at low level (decreased by δ), the counter count downs.

Thus here counter acts as a accumulator that adds and substracts as directed and store. The digital output of counter is converted to the analog signal by Digital to Analog converter.

***(b)* Generation of DM by comparing the signal at quantinization level**

Denote the input signal by $x(t)$ and the staircase approximation to it as $u(t)$. Thus now we find the following relations for delta modulation (Fig. 3.8(*b*)).

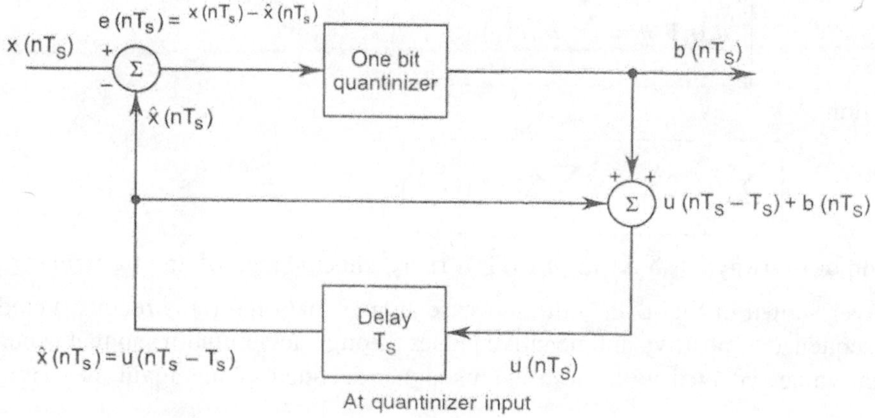

Fig. 3.8 *(b) DM generation*

(*a*) $$e(nT_s) = x(nT_s) - \hat{x}(nT_s) \quad (3.30)$$

But $\hat{x}(nT_s) =$ Delayed version of staircase output.

If the present value of staircase output

$$= u(nT_s) \quad (3.31)$$

Then delayed version $= u(nT_s - T_s)$.

So $$\boxed{e(nT_s) = x(nT_s) - u(nT_s - T_s)} \quad (3.32)$$

Thus the signal generated by the comparison of present value of signal with previous value of sample.

(*b*) Now one bit quantinizer produces only + δ and – δ. Thus the output of one bit quantinizer.

$$\boxed{b(nT_s) = \delta\, Sgn\,[e(nT_s)]} \quad (3.33)$$

Simply $Sgn\ [e\ (nT_s)]$ represents the sign of $e\ (nT_s)$ it would be + ve for + δ and it would be – ve for – δ.

(*c*) Input for the delayed circuit is

$$\boxed{u\ (nT_s) = u\ (nT_s - T_s) + b\ (nT_s)} \tag{3.34}$$

That is generated by just adding the previous value of staircase with the $\pm\ \delta$.

Now discussed, above solve for we assume that accumulator is initially set to zero;

So $\quad u\ (nT_s - T_s) = 0$ so

$$\left.\begin{aligned} e\ (nT_s) &= x\ (nT_s) \\ u\ (nT_s) &= b\ (nT_s) \end{aligned}\right\} \tag{3.35}$$

and

So output $\quad b\ (nT_s) = \delta\ Sgn\ [e\ (nT_s)]$

Now substitute values.

$$u\ (nT_s) = \delta\ Sgn\ [e\ (nT_s)]$$

Now if we take for an *n* number of steps

$$\boxed{\begin{aligned} u\ (nT_s) &= \delta \sum_{i=0}^{n} Sgn\ e\ (iT_s) \\ u\ (nT_s) &= \sum_{i=0}^{n} b\ (iT_s) \end{aligned}} \tag{3.36}$$

Now use equation

$$\boxed{\sum_{i=0}^{n} b\ (iT_s) = \delta \sum_{i=0}^{n} Sgn\ e\ (iT_s)} \tag{3.37}$$

Thus the output is always $\pm\ \delta$ As $[Sgn\ e\ (iT_s)]$ represents the sign of any particular sample.

In the receiver shown in Fig. 3.8(*c*) the staircase approximation $u\ (t)$ is reconstructed by passing the incoming sequence of positive and negative pulses through accumulator: so that accumulator will add the present value received with the previous signal decoded. Thus again the original signal is recovered.

The out off Band quantinization noise in the high-frequency staircase waveform $u\ (t)$ is rejected by passing it through a low-pass filter with a bandwidth equal to the original signal bandwidth.

3.4.2 Quantinization Noise in Delta Modulation

In the delta modulation, the two parameters plays an important role as we have discussed: the step size and the sampling frequency.

These parameter must be choosen very cleverly otherwise we may have two major sources of quantinization noise in DM systems.

(*i*) Slope over load distortion and (*ii*) Gangular noise.

If the rate of change of quantinization is slow in comparison to the slope (rate of change) of message signal, it is called **slope overload**.

If the rate of change of quantinization is fast in comparison to the slope of signal then it is called Granular noise. Granular noise is the noise that results because of approximating the analog signal by step like waveform.

The desired condition to avoid these conditions, the mathematical explanation has been discussed in the example. Due to both these problems, the signal and the approximation don't interact so the quantinization of signal is not achieved. (Fig. 3.8(*c*)).

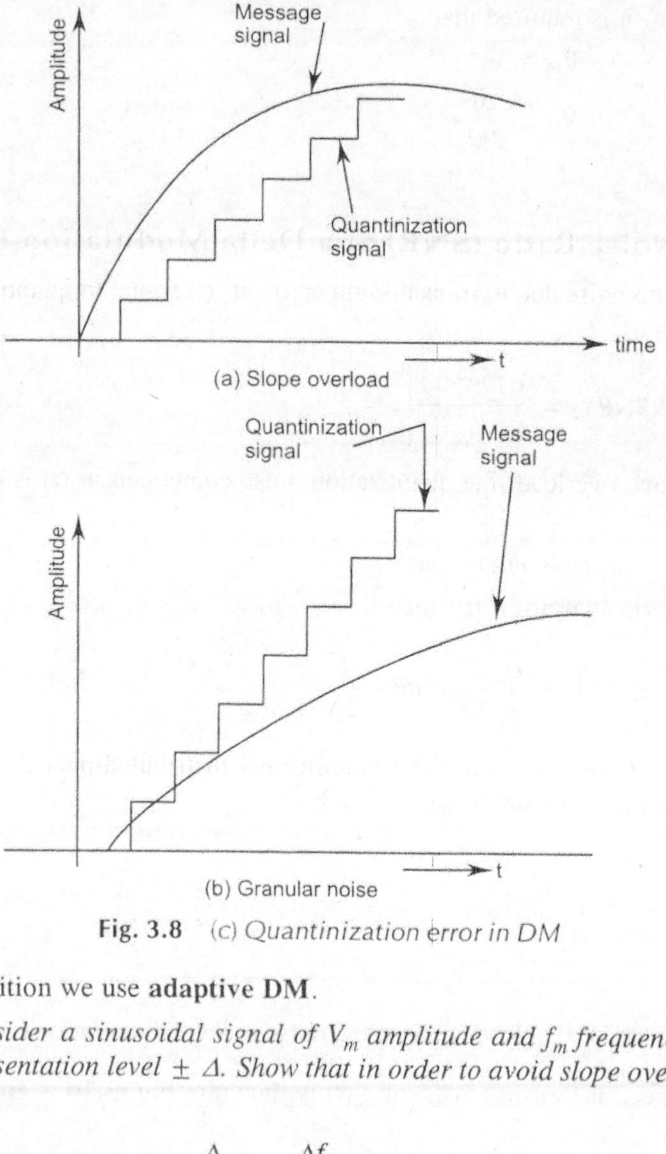

Fig. 3.8 *(c) Quantinization error in DM*

To avoid this condition we use **adaptive DM**.

Example 24. *Consider a sinusoidal signal of* V_m *amplitude and* f_m *frequency applied to a delta modulator with representation level* $\pm \Delta$. *Show that in order to avoid slope overload distortion, it is necessary that*

$$V_m \leq \frac{\Delta}{2\pi f_m T_s} \leq \frac{\Delta f_s}{2\pi f_m}$$

where T_s = *sampling period.*

Solution. We can say that signal is

$$x(t) = V_m \sin 2\pi f_m t$$

The maximum slope of delta modulation may be given as-

$$\text{Maximum slope} = \frac{\text{Step Size}}{\text{Sampling period}} = \frac{\Delta}{T_s}$$

But Maximum slope $= \left| \frac{dx(t)}{dt} \right|$ max. $= V_m \cdot 2\pi f_m$

To avoid slope overload, it is required that

$$V_m \cdot 2\pi f_m \leq \Delta f_s$$

$$V_m \leq \frac{\Delta f_s}{2\pi f_m}$$

To avoid slope overloading.

3.4.3 Signal to Noise Ratio (SNR) of a Delta Modulation System

Assuming that there is no noise due to transmission error, $n_o(t)$ signal to quantization noise ratio of the DM system is given by

$$(\text{SNR})_o = \frac{E\left[m^2(t)\right]}{E\left[m_q^2(t)\right]} \quad (3.38)$$

In the absence of slope overload, the quantization noise component $n_q(t)$ is the response of base band filter to $e_q(t)$ where

$$e_q(t) = m(t) - \hat{m}(t)$$

If we assume a uniform PDF for $e_q(t)$, then

$$E\left[e_q^2(t)\right] = \int_{-\Delta}^{\Delta} \frac{1}{2\Delta} e^2 de = \frac{1}{2\Delta} \frac{e^3}{3} \Bigg|_{-\Delta}^{\Delta} = \frac{\Delta^3}{3} \quad (3.39)$$

The normalized power of the noise waveform is uniformly distributed over the interval 0 to $\pm \Delta$ as shown in Fig. 3.9 where Δ is the step size.

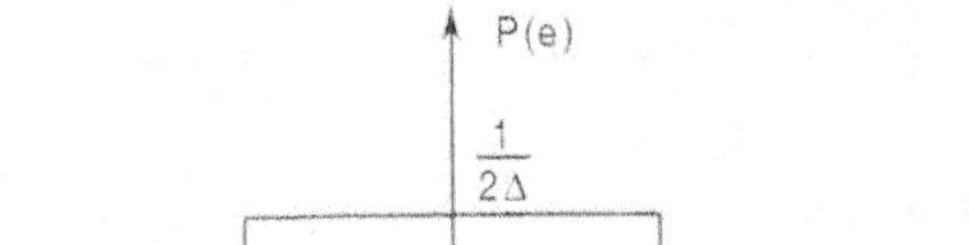

Fig. 3.9 *PDF of quantizing noise in delta modulation*

The PSD (Power Spectral Density) $G_q(f)$ of $e_q(t)$ is shown in Fig. 3.10 is given by

$$G_q(t) = \begin{cases} \frac{\Delta^2/3}{2f_s} = \frac{\Delta^2}{6f_s} & -f_s \leq |f| < +f_s \\ 0 & \text{otherwise} \end{cases}$$

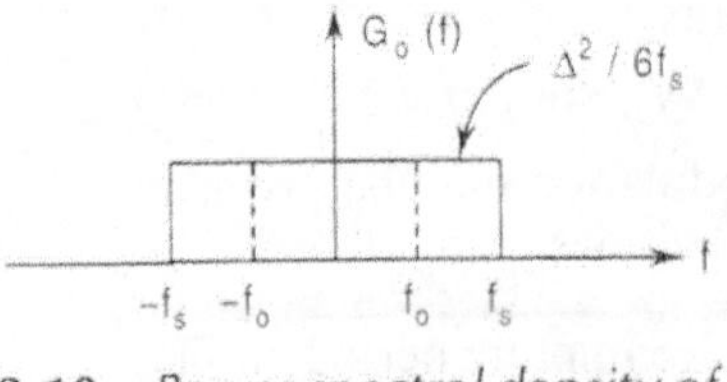

Fig. 3.10 *Power spectral density of $e_q(t)$*

Since $n_q(t)$ is the response of base band filter to $e_q(t)$. The normalized power of the waveform $n_q(t)$ is given by

$$E\left[n_q^2(t)\right] = \int_{-f_o}^{f_o} G_q(f)\, df = \int_{-f_o}^{f_o} \frac{\Delta^2}{6 f_s}\, df = \frac{\Delta^2}{6 f_s} . 2 f_o = \frac{\Delta^2}{3}\left(\frac{f_o}{f_s}\right) \quad (3.40)$$

For the signal $m(t) = v \cos 2\pi f_o t$, the average power of the signal is

$$E\,[m^2(t)] = \frac{V^2}{2}$$

To avoid slope overload, the maximum value of V_m is given by

$$V_m = \frac{\Delta f_s}{2\pi f_o}$$

Signal to noise Ratio, SMR is therefore given by

$$SMR = \frac{E\left[m^2(t)\right]}{E\left[n_q^2(t)\right]} = \frac{\frac{V^2}{2}}{\frac{V^2}{3}\left(\frac{f_o}{f_s}\right)} = \frac{\Delta^2 f_s^2}{8\pi^2 f_o^2} \times \frac{3}{\Delta^2}\left(\frac{f_s}{f_o}\right) \quad (3.41)$$

Therefore,

$$SMR = \frac{3}{8\pi^2}\left(\frac{f_s}{f_o}\right)^3 \quad (3.42)$$

$$SMR \propto f_s^3 .$$

3.5 ADAPTIVE DELTA MODULATION

As we have discussed above there are two types of problems called **slope overload** and **Granular noise**.

So the performance of a delta modulation can be improved significantly by making the step size of the modulator assume a time-varying form. In particular, when input signal is varying fast (called slope overload), the step size must be increased. When signal is varying low (called granular noise), the step size must be reduced. So in this way, the step size is adapted to the level of the input signal. So resulting method is called adaptive delta modulation (ADM).

These are several types of ADM, depending on the type of scheme used for adjusting the step size.

(*a*) In one type, a discrete set of values is provided for the step size.

(*b*) A continuous range for step size variation is provided.

Fig. 3.8 (*b*) shows the block diagram of transmitter and receiver. Basically we need the circuit to perform the operation of variations in step size. In Fig. 3.9 (*b*) we have used a microprocessor that generate a step size that is basic step size. Now if the value of present signal $m(t)$ is increasing then microprocessor provides an addition in the stored step size. If the value of present signal $m(t)$ is decreasing then microprocessor provides an substraction to the stored step size.

The physical realization of concept of adaptive delta modulation is shown here.

Here the step size $\Delta(nT_s)$ 'or' $2\,\delta\,(nT_s)$ is limited in the range of a choosen maximum and minimum values; we may write

$$\delta_{min} \leq \delta\,(nT_s) \leq \delta_{max} \quad (3.43)$$

The upper limit δ_{max} is choosen in such a way that it should be able to control the slope-overload distortion. The lower limit is chosen in such a way that it should be able to control the granular noise.

At the receiver side; the inverse process of step size variation is involved first and after that the process is same as adapted in the case of DPCM.

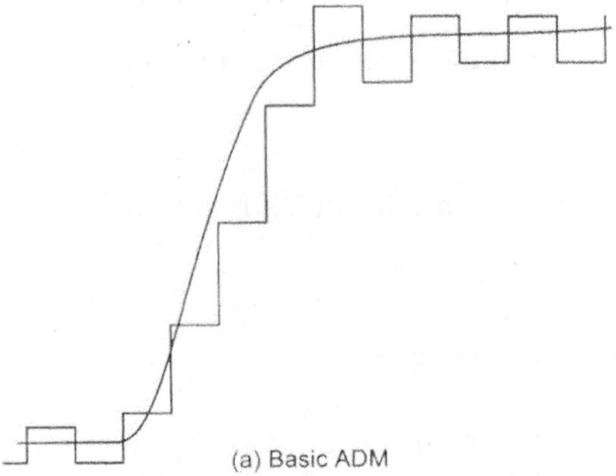

(a) Basic ADM

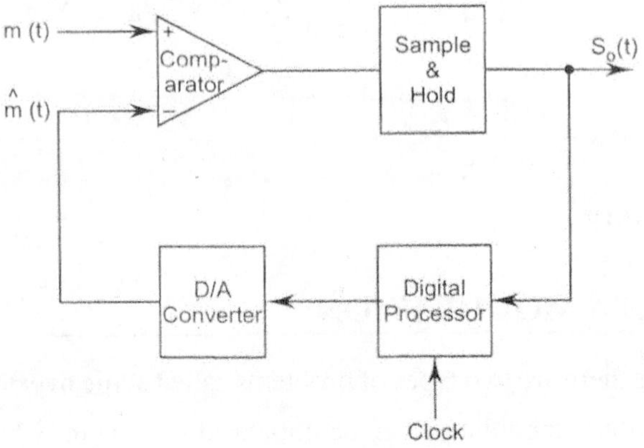

(b) Block diagram

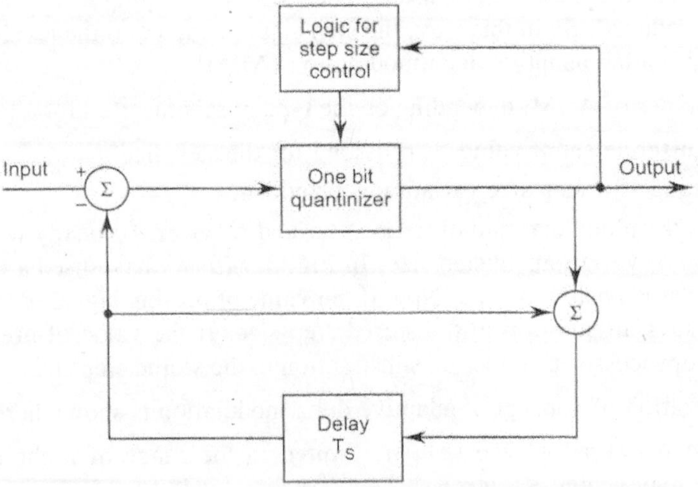

(c) Realization

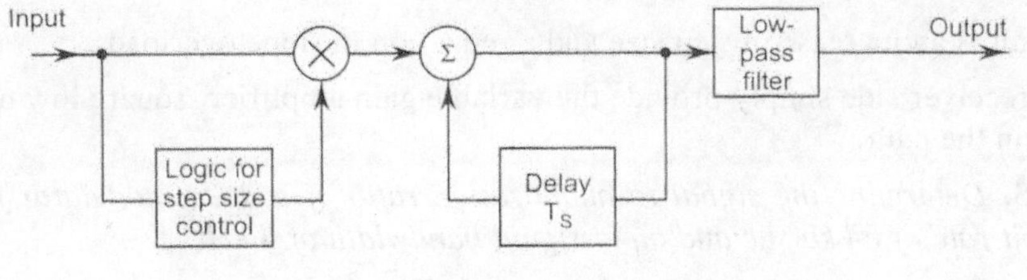

(d) DM demodulation

Fig. 3.11. *ADM generation*

3.6 CONTINUOUSLY VARIABLE SLOPE DELTA MODULATION (CVSD)

ADM system that adjust the step size to any given value is shown. The amplifier has a variable gain. That is, its gain is a function of the voltage applied at its gain control terminal. We assume the characteristics of the amplifier such that when gain control voltage is zero, its gain is low, and the gain increases with increasing positive gain control voltage.

The resistor-capacitor combination serves as an intregrator, the voltage across capacitor being proportional to the applied signal (V_0). Now this voltage is used to control the gain of amplifier. The square device make sure that the input at the amplifier is always positive.

Assume now that *m* (*t*) is making only small variations; so that output so will have very small value and consists of alternate polarity pulses. These pulses, when Integrated, yield an average output of almost zero. So the output of gain-control input is, hence, almost zero, the consequently the step size is reduced.

Now consider the case of slope overload. If *m* (*t*) increases positively 'or' negatively at too rapid a rate that quantinization approximation cannot follow then the output V_0 is then a train of all positive 'or' all negative pulses. The integrator averages and provides a large voltage to increase the gain of amplifier. Because of squaring circuit, the amplifier gain will increase whether the pulses at output are negative 'or' positive.

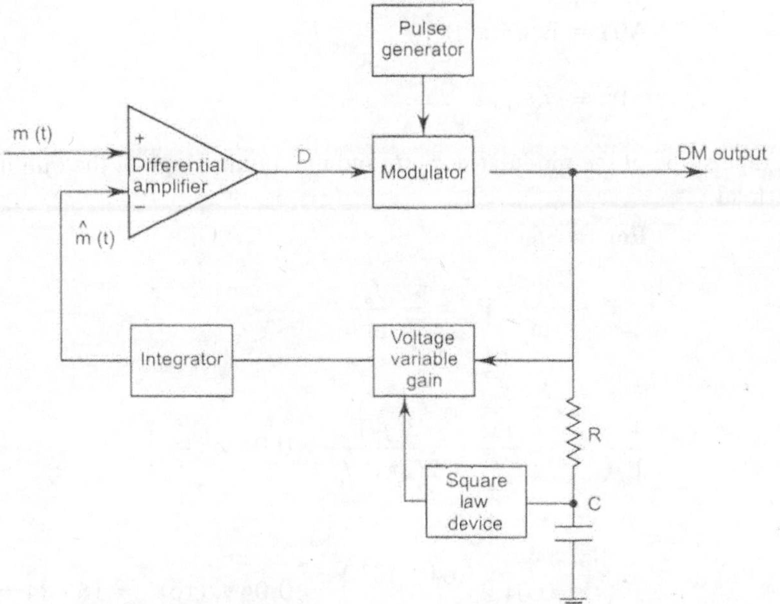

Fig. 3.12 *ADM technique*

The end result is an increase in step size and a reduction in slope overload.

Now at the receiver side simply provide the variable gain amplifier, square low device, RC filter and integrator in the path.

Example 25. *Determine the signal to quantization ratio of a delta modulator for a sinusoidal signal with a bit rate of 64 kb/sec and input signal bandwidth of 4 kHz.*

Solution. Let A (t) represent signal

A (t) represent quantized signal

$$\delta(t) = A(t) - A'(t)$$

Let $\quad S = \text{step size } f(\delta) = \frac{1}{2s} \quad \text{where } -s < \delta < +s$

$$P_q = [\delta(t)]$$

$$P_q = \int_{-s}^{s} \delta^2 f(\delta)\, d\delta;\ P_q = \int_{-s}^{s} \frac{d^2}{2s}\, d\delta = \frac{s^2}{3}$$

Now BP filter at the output of **delta modulator** will eliminate substantial amount of the noise power provided cut off frequency (f_c) of the filter is equal to $f_c = \frac{1}{\tau}$ where τ is the duration of bit.

$$f_c = \frac{1}{\tau}$$

so at filter output $\quad P_{q0} = \frac{s^2}{3} \times \frac{f_c}{f_b}$

f_c = cutoff frequency of LPF and f_b = bit rate

Signal Power : Let us consider sine wave modulator

$$A(t) = E \sin \omega_C t$$

$$P_{s_0} = \bar{A}^2(t) = \frac{E^2}{2}$$

Since the average slope of the modulator = sf_b and maximum slope of the output signal = $E.\omega_c$ to avoid slope overload

$$E\omega_c = Sf_b$$

$$E = \frac{Sf_b}{\omega_c};\ P_{q_0} = \frac{s^2}{2}\frac{f_b^2}{\omega_c^2}$$

$$\frac{P_{s_0}}{P_{q_0}} = \frac{\frac{s^2}{2}\frac{f_b^2}{\omega_c^2}}{\frac{s^2}{3}\frac{f_c}{f_b}} = \frac{3(f_b)^3}{2(2\pi)^2 f_c^3} = 0.04 \times \left(\frac{f_b}{f_c}\right)^3$$

$$= 0.04 \times \left(\frac{64 \times 10^3}{4 \times 10^3}\right)^3 = 0.04 \times (16)^3 = 163.84 = 22 \text{ dB}$$

Example 26. *The pulse rate in a DM system is 56,000 per sec. The input is 5 cos $(2\pi \times 1000\, t)$ + 2 cos $(2\pi \times 2000\, t)$. With t in sec. Find the minimum value of step size which will avoid slope overload distortion. What will be the disadvantages of choosing a value of larger than the minimum ?*

Solution. Given pulse rate 56,000 per sec

$$V_m = 5 \cos (2\pi \times 1000\, t) + 2 \cos (2\pi \times 2000\, t)$$

Find minimum step size or $sf_s > 2\pi fA$

or $$s \geq 2\pi \frac{f}{f_s} \times A \text{ or } s \geq 2\pi \left(\frac{1000}{56000}\right)^5 ; s \geq 2(\pi)\left(\frac{2000}{56000}\right) \times 2$$

or $$s \geq \frac{10\pi}{56} = 0.56v ; s \geq \frac{8\pi}{56} = 0.45v$$

s should not be less than 0.56 v if value is too larger than **granular noise**

Example 27. *A delta modulator system is designed to operate at five times the Nyquist rate for a signal with 3 kHz bandwidth. Determine the maximum amplitude of a 2 kHz input sinusoid for which the delta modulator does not have slope over load. Quantizing step size is 250 mV. Derive the formula that you use.*

Solution. The relation for slope overload distortion which will occur if, $A_m > \dfrac{\delta}{2\pi f_m T_s}$

∴ Slope overload will not occur when, $A_m \leq \dfrac{\delta}{2\pi f_m T_s}$

Maximum frequency in the signal will be W = 3 kHz

∴ Nyquist rate = 2W = 2 × 3 kHz = 6 kHz

Sampling frequency = 5 times Nyquist rate f_s = 5 × 6 kHz = 30 kHz

∴ Sampling interval $T_s = \dfrac{1}{f_s} = \dfrac{1}{30 \times 10^3}$

Step size $\delta = 250 \text{ mV} = 250 \times 10^{-3} = 0.25\text{V}$

Given $f_m = 2 \text{ kHz} = 2 \times 10^3 \text{ Hz}$

∴ Substituting these values in equation for A_m, we have

$$A_m \leq \frac{0.25}{2\pi \times 2 \times 10^3 \times \dfrac{1}{30 \times 10^3}} ; A_m \leq 0.6 \text{ volts which is obtained.}$$

Example 28. *A speech signal is transmitted using delta modulation. The repetition frequency is 56 kbps and the step size is 100 mV. Determine the permission amplitude, of the speech signal, that can be transmitted by the DM system avoiding over distortion.*

Solution. We know that the maximum permissible amplitude for the speech signal equal to the permissible for a sinusoidal signal at 800 Hz. Therefore, we have to evaluate maximum permissible amplitude for a 800 Hz signal. It is given that, f_n = 800 Hz,

$$f_s = 56000 \text{ pulses/second}, \quad E_0 = 100 \text{ mV}.$$

Putting these values in equation, we have

$$\frac{E_0 f_s}{2\pi f_n} = \frac{0.1 \times 56000}{2 \times 3.14 \times 800} = 1.1 \text{ volts.}$$

Example 29. *The pulse rate in DM system is 56,000 per second. The with t in sec. Find the minimum value of step size which will avoid slope overload distortion.*

Solution. Pulse gate given = 56,000 per sec.

$$V_{in} = 5\cos(2\pi\ 1000t) + 2\cos(2\pi\ 2000t) \text{ find minimum step size}$$

$$sf_s > 2\pi fA;\ S \geq 2\pi \frac{f}{f_s} A$$

$$S \geq 2\pi\left[\frac{1000}{56000}\right]5 \text{ or } S \geq 2\pi\left[\frac{1000}{56000}\right]2 \text{ or } S \geq \frac{10\pi}{56} = 0.56\text{V or } S \geq \frac{8\pi}{56} = 0.45\text{V}$$

So S should not be less than 0.56V

Example 30. *A delta modulator system is designed the operate at five times the Nyquist rate for a signal having a bandwidth equal to 3KHz bandwidth. Calculate the maximum amplitude of a 2KHz input sinusoidal for which the delta modulators does not have slope over load. Given that the quantization step size is 250mV.*

Solution. The relation for slope overload distortion which will occur if

$$A_m \geq \frac{\delta}{2\pi f_m T_S} \qquad \ldots(1)$$

Thus, slope overload will not occur if

$$A_m \leq \frac{\delta}{2\pi f_m T_S} \qquad \ldots(2)$$

The maximum frequency in the signal is

$$f_m = 3\text{KHz}$$

Hence Nyquist rate $= 2f_m = 2 \times 3\text{ KHz} = 6\text{ KHz}$

Sampling frequency = 5 times Nyquist rate

$$f_S = 5 \times 6\text{ KHz or } f_S = 30\text{ KHz}$$

Hence sampling interval $T_S = \frac{1}{f_S} = \frac{1}{30 \times 30^3}$ sec

Given that step size $\delta = 250\text{mV} = 250 \times 10^{-3}\text{V}$

$$\delta = 0.25\text{V}$$

Given that $f_m = 2\text{ kHz} = 2 \times 10^3\text{Hz}$

Putting these value in equation (2) we get

$$A_m \leq \frac{0.25}{2\pi \times 2 \times 10^3 \times \frac{1}{30 \times 10^3}} \Rightarrow A_m \leq 0.6\text{V}$$

Example 31. *The pulse rate in a DM system is 56,000 per sec. The input signal is 5 cos $(2\pi \times 1000t)$ V, with t in sec find the maximum value of step size will avoid slope over load distortion what will be the disadvantages of choosing a value of larger than the minimum ?*

Solution. Given pulse rate 56,000 per sec

$$V_m = 5\cos(2\pi \times 10^3 t) + 2\cos(2\pi \times 2 \times 10^3 t)$$

find minimum step size.

$$Sf_S \geq 2\pi fA \qquad \text{or} \qquad S \geq 2\pi \frac{f}{f_S} A$$

$$S \geq 2\pi\left(\frac{(10)^3}{56\times10^3}\right)\times 5; \qquad S \geq 2\pi\left(\frac{2\times10^3}{56\times10^3}\right)\times 2; S > \frac{10\pi}{56} = 0.56\,\text{V}$$

$$S \geq \frac{8\pi}{56} = 0.45\,\text{V}; \qquad \text{S should not be less than 0.56V}$$

Its value is too larger than granular noise.

Example 32. *A DM system is designed to operate at 3 times the Nyquist rate for a signal with a 3-kHz bandwidth. The quantizing step size is 250 mV.*

(a) Determine the maximum aplitude of a 1-kHz input sinusoid for which the delta modulator does not show slope overload.

(b) Determine the postfiltered output signal-to-quantizing-noise ratio for the signal of part (a).

Solution. *(a)* $m(t) = A\cos\omega_m t = A\cos 2\pi(10^3)t$; $\left|\frac{dm(t)}{dt}\right|_{max} = A(2\pi)(10^3)$

By equation $A > \frac{\Delta}{\omega_m T_s} = \frac{\Delta}{2\pi}\left(\frac{f_s}{f_m}\right)$, the maximum allowable amplitude of the input sinusoid is

$$A_{max} = \frac{\Delta}{\omega_m T_s} = \frac{\Delta}{\omega_m} f_s = \frac{250}{2\pi(10^3)} 3(2)(3)(10^3) = 716.2\,\text{mV}$$

(b) From equation $(\text{SNR})_0 = \left(\frac{S}{N_q}\right)_0 = \frac{3f_s^2}{8\pi^2 f_m^2 f_M}$, and assuming that the cutoff frequency of the low-pass filter is f_m, we have

$$(\text{SNR})_0 = \left(\frac{S}{N_q}\right)_0 = \frac{3\left[(3)(6)(10^3)\right]^3}{8\pi^2(10^3)^3} = 221.6 = 23.5\,dB$$

Example 33. *Compare the overall output S/N ratio for 8 bit PCM and DM system used for transmitting a base band signal whose spectrum is confined from 300 to 3000 Hz. Assume that both system operates at 9 bit rate of 64 kbit/sec and use a PSK signaling scheme with* $\left(\frac{S_{av}}{\eta f_m}\right) = 20\,dB$.

Solution. *(a)* PCM system, we have, $\frac{1}{T_b} = 64{,}000 \left(\frac{S_{av}}{\eta f_m}\right) = 100$

and

$$P_e = Q\left(\sqrt{\frac{2S_{av}T_b}{\eta}}\right) = Q\sqrt{9.375} = 10^{-3}$$

Hence

$$\frac{S_O}{N_O} = \frac{2^{2N}}{1+4P_e 2^{2N}} = 24\,\text{dB}$$

(*b*) DM system $f_1 = 300, f_m = 3000, f_s' = 64{,}000$ and $P_e = Q\left(\sqrt{9.375}\right) = 10^{-3}$

$$f_s' = 64{,}000, f_1 = 300 \text{ and } f_m = 3000 \text{ we have } \frac{S_O}{N_O} = 20\,\text{dB}$$

3.7 COMPARISING BETWEEN PCM AND DM

(*i*) DM needs a single circuit as compared to PCM. But the signal to quantization noise ratio is less in DM than in PCM

(*ii*) For a good quality transmission the bandwidth needed by DM is more than PCM.

The use of DM can be recommended for the following two situations :

(*a*) When simple circuitry is most importance and the allowable bandwidth is large.

(*b*) When the bandwidth conservation is desirable at the cost of the quality of transmission.

Table 3.2 Comparison between PCM, adaptive delta modulation and differential pulse code modulation

S. No.	*Parameter Differential Pulse of comparison Code Modulation (DPCM)*	*Pulse Code Modulation (PCM)*	*Delta Modulation (DM)*	*Adaptive Delta Modulation (ADM)*
1.	Number of bits.	It can use 4, 8 or 16 bits per sample.	Its uses only one bit for one sample.	Only one bit is used Bits can be more than one to encode one sample. but are less than PCM.
2.	Levels and step size	The number of levels depend on number of levels size is fixed.	Step size is kept fixed and cannot be varied.	According to the signal Here, Fixed number of variation, step size levels are used. varies (*i.e.*, Adapted).
3.	Quantization error and distortion	Quantization error depends on number of levels size is fixed.	Slope overload distortion and granular noise are present.	Quantization noise is Slope overload distortion present but other errors and quantization noise are absent. are not encountered.
4.	Transmission bandwidth	Highest bandwidth is required since number of bits are high.	Lowest bandwidth is required.	Lowest bandwidth is Bandwidth required is required. lower than PCM.
5.	Feedback	There is no feedback in transmitter or receiver.	Feedback exists in transmitter	Feedback exists. Here, Feedback exists.
6.	Complexity of notation	System complex.	Simple.	Simple. Simple.

REVIEW QUESTIONS

1. What do you understand by pulse code modulation? What are the basic steps of PCM?
2. What is Uniform Quantinization? Explain different characteristics of a Uniform Quantinization.
3. Why do you use Non-Uniform Quantinization? Explain A-law and μ-law of Quantinization.
4. What are the advantages of DPCM over PCM? Explain modulation and Demodulation of DPCM.
5. What is DM? Explain modulation and Demodulation of DM.
6. What are slope overload and granular noise?
7. What is adaptive DM? Why is it better than DM?
8. Explain CVSD.
9. (*a*) Draw block diagram and explain working of DPCM. Compare its performance w.r.t. PCM technique briefly.
10. (*a*) Draw the circuit diagram of PAM modulator and demodulator. Explain their working.
    (*b*) A 6 bit single channel PCM system gives an output of 60 Kbps. Determine also the highest possible modulation. Frequency with DM modulator with same O/P data rate.
    (*c*) What are the advantages of PCM over all other analog methods of modulation ?
11. (*a*) Consider a sine wave of frequency f_m and amplitude A_m, applied to a delta modulation of step size Δ. Show that the slope overload distortion will occur if $A_m > \frac{\Delta}{2\pi f_m T_s}$ where T_s is the sampling period. What is the maximum power that may be transmitted without slope-covered distortion ?
    (*b*) With the help of the block diagrams of a transmitter and a receiver, explain the differential pulse-code modulation system.
    (*c*) Compare PCM and DPCM on the basis of bit transmission rate.
12. (*a*) A waveform 20 + 20 sin (500*t* + 30°) is to be sampled periodically and reproduced from these sample values.
    (*i*) Find the maximum allowable time interval between sample values.
    (*ii*) How many sample values need to be stored in order to reproduce 1sec of this waveform ?
    (*b*) Explain how companding improves S/N ratio in PCM. What are the advantages of delta modulation over conventional PCM ?
    (*c*) Explain the terms: frequency deviation, pre-emphasis and use of limiters in FM.
13. (a) In a binary PCM system the output signal to quantizing noise ratio is to be held to a minimum of 40 dB. Determine the number of required levels and find the corresponding output signal to quantizing noise ratio.
    (*b*) Adaptive delta modulation.
14. Explain the following Digital Communication Techniques :
    (*a*) DPCM (*b*) ADM
15. (*a*) State and prove uniform sampling theorem for a band limited signal.
    (*b*) Explain the working of a delta modulator. What are its limitations due to fixed step size, how are they overcome ?
16. Describe the following modulation techniques of digital communication:
    (*a*) PCM (*b*) ADM
17. Write short notes on any three of the following:
    (*i*) Time division multiplexing (*ii*) EYE diagram
    (*ii*) Comparison of digital communication systems.

18. (*a*) Explain briefly how differential PCM and delta modulation exploit the redundancies in speech waveform to encode it in fewer bits.

(*b*) Why is adaptive delta modulator an improvement over linear delta modulator ? Give a schematic diagram of adaptive delta modulator with digitally controlled companding and explain its working briefly.

(*c*) If a 25 dB signal-to-quantization noise ratio should be maintained in a delta-modulator circuit, determine the quantization bit rate for an analog input signal with bandwidth of 3.4 kHz.

19. (*a*) Clearly differentiate between the process and features of –

(*i*) Natural sampling and flat-top sampling

(*ii*) Continuous companding and segmented companding

(*iii*) Analog companding and digital companding

(*b*) A signal $x(t) = 2\cos 400\pi t + 6\cos 640\pi t$ is ideally sampled at $f_s = 500$ Hz. If the sampled signal is passed through an ideal low pass filter with a cut-off frequency of 400 Hz, what frequency components will appear in the output ?

20. (*a*) What is a frame-alignment-word (FAW) in PCM transmission ? What parameters are considered in choosing a proper FAW ? How is frame synchronization achieved with the help of FAW ? Give a state-transitition diagram for the process.

(*b*) What important functions are performed by the regenerator circuit of a receiver ? Why is equalization carried out ?

(*c*) Give a circuit and waveforms to explain clock extraction process in a PCM system using AMI line coding.

4

DIGITAL BASE BAND TRANSMISSION

4.1 DIGITAL DATA AND DIGITAL SIGNAL

Digital to digital encoding or conversion is the representation of digital information by a digital signal. For example, when you transmit data from your computer to your printer both the original data and the transmitted data are digital.

Data encoding schemes plays an important role to improve system performance. A variety of encoding schemes are in use. These schemes can be compared on basis of following factors.

(*i*) **Power Spectral Density :** It is desirable to have zero PSD (power spectral density) at $\omega = 0$ (no D.C. component) because AC coupling and transformers are used at the repeaters.

(*ii*) **Transmission Bandwidth :** A lack of high frequency components means that less bandwidth is required for transmission. In addition, lack of a Direct Current (DC) component is also desirable. Finally, the magnitude of the effects of signal distortion and interference depends on the spectral properties of the transmitted signal.

(*iii*) **Error Detection :** Error detection is the responsibility of a layer of logic above the signalling level known as data link layer. However, it is useful to have some error-detection capability build into the physical signaling encoding scheme, this permits errors to be detected more quickly.

(*iv*) **Clocking :** It is important to maintain synchronization between transmitter and receiver so as to determine the beginning and end of each bit position at receivers. One expensive approach is to provide a separate clock-lead to synchronize the transmitter and receiver.

(*v*) **Cost and Complexity :** Although digital logic continuous to drop in price, expense should not be ignored. In particular, the higher the signaling rate to achieve a given data rate, the greater the cost.

(*vi*) **Transparency :** It should be possible to transmit a digital signal correctly regardless of the pattern of 1's and 0's. We will see in coming section that a long string of 0's could cause errors in timing extraction in on–off and bipolar cases. If the data are so coded that for every possible sequence of data, the coded signal is received faithfully, the code is transparent.

(*vii*) **Signal Interference and Noise Immunity :** A good signal design should be noise immuned. Its performance should not be affected in presence of noise. This ability is usually expressed in terms of bit error rate.

Various Signal Encoding Techniques are shown in Fig. 4.1.

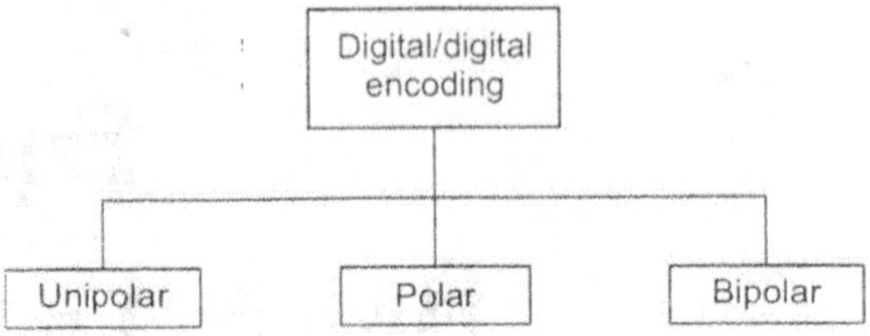

Fig. 4.1

4.2 UNIPOLAR

Unipolar encoding is very simple and very primitive. It is almost obsolete today. The polarity of a pulse refers to whether it is positive or negative. Unipolar encoding is so named because it uses only one polarity. This polarity is assigned to one of the two binary states, usually the 1. The other state, usually the 0 is represented by zero voltage.

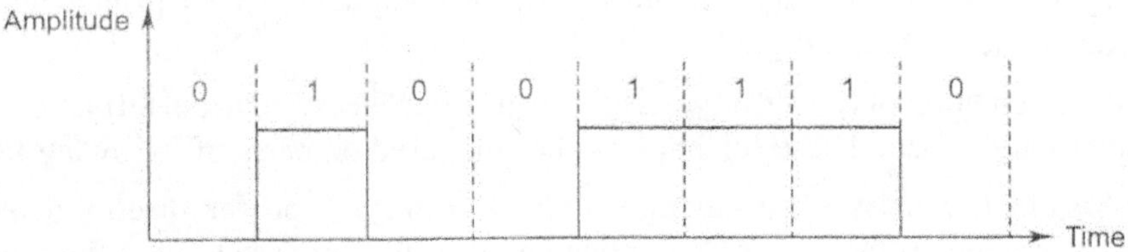

Fig. 4.2 *Unipolar encoding*

4.3 POLAR

Polar encoding uses two voltage levels : one positive and one negative. Various polar encoding methods are used, among them important are Non-return to zero (NRZ) return to zero (RZ) and biphase. NRZ encoding include two method : Non-return to zero level (NRZ–L), and Non-return to zero, inverted (NRZ–1). Biphase also refers to two methods. The first, Manchester, is the method used by ethernet LANs. The second, differential Manchester, is the method used by Token Ring LANs.

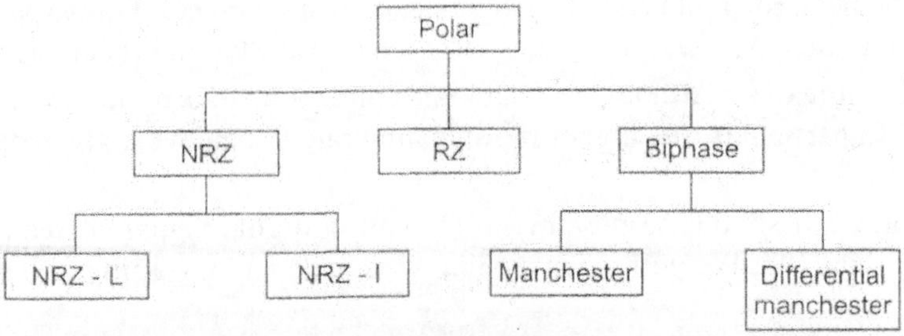

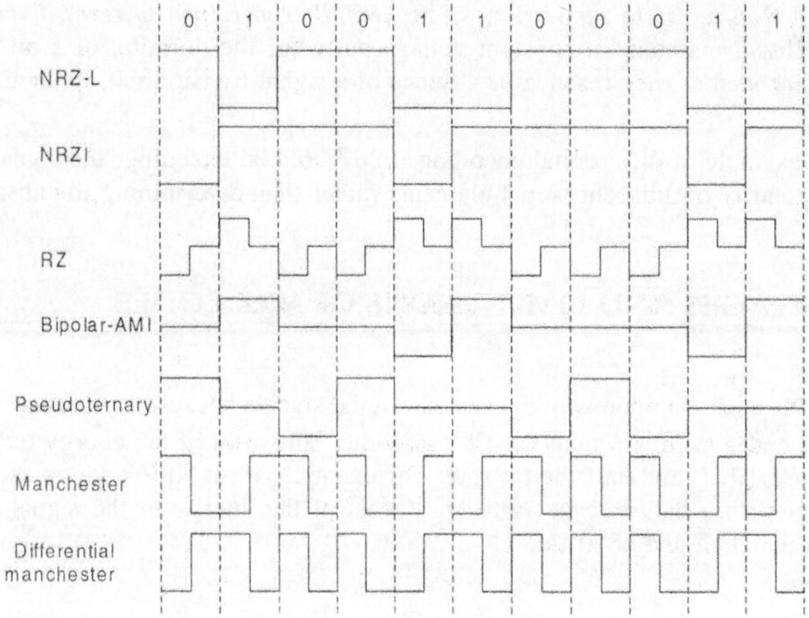

Fig. 4.3 *Digital signal encoding formats*

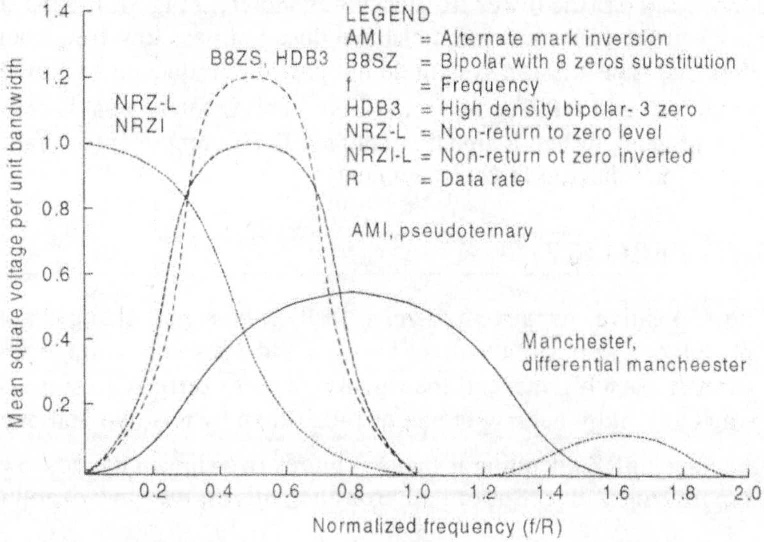

Fig. 4.4 *Spectral density of various signal encoding schemes*

4.4 NON-RETURN TO ZERO (NRZ)

The most common, and easiest, way to transmit digital data is to use two different voltage levels for the two binary digits (0 and 1). Codes that follow this strategy share the property that the voltage level is constant during a bit interval; there is no transition during that interval. This is known as ***Non-return to zero level (NRZ-L)***. For example, the negative voltage can be used to represent binary 0, with a constant positive voltage used to represent binary 1. In NRZ-L, a negative voltage is used to represent one binary value and a positive voltage is used to represent the other. NRZ-L is illustrated in Fig. 4.3.

A variation of Non-return to zero is known as ***NRZ–I (Non-return to zero, invert on ones).*** As with NRZ–L, NRZ–I maintains a constant voltage pulse for the duration of a bit time. The data themselves are encoded as the presence or absence of a signal transition at the beginning of the bit time.

NRZ–I is an example of differential encoding. In differential encoding, the signal is decoded by comparing the polarity of adjacent signal elements rather than determining the absolute value of a signal element.

4.5 ADVANTAGES AND LIMITATIONS OF NRZ CODES

1. *Advantage*
   (*i*) NRZ is most common way to transmit digital signals because it is easiest for engineers.
   (*ii*) NRZ codes make efficient use of bandwidth. The most of the energy in NRZ signals lie between D.C. and half the bit rate. For example, if an NRZ code is used to generate a signal with a data rate of 9600 bps. Most of the energy in the signal is concentrated between D.C. and 4800 Hz.
2. *Limitations*
   (*i*) In NRZ codes strings of consecutive 0's and 1's lead to periods where the signal remains constant. These strings occurs frequently enough to produce a spectrum that has its components concentrated at the lower frequencies as shown in Fig. 4.4. This situation presents a problem when the communication channel does not pass low frequencies. For example, most telephone transmission system do not pass the frequencies below than 200 Hz.
   (*ii*) The average value of NRZ codes is non–zero. This creates, what is called a direct current (D.C.) component. When a signal contains a D.C. component, it cannot travel through media that cannot handle D.C. components.

4.6 RETURN TO ZERO (RZ)

RZ uses three values : positive, negative and zero. In RZ, the signal changes not between bits but during each bit. Like NRZ–L, a positive voltage means 1 and a negative voltage means 0. But, unlike NRZ–L, halfway through each bit interval, the signal returns to zero. A 1 bit is actually represented by positive to zero and a 0 bit by negative to zero, rather than by positive and negative alone.

The main disadvantage of RZ encoding is that it requires two signal changes to encode one bit and therefore occupies more bandwidth. But of the three alternatives, we have examined so far, it is the most effective.

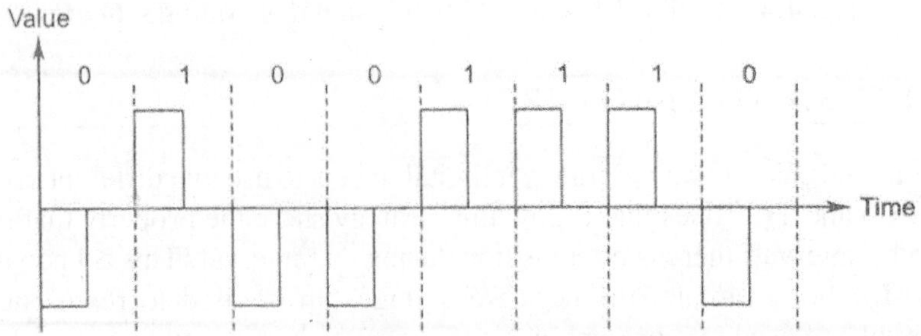

Fig. 4.5 *RZ encoding (Transitions at middle of each bit can be used for synchronization)*

4.7 BIPHASE

Biphase encoding technique is a good solution to the problem of synchronization in NRZ technique. In this method, the signal changes at the middle of the bit interval but does not return to zero. Instead, it continues to the opposite pole. Due to these frequent transitions, timing circuit at receiver can maintain synchronization with transmitting end.

There are two types of biphase encoding techniques in use today :

1. *Manchester* 2. *Differential Manchester.*

1. *Manchester:* Manchester encoding uses the inversion at the middle of each bit interval for both synchronization and bit representation. A negative to positive transition represents binary 1 and a positive to negative transition represents binary 0.
2. *Differential Manchester.* In differential Manchester, the inversion at the middle of the bit interval is used for synchronization, but the presence or absence of an additional transition at the beginning of the interval is used to identify the bit.

Advantages of biphase techniques are :

(*i*) *Synchronization :* Because there is a predictable transition during each time, the receiver can synchronize on that transition. For this reason, the biphase codes are known as self-clocking codes.

(*ii*) *No D.C. Component :* Biphase codes have no D.C. component, yielding the benefits described earlier.

(*iii*) *Error Detection :* The absence of an expected transition can be used to detect errors. Noise on the line would have to invert both the signal before and after the expected transition to cause an undetected error.

A significant disadvantage of biphase technique is that it requires at least one transition per bit time and may have as many as two transitions. Thus, the maximum modulation rate is twice that for NRZ; this means that the bandwidth required is correspondingly greater.

4.8 BIPOLAR

Bipolar encoding uses three voltage levels, negative positive, and zero (–A , +A, and 0). Zero level in bipolar encoding is used to represent binary 0. The 1s are represented by alternating positive and negative voltage.

Several advantages of this approach are :

(*i*) Bandwidth of the resulting signal is considerably less than the bandwidth of NRZ signals.

(*ii*) There will be no loss of synchronization if a long string of 1's occur. Each 1 introduces a transition, and the receiver can resynchronize on that transition.

(*iii*) The pulse-alteration property provides a simple means of error detection. Any isolated error, whether it detects a pulse or adds a pulse, causes a violation of this property.

(*iv*) Because the 1 signals alternate in voltage from positive to negative, there is no net D.C. component.

Limitations of bipolar encoding techniques :

(*i*) In bipolar encoding, the line signal may take on one of three levels and the receiver has to distinguish between these three levels instead of just two levels as in the other signaling formats previously discussed.

(*ii*) Although a degree of synchronization is provided with these codes, a long string of 0's is still a problem. A binary 0 is represented by no line signal and there will be no transition for a long string of 0. This may lead to loss of synchronization between transmitter and receiver.

Because of three voltage levels used in multilevel binary signal, it requires approximately 3 *d*B more signal power than a two-valued signal for the same probability of bit error.

Three types of bipolar encoding are in popular use by the data communications industry :

(*i*) AMI, (*ii*) B-8ZS, (*iii*) HDB3.

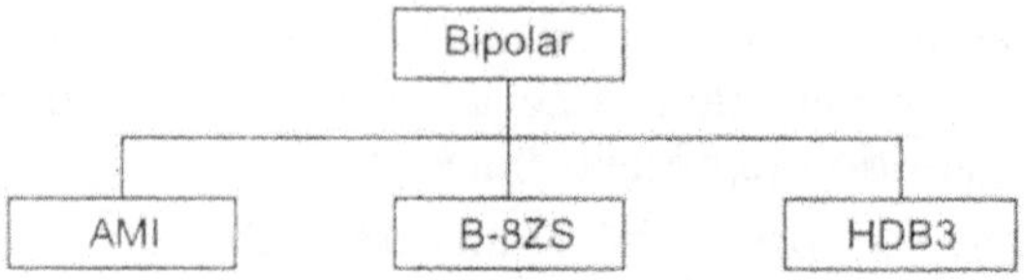

Fig. 4.6

4.9 BIPOLAR ALTERNATE MARK INVERSION (AMI)

A simple bipolar encoding scheme is **Alternate Mark Inversion (AMI)**. In the name alternate mark inversion, the word mark comes from telegraphy and means 1. So AMI means alternate 1 inversion. A binary 0 is represented by no line signal. Fig. 4.7 gives an example.

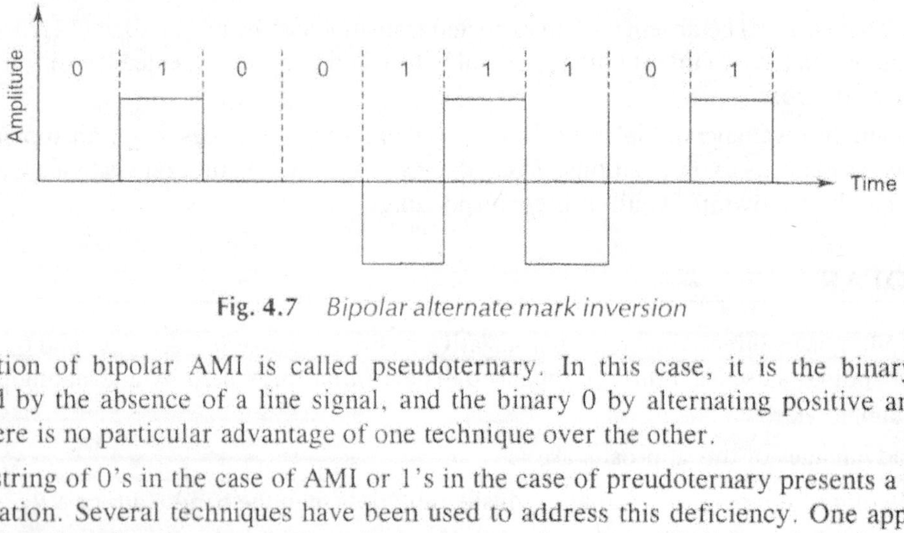

Fig. 4.7 *Bipolar alternate mark inversion*

A variation of bipolar AMI is called pseudoternary. In this case, it is the binary 1 that is represented by the absence of a line signal, and the binary 0 by alternating positive and negative pulses. There is no particular advantage of one technique over the other.

A long string of 0's in the case of AMI or 1's in the case of preudoternary presents a problem of synchronization. Several techniques have been used to address this deficiency. One approach is to insert additional bits that force transitions. This technique is used in ISDN for relatively low data rate transmission. At a high data rate, this scheme is expensive, as it results in an increase in already high signal transmission rate. To cope with this problem at high data rates, a technique that involves scrambling the data is used.

Idea behind scrambling is to use sequences that would result in a constant voltage level on the line as filling sequences to provide sufficient transitions for the receiver's clock to maintain synchronization.

4.10 BIPOLAR WITH 8–ZEROS SUBSTITUTION (B-8ZS)

The difference between B-8ZS and bipolar AMI occurs whenever eight or more consecutive 0's are encountered in the data stream. The solution provided by B-8ZS is to force artificial signal changes, called violations, within the 0 string. Anytime eight 0's occur in succession, B-8ZS introduces changes in the pattern based on the polarity of the previous 1 (the 1 occurring just before the 0's). The changes in pattern are done with the following rules.

(*i*) If an octet of all zeros occurs and the last voltage pulse preceding this octet was negative, then the eight zeros of the octet are encoded as 000 – + 0 + –.

(*ii*) If an octet of all zeros occurs and the last voltage pulse preceding this octet was positive, the eight zeros of the octet are encoded as 000 + – 0 – +.

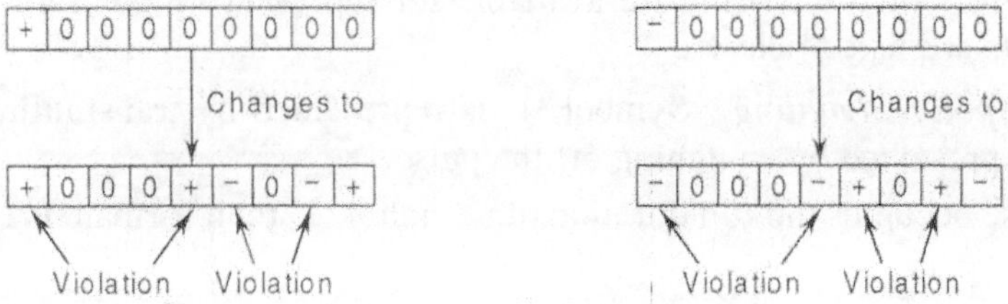

(a) Positive polarity of previous bit (b) Negative polarity of previous bit

Fig. 4.8 *B-8ZS encoding*

This technique forces two code violations (signal patterns not allowed in AMI) of the AMI code, an event unlikely to be caused by noise or other transmission impairment. The receiver recognizes the pattern and interprets the octet as consisting of all zeros.

4.11 HIGH–DENSITY BIPOLAR 3 (HDB3)

HDB3 is the convention adopted in Europe and Japan. HDB3 introduces changes into the bipolar AMI pattern every time four consecutive 0's are encountered instead of waiting for the eight zeros as by B8ZS. Although the name is HDB3, the pattern changes whenever there are four 0's in succession.

The pattern of violation in HDB3 is based on the number of 1's that have occurred in the bit stream since the last substitution as well as the polarity of the previous 1 bit. In this scheme fourth zero is replaced with a code violation. It is also ensured that successive violations are of alternate polarity so that no D.C. component is introduced. This condition is tested for by knowing whether the number of pulses since the last violation is even or odd and the polarity of the last pulse before the occurrence of the four zeros.

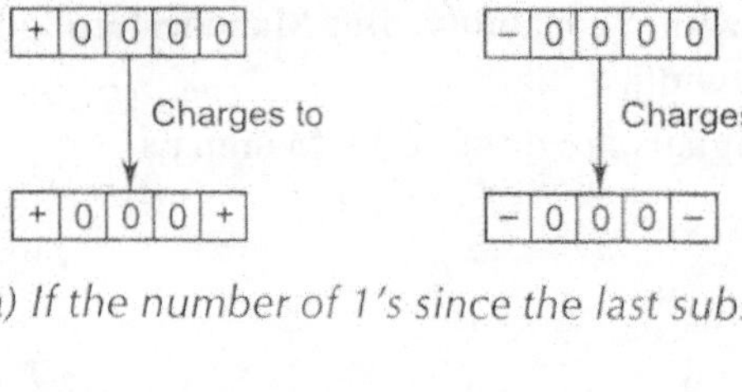

Fig. 4.9 *(a) If the number of 1's since the last substitution is odd*

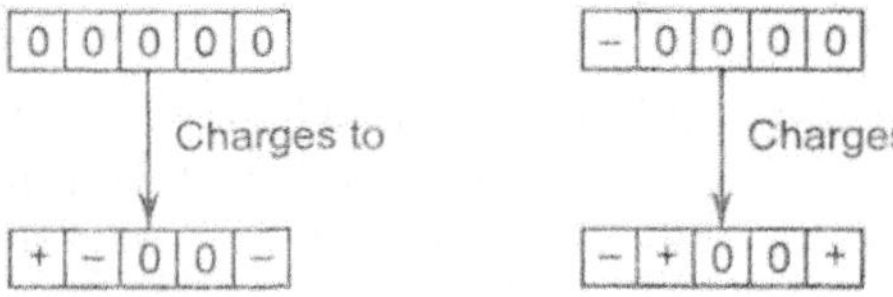

Fig. 4.9 *(b) If the number of 1's since the last substitution is even*

4.12 LINE CODING

In order to send the encoded digital data over a channel, it is necessary to translate the data in the form of electrical waveforms.

A number of formats or waveforms are available for representing the digital data. Some of the useful digital formats are listed below :

(*i*) *On-Off or Unipolar Signaling* : Symbol '1' is represented by transmitting a pulse, where as symbol 0 is represented by switching off the pulse.

When the pulse occupies the full duration of a symbol '1' then formate is called Non-Return to zero (NRZ) type.

When the pulse occupies usually for half of the symbol duration, it is said to be return-to-zero (RZ) type.

It is simple to implement but contains a D.C. component that is often found to be objectionable.

(*ii*) *Polar Signaling* : In the polar formate, the positive pulse is transmitted for symbol '1' and a negative pulse for symbol 0. It can be of the NRZ or RZ type.

As it has not D.C. component (due to positive and negative pulses) so better than unipolar.

(*iii*) *Bipolar Signaling* : In bipolar format of signaling, positive and negative pulses are used alternately for the transmission of 1s and no pulses for the transmission of 0's. It can also be of NRZ or RZ type.

It does not contain any D.C. component even if this is long string of 0's and 1's. It is also good for **error detecting** because due to noise, the alternative pulses (positive and negative pulses) will change the pattern and thus if we are getting alternative positive and negative pulses, then it is error.

So it is adopted by the T_1 carrier system for digital telephony.

(*iv*) *Manchester Signaling* : It is also known as *biphase baseband signaling*. In this format of signaling, symbol '1' is represented by the positive pulse for one half of the symbol duration, followed by a negative pulse for the remaining half of the symbol duration for symbol '0', these two pulses are transmitted in reverse order.

The NRZ versions of unipolar, polar and unipolar format make efficient use of bandwidth. But they don't have synchronization capability. But Manchester Coding has this capability but on the basis of extended bandwidth.

Remember that all these signals are discrete PAM signals.

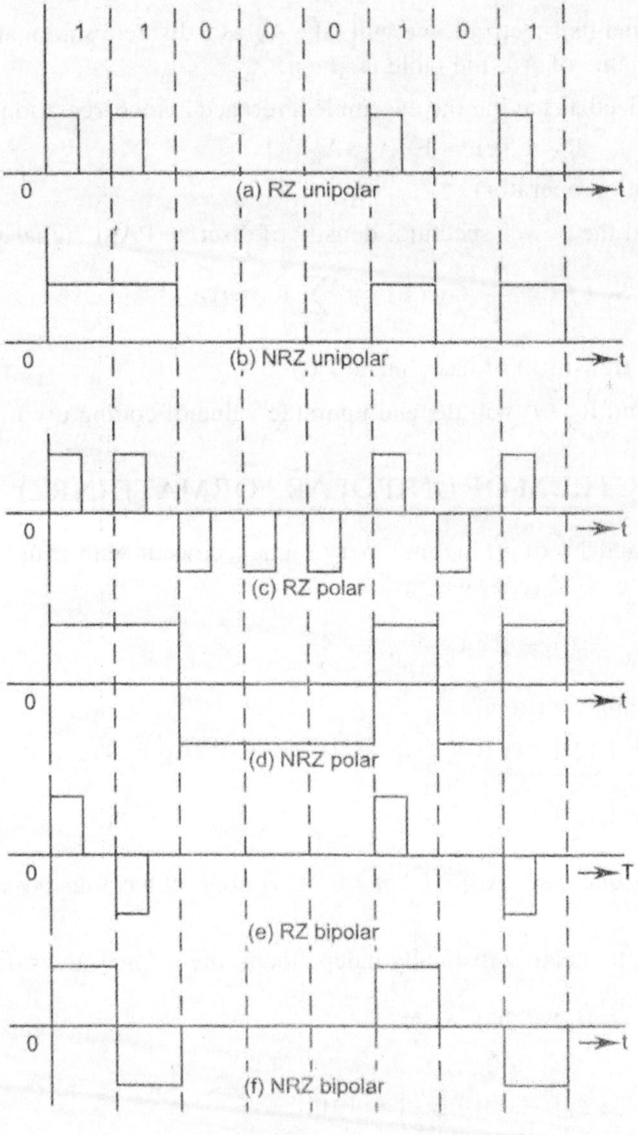

Fig. 4.10 *Line coding.*

4.13 POWER SPECTRUM OF LINE CODING SIGNALS

As we have drawn the signal format in Fig (4.1), we can say that these are example of a discrete amplitude-modulated pulse train. That can be given by :

$$X(t) = \sum_{K=-\infty}^{\infty} A_K \, v(t - KT) \tag{4.1}$$

where coefficient A_K is a discrete random variable, $u(t)$ is the basic pulse shape, and T is the symbol duration. The Basic Pulse $u(t)$ is centered at the origin, $t = 0$ and normalised such that

$$v(0) = 1.$$

Now if we want to find its spectrum, we will take A_K as a discrete stationary random source. For different formates the value of A_K, the table is given.

The source is characterized as having the ensemble-averaged autocorrectation function.

$$R_A = (n) = E[A_K, A_{K-n}] \tag{4.2}$$

(where E is the expectation operator).

Accordingly, we find the power spectural density of discrete PAM signal $x(t)$ by the equation :

$$S_X(f) = \frac{1}{T} | v(f) |^2 \sum_{n=-\infty} R_A(n) e^{-j2\pi n f T} \tag{4.3}$$

where $v(f)$ = Fourier transform of basic pulse $v(t)$.

The value of $v(f)$ and $R_A(n)$ will depend upon the value of coding used.

4.14 POWER SPECTRUM OF UNIPOLAR FORMATE (NRZ)

Let us assume that 0's and 1's of a random binary sequence occur with equal probability. Then, for a unipolar formate of the NRZ type, we have :

$$P(A_K = 0) = P(A_K = a) = \frac{1}{2}. \tag{4.4}$$

Hence for $n = 0$, we may write

$$E[A_K^2] = (0)^2 P(A_K = 0) + (a)^2 P(A_K = a)$$

$$\boxed{E[A_K^2] = \frac{a^2}{2}} \tag{4.5}$$

Consider next the product $A_K . A_{K-n}$ for $n \neq 0$. This product has four possible values, namely 0, 0, 0, and a^2.

Assuming that all symbols are statistically independent, these four values occur with a probability of $\frac{1}{4}$ each. Hence, for $n \neq 0$, we may write

$$E[A_K . A_{K-n}] = 3(0)\left(\frac{1}{4}\right) + (a)^2\left(\frac{1}{4}\right) \tag{4.6}$$

$$= \frac{a^2}{4}, \; n \neq 0. \tag{4.7}$$

We may thus express the autocorrelation function $R_A(n)$ as follows :

$$R_A(n) = \begin{Bmatrix} \frac{a^2}{2} & ; & n = 0 \\ \frac{a^2}{4} & ; & n \neq 0 \end{Bmatrix} \tag{4.8}$$

For a basic pulse $v(t)$, we have a rectangular pulse of unit amplitude and duration T_b. Hence, the fourier transform of $v(t)$ equals

$$v(f) = T_b \sin c\,(f T_b) \tag{4.9}$$

So substituting this value in equation

$$S_X(f) = \frac{a^2 T_b}{4} \sin^2 c\,(f T_b) + \frac{a^2 T_b}{4} \sin c^2 \sum_{n=-\infty}^{\infty} e^{-j2\pi n f T_b} \tag{4.10}$$

But by poission's equation ; the equation

$$\sum_{n=-\infty}^{\infty} e^{-j2\pi f T_b} = \frac{1}{T_b} \sum_{m=-\infty}^{\infty} \delta\left(f - \frac{m}{T_b}\right) \tag{4.11}$$

where $\delta(f)$ denotes a Dirac delta function at $f = 0$;

As the sinc function sinc $(f T_b)$ has null values at $f = \pm \frac{1}{T_b}, \pm \frac{2}{T_b}$, so we can write,

$$S_X(f) = \frac{a^2 T_b}{4} \operatorname{sinc}^2 \left(f T_b\right) + \frac{a^2}{4} \delta(f) \tag{4.12}$$

(Remember that $T = T_b$ called Nyquist sample rate)

Thus, it contains a D.C. part also, it has been drawn by curve a.

4.15 POWER SPECTRUM OF POLAR FORMATE (IN NRZ FORMATE)

Again for polar formate

$$R_A(n) = \begin{cases} a^2 & ; \quad n = 0 \\ 0 & ; \quad n \neq 0 \end{cases} \tag{4.13}$$

The basic pulse is same as that of used for unipolar formate.

Again $\qquad S_X(f) = a^2 T_b \operatorname{sinc}^2 (f T_b) \qquad (4.14)$

again it has drawn by curve b.

4.16 POWER SPECTRUM OF BIPOLAR FORMATE (IN NRZ FORMATE)

We note that the bipolar formate ; it has three levels : a, 0 and $-a$, then for event of equal propability; we have probability of occurrence of these three levels as follow :

$$\left.\begin{aligned} P(A_K = a) &= \frac{1}{4} \\ P(A_K = 0) &= \frac{1}{2} \\ P(A_K = -a) &= \frac{1}{4} \end{aligned}\right\} \tag{4.15}$$

Hence for $n = 0$; we may write

$$E[A_K^2] = (a)^2 P(A_K = a) + (0)^2 P(A_K = 0) + (-a)^2 P(A_K = -a)$$

$$= \frac{a^2}{2}. \tag{4.16}$$

Now for $n = 1$, we can calculate the (A_K, A_{K-1}) as we have totally four possible form :

$$\left.\begin{aligned} &= 0 . 0 = 0 \\ &= 0 . a = 0 \text{ ('or' } 0 . -a = 0) \\ &= a . 0 = 0 \text{ ('or' } -a . 0 = 0) \\ &= a . -a = -a^2. \end{aligned}\right\} \tag{4.17}$$

So we have totally four events so have probability of $\frac{1}{4}$. So,

$$E[A_K, A_{K-1}] = (3)(0)\left(\frac{1}{4}\right) + (-a)^2\left(\frac{1}{4}\right)$$

$$= -\frac{a^2}{4}. \tag{4.18}$$

It can be seen that for $n > 1$, we have :

$$E[A_K . A_{K-n}] = 0$$

Again, we have :

$$R_A(n) = \begin{cases} \frac{a^2}{2} & ; \quad n = 0 \\ -\frac{a^2}{4} & ; \quad n = \pm 1 \\ 0 & ; \quad \text{otherwise} \end{cases} \tag{4.19}$$

The basic pulse remained same as previous cases. So power spectrum density of the NRZ formate is given by :

$$S_X(f) = T_b \sin c^2 (f T_b) \left[\frac{a^2}{2} - \frac{a^2}{4}\left(e^{j2\pi f T_b} + e^{-j2\pi f T_b}\right)\right]$$

$$= \frac{a^2 T_b}{2} \sin c^2 (f T_b) [1 - \cos (2\pi f T_b)$$

$$\boxed{S_X(f) = \frac{a^2 T_b}{2} \sin c^2 (f T_b) \sin^2 (\pi f T_b)} \tag{4.20}$$

This is shown in curve C.

4.17 POWER SPECTRUM OF MANCHESTER CODING (IN NRZ FORMATE)

The autocorrelation function $R_A(n)$ for the manchester formate is same as used for polar formate.

The basic pulse $v(t)$ for a manchester formate consists of a doublet pulse amplitute and total duration T_b ; Hence the fourier transform of pulses equals.

$$v(f) = jT_b \text{ sinc}\left(\frac{fT_b}{2}\right) . \sin\left(\frac{\pi f T_b}{2}\right). \tag{4.21}$$

Again by the same method as adopted above; the PSD of Manchester formate is given by :

$$S_X(f) = a^2 T_b \text{ sinc}^2\left(\frac{fT_b}{2}\right) . \sin^2\left(\frac{\pi f T_b}{2}\right). \tag{4.22}$$

The plot of PSD has been drawn in curve *d*.

It can be seen that for Manchester coding, the bandwidth requirement is high.

Table 4.1 Data formates for line coding

NRZ Line Coding	***Coefficient***	***Remark***
Unipolar	$A_K = \begin{cases} a & ; \text{ Symbol 1} \\ 0 & ; \text{ Symbol 0} \end{cases}$	Rectangular pulse of unit amplitude and duration T_b.
Polar	$A_K = \begin{cases} a & ; \text{ Symbol 1} \\ -a & ; \text{ Symbol 0} \end{cases}$	
Bipolar	$A_K = \begin{cases} a, -a & ; \text{ is alternating 1's} \\ 0 & ; \text{ Symbol 0} \end{cases}$	
Manchester	$A_K = \begin{cases} a & ; \text{ Symbol 1} \\ -a & ; \text{ Symbol 0} \end{cases}$	Doublet pulse of height $\pm$ 1 and total duration T_b

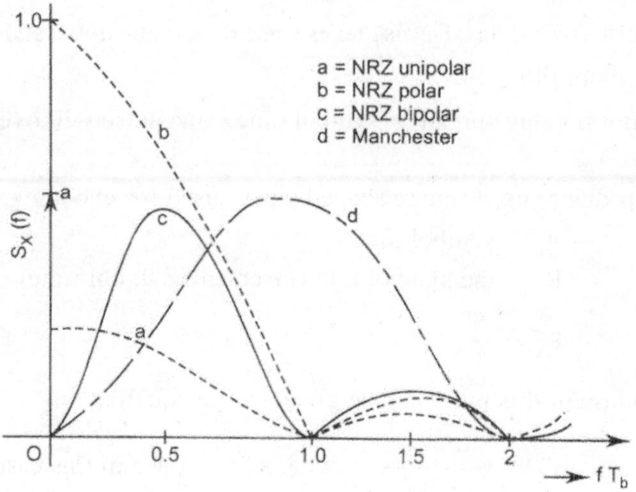

Fig. 4.11 *PSD of line coding.*

4.18 INTERSYMBOL INTERFERENCE (ISI)

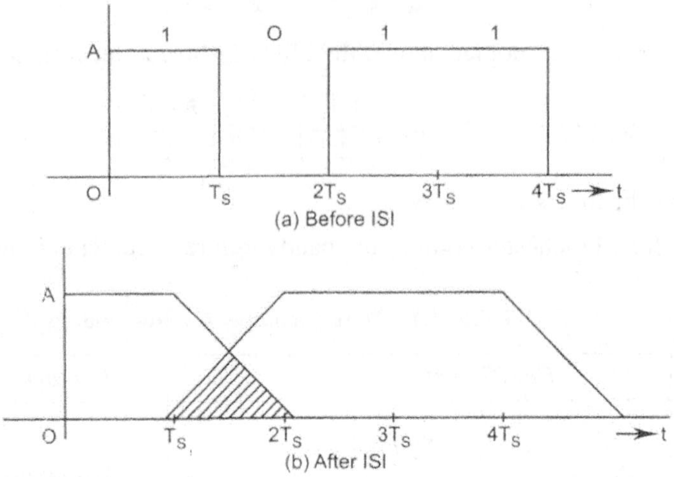

Fig. 4.12 *Inter symbol interference*

Till now we have considered that the signal is distorted due to noise only. But practically we have another problem that is called dispersion of the pulse shaping. And the name of this problem is intersymbol interference.

The effect of this practically can be understood in Fig. 4.3 (*a*) and Fig. 4.3 (*b*).

Due to channel, basically we get a time delay introduced in the pulse. That can be represented by :

$$y(t) = u \sum_{K=-\infty}^{\infty} A_K \, P\left(t - KT_b - t_d\right) \tag{4.23}$$

where t_d = time delay introduced by the channel. Thus, we can say that :

(*a*) Basic pulse (without interference and time delay T_d) takes some time for die down and that is called symbol time (T_b).

(*b*) Now modified pulse (with delay T_d) also takes some time to die down and it is called *delay spread.*

If delay spread ≤ Symbol time

No ISI will occur. But if delay spread > symbol time ; one pulse will overlap to other and that is called ISI.

The basic problem is due to used square shaped pulses used for encoding of signal.

Here I_s = symbol time

R_s = the symbol rate (inverse of symbol time)

where, $R_s = \dfrac{1}{T_s}$

The frequency response of this pulse can be given by the equation.

$$h(f) = \frac{\sin 2\pi t / T_s}{2\pi t} \quad \text{where } T_s = 1 \text{ sec in this case.} \tag{4.24}$$

In this case, we can see that bandwidth of signal is equal to the symbol rate of signal.

But square pulse has some disadvantages and they are :

(*i*) The square pulse is difficult to create in time domain because of rise time and a decay time.

(*ii*) The main disadvantage is that frequency response goes on forever and decays slowly. That is the reason that why it is very sensitive to the intersymbol interference.

4.18.1 Ideal Solution of ISI

Now it is very practical, if we use a (sin *c*) function in the time domain that will cause a pulse in frequency domain. Here we have drawn the figure, we can see that by using a sin *c* function, we can reduce the bandwidth requirement just half of the bit rate. And now it is very less sensitive to the ISI.

This sin *c* function is given by :

$$P(t) = \frac{\sin 2\pi B_0 t}{2\pi B_0 t} = \sin c\,(2B_0 t) \tag{4.25}$$

where, $B_0 = \frac{1}{2T_b}$ as discussed above.

In the frequency domain, P (*f*), the fourier transform of P (*t*) is given by :

$$P(f) = \frac{1}{2B_0},\ 0 \le |f| < B_0 \tag{4.26}$$

$$= 0\ B_0 < |f|.$$

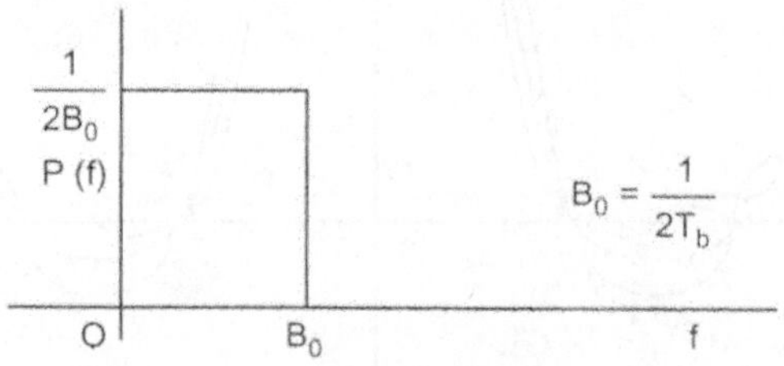

(a) Ideal pulse spectrum in time domain

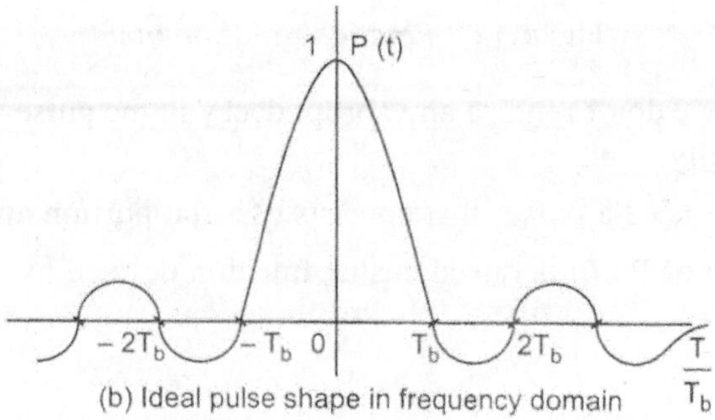

(b) Ideal pulse shape in frequency domain

Fig. 4.13 *Ideal solution of ISI*

Thus, the P (f) signal minimize the problem of ISI ; but the main disadvantages of this method is that:

(i) The ideal frequency response P (f) is not possible and followed.

(ii) The time function P (t) also decreases as $1/|t|$ for large $|t|$ resulting in a slow rate of decay. This is due to discontinuity of P (f) at $f = \pm B_0$.

4.18.2 Practical Solution

So now we will try to attend the practical solution very near to the ideal one.

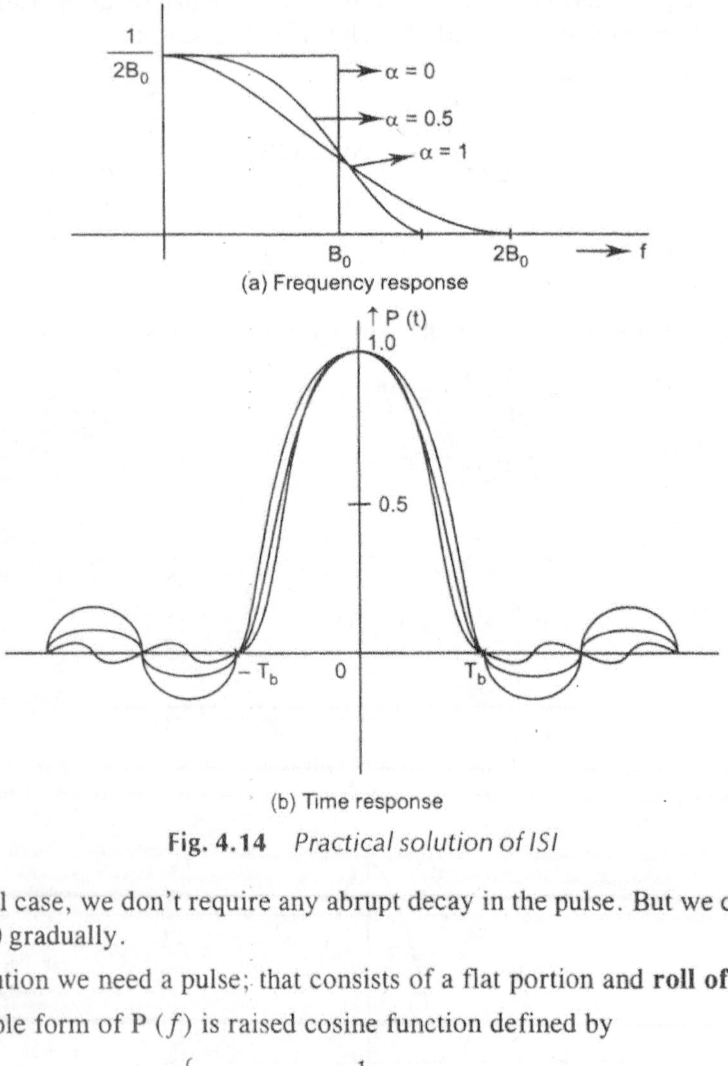

Fig. 4.14 *Practical solution of ISI*

Now in practical case, we don't require any abrupt decay in the pulse. But we can choose P (f) to decrease towards 0 gradually.

In practical solution we need a pulse; that consists of a flat portion and **roll off portion**.

One such possible form of P (f) is raised cosine function defined by

$$P(f) = \begin{cases} \dfrac{1}{2B_0} & ; \quad 0 \le |f| < f_1 \\[2ex] \dfrac{1}{2B_0}\left\{1 + \cos\left[\dfrac{\pi(|f| - f_1)}{2B_0 - 2f_1}\right]\right\} & ; \quad f_1 < |f| < 2B_0 - f_1 \\[2ex] 0 & ; \quad 2B_0 - f_1 < |f| \end{cases} \qquad (4.27)$$

The frequency f_1 is given by :

$$\alpha = 1 - \frac{f_1}{B_0} \tag{4.28}$$

where α = roll of factor for $\alpha = 0$;

we have $f_1 = B_0$ and now P (f) has the ideal form.

The pulse shape P (t) corresponding to the frequency response P (f) can be obtained by taking inverse fourier transform on the both sides of equation -

Thus,

$$P(t) = \text{sin}\,c\,(2B_0 t)\frac{\cos(2\pi\alpha B_0 t)}{1 - 16\alpha^2 B_0^2 t^2} \tag{4.29}$$

The time response P (t) consists of two factors :

(i) The factor sin c ($2B_0t$) which is same as that associated with the ideal solution.

(ii) A factor $\frac{\cos(2\pi\alpha B_0 t)}{1 - 16\alpha^2 B_0^2 t^2}$ which decreases as $\frac{1}{|t|^2}$ for large $|t|$.

Due to these roll of factor causes the practical solution of that practical difficulty.

Bandwidth requirement for raised cosine function :

We can see from figure, that now bandwidth requirement for the spectrum is :

$$B = 2B_0 - f_1 \tag{4.30}$$

Now again roll off factor

$$\alpha = 1 - \frac{f_1}{B_0}. \tag{4.31}$$

So, we obtain

$$B = B_0\,(1 + \alpha) \tag{4.32}$$

But the transmission bandwidth of ideal case = B_0.

So, we get the exceed in the bandwidth requirement = αB_0.

Example 1. *An NRZ-L signal is passed through a filter with r = 0.6 and then modulated onto a carrier. The data rate is 2400 bits/sec. Evaluate the bandwidth for ASK and FSK. Assume Δf = 5 KHz for FSK.*

Solution. Bandwidth for ASK = $(1 + r)$ R

Here we have $r = 0.6$

$R = 2400$

Therefore for bandwidth = (1 + .6) 24000 = 1.6 × 2400 = 3840 Hz.

Banwidth for FSK = $2\Delta F + (1 + r)\,R = 2 \times 5 \times 10^3 + (1 + .6)\,2400$ = 13.84 KHz.

Example 2. *Binary data is transmitted over a certain channel at a rate of R_b bits/S. To reduce the transmission bandwidth, it is decided to transmit this data using 8–ary multiamplitude signaling.*

(a) *By what factor is the bandwidth reduced ?*

(b) *By what factor is the transmitted power increased, assuming the minimum separation between pulse amplitudes to be the same in both cases ?*

[**Hint** : Take the pulse amplitudes to be ± A/2, ± 3A/2, ± 5A/2 and ± 7A/2 so that the minimum separation between various amplitude levels in A, the same as in the binary case pulses ± A/2. Assume all the eight levels equally likely].

Solution. (*a*) For an 8–array system bandwidth will reduce by a factor of $\log_2 8 = 3$

(*b*) Let the pulse amplitudes for 8 array system be $\pm\frac{A}{2}, \pm\frac{3A}{2}, \pm\frac{5A}{2}$ and $\pm\frac{7A}{2}$ and pulse amplitude for binary system be $\pm\frac{A}{2}$.

Since power is directly proportional to square of the maximum amplitude

$$\therefore \quad \frac{P_{binary}}{P_{8\,array}} = \left(\frac{\frac{A}{2}}{\frac{7A}{2}}\right)^2 = \frac{1}{49}$$

Example 3. *Sketch the waveform for the bit stream 0 1 0 0 1 1 1 0 using pseudoternary.*

Solution. Sketch 0 1 0 0 1 1 1 0 using pseucloternary

In psuedotevnary a 1 in encoded by no pulse while a 0 in encoded by a pulse $p(t)$ or $-p(t)$ depending on whether the previous 0 is encoded by a pulse $p(t)$ or $-p(t)$

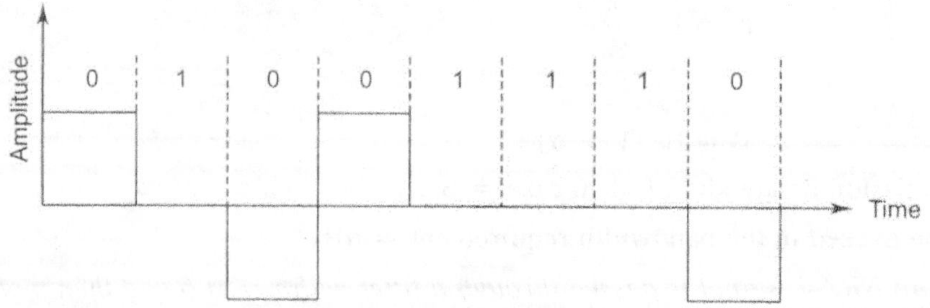

Fig. 4.15 *Psuedototernary*

REVIEW QUESTIONS

1. Define and draw the different types of line codings.
2. For digital data

   1 1 0 1 0 1 0

   Draw :

   (*a*) RZ and NRZ Unipolar

(*b*) RZ and NRZ Polar
(*c*) RZ and NRZ Bipolar
(*d*) RZ and NRZ Manchester Coding.

3. What is ISI ? Explain the ideal and practical solution of this problem.

4. Explain explanatory notes on
(*i*) B-8ZS signaling
(*ii*) HDBZ coding

5. Enumerate the need for line coding list its advantages and disadvantages.

6. List the important properties, a digital signaling format must posses.

7. Sketch the various digital signaling formats for the bit stream $b(t)$ = 100101101001.

5

Digital Multiplexing

5.1 MULTIPLEXING

Multiplexing is the set of techniques that allows simultaneous transmission of multiple signals across a single data link. It is method for efficient utilization of system resources.

As data and telecommunications usage increases, so does traffic. We can accommodate this increase by continuing to add individual lines each time a new channel in needed, or we can install higher capacity links and use each to carry multiple signals.In a multiplexed system, *n* devices share the capacity of one link. Signals are multiplexed using three basic techniques :

(*i*) Frequency division multiplexing (FDM)

(*ii*) Time division multiplexing (TDM)

(*iii*) Code division multiplexing (CDMA).

The time division multiplexing is subdivided into synchronous (usually just called TDM) and asynchronous TDM, also called statistical TDM or concentrator.

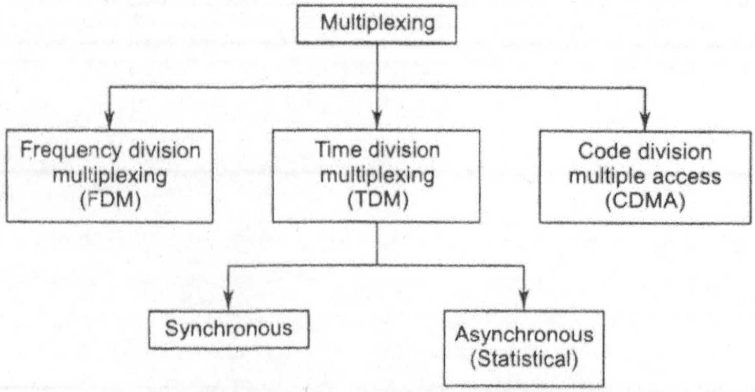

Fig. 5.1 *Classification of multiplexing techniques*

5.2 FREQUENCY DIVISION MULTIPLEXING

Frequency division multiplexing is an analog technique that can be applied when the bandwidth of a link is greater than the combined bandwidths of the signals to be transmitted. In FDM the bandwidth is divided into a number of frequency slots, each of which can accommodate the signal of an individual connection.

Fig. 5.2 gives a conceptual view of FDM.

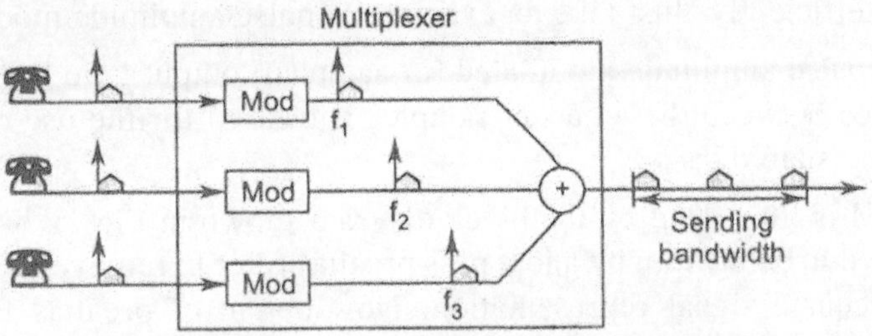

Fig. 5.2 *(a) FDM at transmitting end*

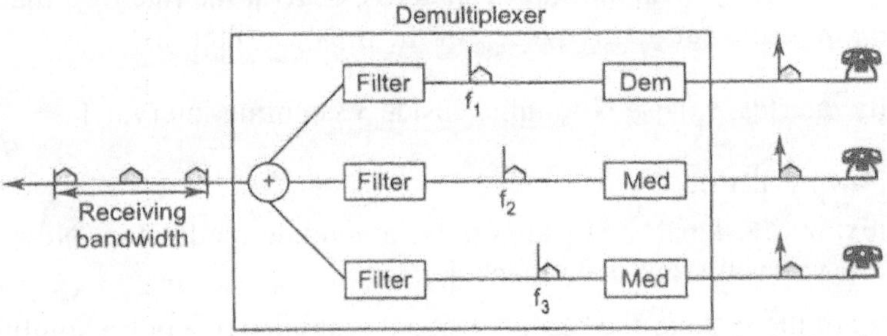

Fig. 5.2 *(b) FDM at receiving end*

5.3 FEATURES OF FDM

(*i*) FDM requires tight RF filteing to minimize adjacent channel interference.

(*ii*) FDM system have higher system costs as compared to TDM systems, because of single channel per carrier design and the need to use costly band pass filters to eliminate spurious radiation at the base station.

(*iii*) Since FDM is a continuous transmission scheme, fewer bits are needed for overhead purposes (such as synchronization and framing bits) as compared to TDM.

(*iv*) The complexity of FDM system is lower when compared to TDM system, though this is changing as digital signal processing methods improve for TDM.

(*v*) The bandwidth of FDM channels are relatively narrow as each channel supports only one circuit per carrier. That is, FDM is usually implemented in narrow band systems.

(*vi*) The symbol time of a narrowband signal is larger as compared to the average delay spread. This implies that the amount of inter symbol interference is low and, thus, little or no equalization is required in FDM narrowband systems.

(*vii*) If an FDM channel is not in use, then it sits idle and cannot be used by other users to increase or share capacity. It is essentially a wasted resource.

(*viii*) The FDM channel carries only one phone circuit at a time.

(*ix*) After the assignment of a frequency channel to user, transmission is continuous.

5.4 TIME DIVISION MULTIPLEXING (TDM)

Multiplexing may be defined as a technique which allows many users to share same communication channel simultaneously. The sharing of channel generally is based on the basis of time and frequency and they are called **Time Division Multiplexing** and **Frequency Division Multiplexing** respectively. Here we will discuss time division multiplexing. As we know that in digital signals, we have space between the successive digital code word that spacing can be utilized by the other signals. Thus, we can utilize the channel efficiently. Just take an example of pulse amplitude modulation.

As we know that in pulse amplitude modulated (or sampled) output ; we have some samples in a time period so the space between the adjacent samples is cleared for the use of other independent message signal on a time-shared basis.

The concept of TDM is illustrated by the block diagram shown in Fig. 5.3. Each input message signal in firstly restricted in bandwidth by a low pass prealias filter to remove the frequencies that are non-essential to an adequate signal representation. Now output of prealias filter is applied to a commutator. The commutator is a electronic switch. The function of commutator is :

(1) To take a narrow sample of each of the N input message at the rate of f_s that is slightly higher than 2W where W is the cut off frequency of the prealias filter and

(2) To sequentially interleave these N samples inside a sampling interval $T_s = \frac{1}{f_s}$. This operation is basically showing TDM.

After the multiplexing the signals ; signal is pulse amplitude modulated. Now after modulation, the signal is suitable for transmission over channel.

At the receiver end of the system, the received signal is applied to a pulse amplitude demodulator. This demodulator performs the reverse process to that of pulse amplitude modulator.

Again we have decommutator that redistribute the signals at the N channel. For proper redistribution; this decommutator should be sychronized with commutator switch.

Analysis :

In a time period T_s total pulses = N

(*i*) Number of pulses in 1 one second $= \frac{N}{T_s}$

or Signaling rate $= Nf_s$

or signaling rate

$$\boxed{r = Nf_s \geq 2Nf_H} \qquad (5.1)$$

(*ii*) Transmission Bandwidth of TDM channel

$$= \text{Bandwidth of LPF}$$

$$= \frac{1}{2} Nf_s$$

$$(f_s)_{min} = 2f_H$$

So $$\boxed{(B.W.)_{min} = Nf_H.} \qquad (5.2)$$

Here the multiplexed signal have some bandwidth f_H.

Limitations :

1. The TDM is highly sensitive to the dispersion in the common channel. So that we can see inter symbol interference in the TDM.
2. TDM also have crosstalk due to non-linearity of channel.

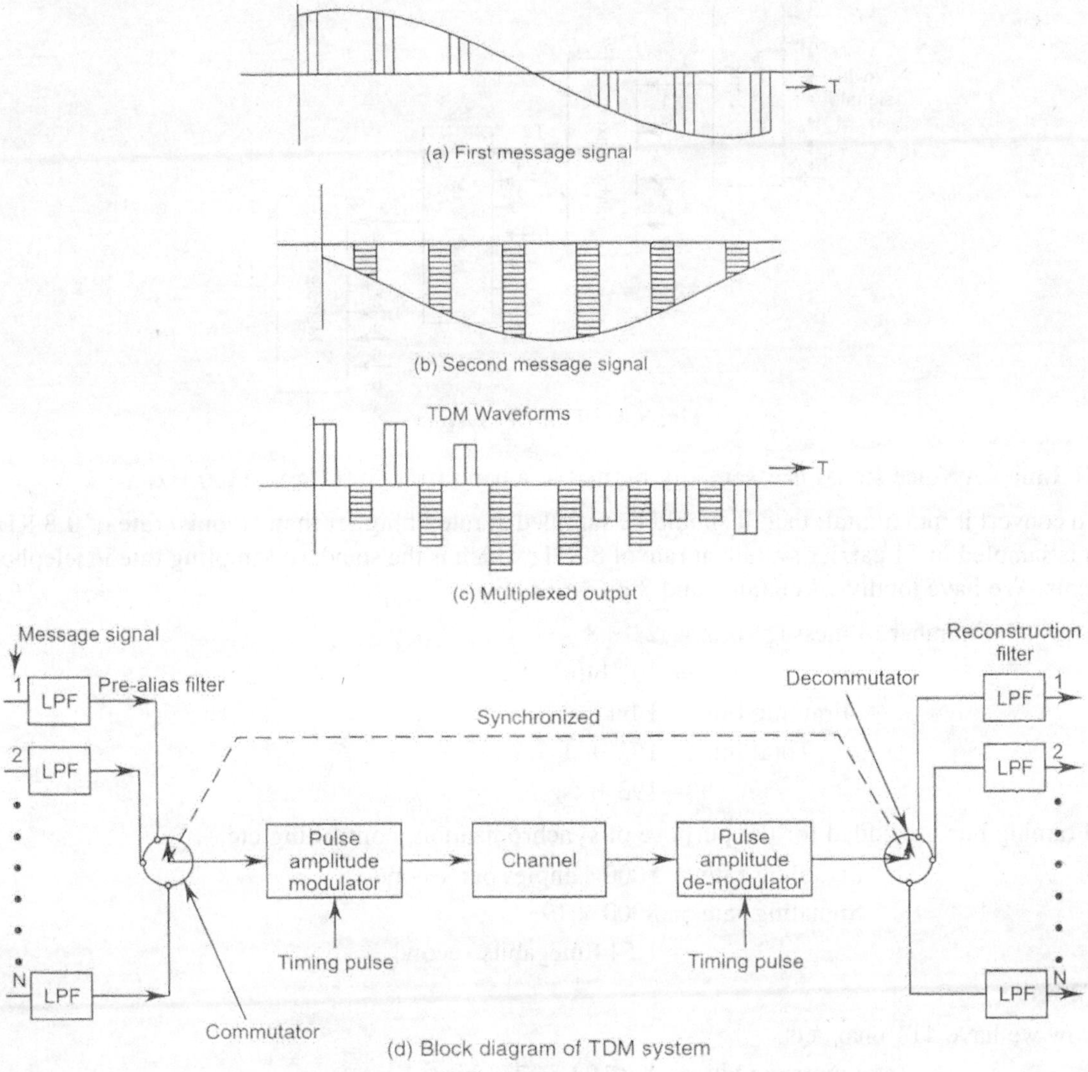

(d) Block diagram of TDM system

Fig. 5.3

5.5 DIGITAL MULTIPLEXER

In TDM ; we have seen that a group of analog signals are sampled sequentially in time at a common sampling rate and then multiplexed for transmission over a common line. Now we will consider the multiplexing of digital signals at different bit rate. Due to this type of service, we are able to combine several digital signals such as computer output, digitalized voice signal television signals into a single data stream.

In digital multiplexing system, a low speed data is taken and then with the help of a selector switch interleaving of bits take place and then signal is transmitted on a high speed channel.

5.6 THE T1 DIGITAL SYSTEM

The basic time division multiplexing scheme called the T1 digital system developed by AT and T Bell labs is shown in Fig. 5.4. It uses multiple signals over telephone line using wideband coaxial cable.

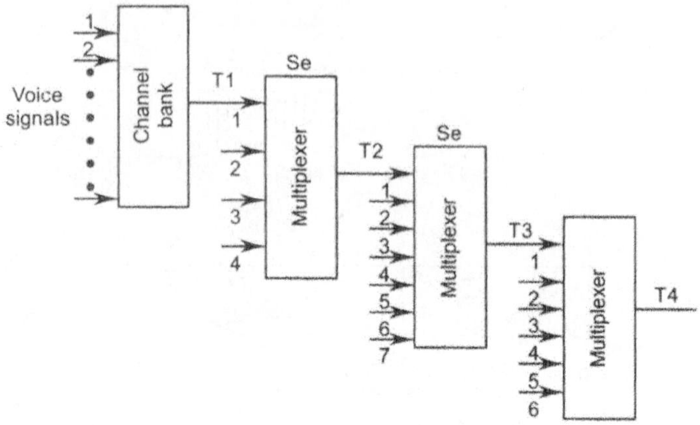

Fig. 5.4 *T carrier system*

T1 Line : A voice signal is essentially limited to a band from 300 Hz to 3400 Hz.

To convert it into digital; then it should be sampled at rate of higher than Nyquist rate of 0.8 KHz. So it is sampled in T1 carrier system at rate of 8 KHz which is the standard sampling rate in telephone systems. We have totally 24 channel and 8 bit word length.

Thus total number of message bits $= 24 \times 8$

$= 192$ bits.

Framing bits $= 1$ bit

Total bits $= 192 + 1$

$= 193$ Bits.

(Framing bits are added for the purpose of synchronization, controlling etc.)

Sampling rate $= 8000$ samples per second

Signaling rate $= 8000 \times 193$

$= 1.544$ megabits/second.

T2 line :

Now we have 4T1 channels.

So total message bits $= 4 \times 193 = 772$ bits.

Added frame bits $= 17$ bits

So total bits $= 772 + 17$

$= 789$ bits

Signaling rate $= 789 \times 8000$

$= 6.312$ Mb/second.

T3 line :

This stage has 7 channels each having 789 bits.

Total message bits $= 789 \times 7$

$= 5523$ bits

Added frame bits = 69 bits

So total bits = 5523 + 69

= 5592 bits.

Signaling rate = 5592 × 8000

= 44.736 Mb/sec.

T4 line :

This stage has 6 channels each having 5592 bits.

Total message bits = 5592 × 6

= 33552 bits

Added frame bits = 720 bits

Total bits = 34272 bits.

So signaling rate = 34272 × 8000

= 274.176 Mb/sec.

5.6.1 M12 Multiplexer

Four T1 lines are multiplexed in a M12 multiplixer to generate T2 transmission system. (Similarly seven T2 lines are multiplexed in a M23 multiplexer to generate T3 etc). But all the T1 lines cannot be multiplexed without any arrangement. They must include some form of framing, so that its individual components can be identified at the receiver.

M_0 (48)	C_I (48)	F_0 (48)	C_I (48)	C_I (48)	F_1 (48)
M_1 (48)	C_{II} (48)	F_0 (48)	C_{II} (48)	C_{II} (48)	F_1 (48)
M_1 (48)	C_{III} (48)	F_0 (48)	C_{III} (48)	C_{III} (48)	F_1 (48)
M_1 (48)	C_{IV} (48)	F_0 (48)	C_{IV} (48)	C_{IV} (48)	F_1 (48)

Fig. 5.5 *M12 MUX*

Fig. 5.5 illustrates the signal formate of M12 multiplexer. Total bits (789 bits) are not transmitted in bulk but they are grouped into 48 bits each. Now this complete frame is again divided into smaller subframes each having 6 packets of 48 bits. The first subframe is transmitted, then the second, the third and the fourth in that order. 48 bits are not taken by single channel. Bit-by-bit interleaving of the incoming four T1 bit streams is used to accumulate a total of 48 bits, 12 from each input. A control bit is then inserted by the multiplexer. Each frame contains total 24 control bits, separated by sequences of 48 data bits. Three types of control bits are used in the M12 Multiplexer to provide synchronization and frame indication and to identify which of the 4 input signals has been stuffed. These control bits are labeled as F, M, and C as shown in Fig. 5.5. These functions are as follows :

1. The F-control bits, two per subframe, constitute the main framing pulses. The subscripts on the F-control bits denote the actual bit (0 or 1) transmitted.
2. The M-control bits, one per subframe, form secondary framing pulses to identify the four subframes. Here again the subscript on the M-control bits denote the actual bit (0 or 1) transmitted.
3. The C-control bits, three per subframe are stuffing indicators. In practicular

C_I refers to input channel I,

C_{II} refers to input channel II, and so on.

The insertion of a stuffed bit in this T1 signal is indicated by setting all three C-control bits to '1'. To indicate no stuffing three are set '0'.

The demultiplexer at the receiving M12 unit first search for the main framing sequence $F_0F_1F_0F_1F_0F_1F_0F_1$. This establishes identify for the four input T1 signals and also for the M and C-control bits. From the M_0M_1 M_1M_1 sequence, the correct framing of the C-control bit is verified. Finally the four T1 signals are properly demultiplexed and destuffed.

The signal format just described has two safeguards.

1. It is possible although unlikely, that with just the F_0F_1 F_0F_1 F_0F_1 F_0F_1 sequence, one of the incoming T1 signals may contain a similar sequence. This could then cause the receiver to decode a wrong code that is why $M_0M_1M_1M_1$ provide verification of proper demultiplexing.
2. The single error correction capability built into the C-control bit ensures that the four T1 signals are properly demultiplexed.

5.6.2 Signaling

In addition to the voice signal, a telephone system must also pass special supervisory signals at the far ends such as for dial tone, for telephone off-hook/on-hook. In the *T1 carrier system* this requirement is accomplished as follows.

For five frame, 8 bits are sended for information in a frame ; but every sixth frame, the least significant bit of each voice channel is deleted and a signaling bit is inserted in its place. This is called $7\frac{5}{6}$ framing.

This is repeated for every sixth frame. The sequence of signaling bits is thus transmitted at a rate equal to the sampling rate divided by six that is $8000 \times \frac{1}{6} = 1333$ Kb/sec.

This type of signaling is called channel associated signaling.

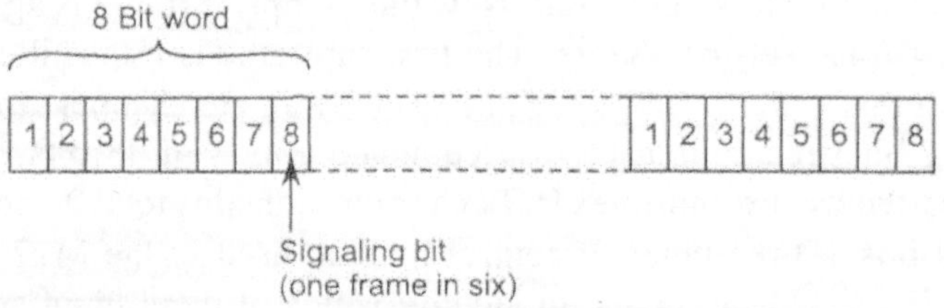

Fig. 5.6 *T1 carrier system*

5.7 SYNCHRONOUS TDM

In synchronous time division multiplexing, The term synchronous has a different meaning from that used in other areas of telecommunications. Here synchronous means that the multiplexer allocates exactly the same time slot to each device at all times, whether or not a device has anything to transmit. Time slot A, for example, in assigned to device A alone and cannot be used by any other device. Each time its allocated time slot comes up, a device has the opportunity to send a portion of its data. If a device is unable to transmit or does not have data to send, its time slot remains empty.

Time slots are grouped into frames. A frame consists of one complete cycle of time slots, including one or more slots dedicated to each sending device (as shown in Fig. 5.7). In a system with *n* input lines, each frame has at least *n* slots, with each slot allocated to carrying data from a specific input line. If all the input devices sharing a link are transmitting at the same data rate, each device has one time slot per frame. However, it is possible to accommodate varying data rates. A transmission with two slots per frame will arrive twice as quickly as one with one slot per frame. The time slots dedicated to a given device occupy the same location in each frame and constitute that devices channel. In Fig. 5.7 we have shown 5 input lines multiplexed onto a single path using synchronous TDM. In this example, all of the inputs have the same data rate, so the number of time slots in each frame is equal to the number of input lines.

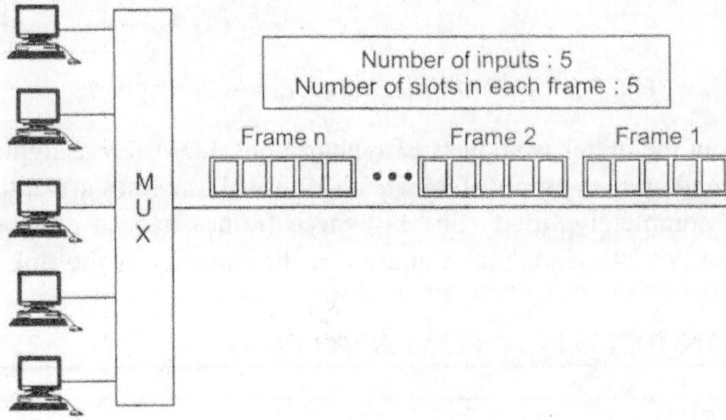

Fig. 5.7 *Synchronous TDM*

Synchronous TDM can be compared to a very fast rotating switch. As the switch opens in front of a device, that device has the opportunity to send a specified amount (x bits) of data onto the path. The switch moves from device to device at a constant rate and in a fixed order. This process is called interleaving.

Interleaving can be done by, bit by byte or by any other data unit. In other words, the multiplexer can take one byte from each device, then another byte from next device and so on. In a given system, the interleaved units will always be of the same size.

Fig. 5.8 shows interleaving and frame building. In the example, we interleave the various transmission by character (equal to one byte each), but the concept is same for data units of any length. As we can see, each device is sending a different message. The multiplexer interleaves the different messages and forms them into frames before putting them onto the link.

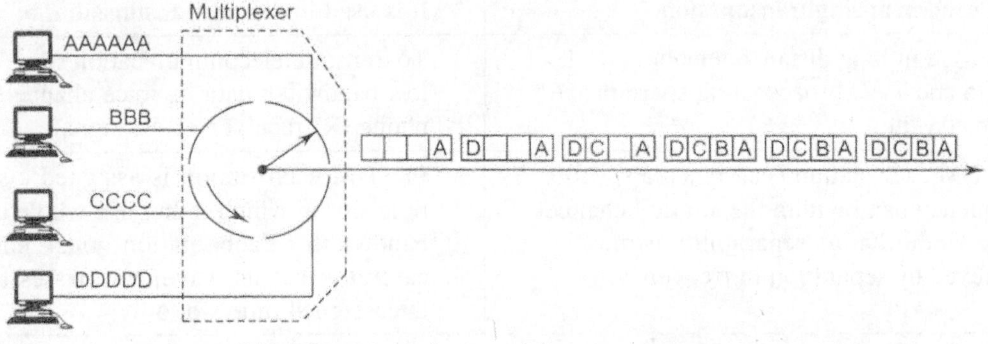

Fig. 5.8 *Multiplexing of frames at transmiter*

At the receiver, the demultiplexer decompose each frame by extracting each character in turn. As a character is removed from a frame, it is passed to the appropriate receiving device.

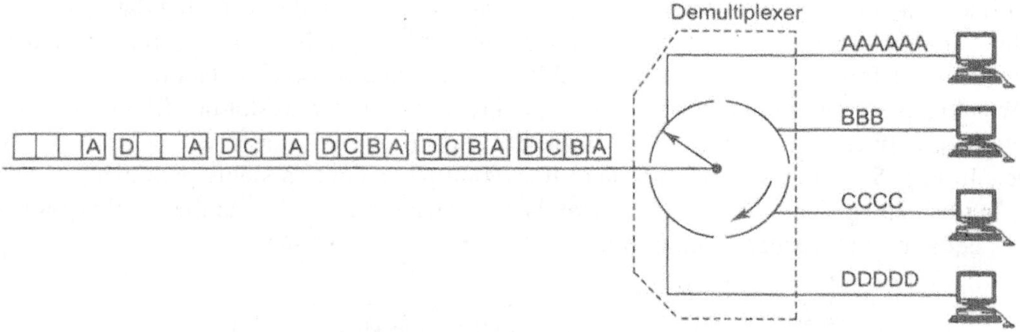

Fig. 5.9 *Demultiplexing of frames at receiver*

Fig. 5.9 also point out the major weakness of synchronous TDM. By assigning each time slot to a specific input line, we end up with empty slots whenever not all the lines are active. In figure only the first there frames are completely filled. The last three frames have a collective six empty slots. Having 6 empty slots out of 24 means that a quarter of the capacity of the link is being wasted.

5.8 COMPARISON BETWEEN TDM AND FDM

FDM	*TDM*
In FDM the available channel band width is divided into slots of frequency.	In TDM time frame is divided into slots.
In FDM the signals are separated in frequency domain but mixed together in time domain.	In TDM the analog signals are separated in time but mixed together in frequency domain.
FDM equipment consists of analog circuit for modulator, carrier generator BPF, demodulator for each channel.	TDM circuitry is similar because of digital consister of a commutator distributor.
FDM does not provide reliable and efficient operation	Here circuitry provide reliable and efficient operation.
FDM uses SSB modulation.	TDM uses PCM modulation.
It is used in analog transmission.	It is used in digital transmission.
It is used in long distance telephone FM Radio and T.V. broadcasting spare probe telemetry etc.	To transmit telecommunication slow, change low bandwidth data in voice channel picture phone (R) receiver, T.V. signal.
In FDM each station is assigned a specific frequency band within the allotted channel bandwidth. Station separability is thus achieved by separation in frequency.	In TDM each station is assigned a specific time slot in which it can use whole channel bandwidth. Each station must ensure it carefully that its waveform passes through the assigned time slot only

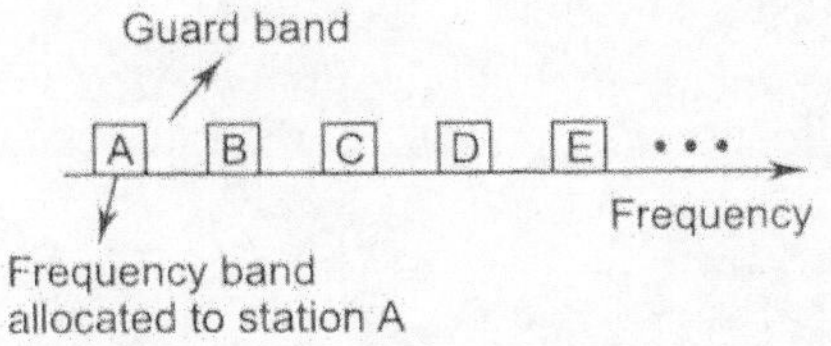 *Frequency separation in FDM*	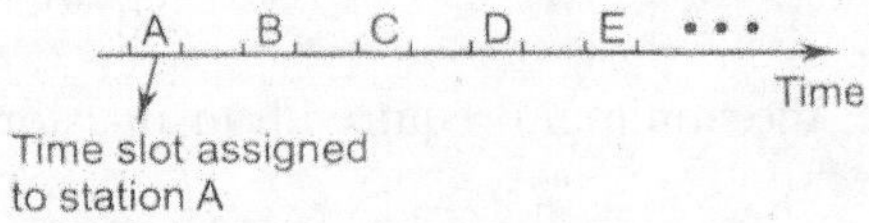 *Time separation in TDM*
FDM is the simplest and most basic format to implement since it requires station configuration most compatible to existing hardware.	TDM involves more complicated station operations, including some form of precise time synchronisation among all the users (stations).
Because of above disadavantage the transmitter amplifier cannot be operated at maximum power output or saturation level.	The above point leads to one of the most significant advantages of TDM, which is that the transmitter can be operated at maximum power output or saturation level.
In FDM Network Management is done by hardware.	In TDM Network Management is done by software.
Carrier can be digitally or analogically modulated.	A reference station transmits repetitive bursts that define the time frames and the transmitting station transmits the traffic bursts during the assigned slots within the frames. Because of the bursty nature of the transmissions. TDM is only suited to digitally modulated signals.
FDM networks can be operated with pre-assigned and demand assigned channels.	TDM networks can also be operated with both channels.
If has small channel cross talk due to non linearities in the circuit that handle the signal in transmitter and receiver.	No crosstalk due to circuit non linearities if the pulses are completely isolated non over lapping since signal from different channel are not handled.

SOLVED EXAMPLES

Example 1. *Voice signal band limited to 3.4 KHz is sampled at 8 KHz and pulse code modulated using 64 quantization levels. Ten such signals are TDM using one 5 bit synchronization word every alternate frame. Calculate the minimum channel bandwidth required for transmission of the TDM signal.*

Solution. With a sampling rate of 8 KHz each frame of the multiplexed signal occupies a period of

$$\frac{1}{8000} = 0.000125 \ \mu \text{ sec}$$

Since each frame consists of ten 5 bit words it contains a total of $10 \times 5 = 50$ bit

Then, duration of each bit is $T_b = \frac{125}{50}\ \mu\text{ sec} = 2.5\ \mu\text{ sec}$

Hence, the minimum required transmission bandwidth is

$$f_T = \frac{1}{2T_b} = \frac{1}{2 \times 2.5 \times 10^{-6}} = 200\text{ KHz}$$

Example 2. *Five signal channel are sampled and time multiplexed. The multiplexed signal is then passed through a lowpass filter. Three of the channels handle signals covering the frequency range 300 to 3300 Hz. Other two carry signals covering the range 50 Hz. to 10 KHz include 0 is a synchronizing signal.*

(a) *What is the minimum sampling ratio?*

(b) *For this sampling rate what is the minimum bandwidth of the low pass filter?*

(c) *What would be the minimum bandwidth if these channels were frequency multiplexed? Using normal AM techniques.*

Solution. (*a*) Sampling rate must be $\geq 2f_m = 2 \times 10\text{ KHz} = 20\text{ KHz}$

$T_s = \frac{1}{f_s} = \frac{1}{2\text{ KHz}} = 50\ \mu\text{ sec}$ in this time 5 channel are multiplexed. Thus $\tau = \frac{50}{5} = 10\ \mu\text{ sec}$

(*b*) Bandwidth $(\text{B.W.}) = \frac{1}{\tau} = \frac{10^6}{10} = 100\text{ KHz}$

(*c*) DSB amplitude modulation require

3 channel require $3 \times 3.3\text{ KHz} = 10\text{ KHz}$

2 channel require $2 \times 10\text{ KHz} = 20\text{ KHz}$

Minimum B.W. requires therefore $= 2 \times 20\text{ KHz} = 40\text{ KHz}$

Thus example illustrate the digital techniques require much larger bandwidth per channel making TDM unavailable.

Example 3. *Ten 9600 bps lines are to be multiplexed using TDM. Ignoring overhead bits, what is the total capacity required for synchronous TDM ? Assuming that we wish to limit average line utilization of 0.8 and assuming that each line is busy 50% of the time, what is the capacity required for statistical TDM ?*

Solution. Capacity for synchronous TDM $= 10 \times 9600 = 9600$ Kbps $= 96$ Kbps

Capacity for asynchronous TDM $= 10 \times 9600 \times \frac{\text{line busy time}}{\text{line utilization}}$

$$= \frac{10 \times 9600 \times \cdot 5}{\cdot 8} = 60\text{ Kbps}$$

Example 4. *A character interleaved time division multiplexer is used to combine the data streams of a number of 110 bps asynchronous terminals for data transmission over a 2400 bps digital line. Each source sends asynchronous characters consisting of 7 data bits, 1 parity bit, 1 start bit and 2 stop bits. Assume that one synchronisation character is sent every 19 data characters and in addition at least 3 percent of the line accommodate speed variations:*

(i) Determine the number of bits per character.
(ii) Determine the number of terminals that can be accommodated by the multiplexer.
(iii) Sketch a possible framing pattern for the multiplexer.

Solution. (*i*) Number of bits per characters = Data bits + parity bit + start bit + stop bit.

$$= 7 + 1 + 1 + 2$$
$$= 11 \text{ bits/character.}$$

(*ii*) Since 3% of the line capacity is reserved for plus stuffing to accommodate speed variations, the available capacity will be

$$= 2400 \times 0.97$$
$$= 2328 \text{ bps}$$

If we use 20 terminals sending one character at a time in TDM plus a synchronization character, the total capacity used is

$$= 21 \times 110 \text{ bps}$$
$$= 2310 \text{ bps}$$

Which is nearly equal to available capacity. Hence 20 terminals can be accommodated by the multiplexer.

(*iii*) The frame format will consist of one synchronization character, followed by 20, 11-bit terminal characters, followed by stuff bits.

Example 5. *Assume that two 600 bps terminals, five 300 bps terminals, and a number of 150 bps terminals are to be time multiplexed in a character interleaved format over a 4800 bps digital line. The terminals send 10 bits/character, and one synchronization character is inserted for every 99 data characters. All the terminals are asynchronous, and 3% of the line capacity is allocated for pulse stuffing to accommodate variations in the terminal clock rates.*

Determine the number of 150 bps terminals that can be accommodated.

Solution. Since 3% of the line capacity is reserved for pulse stuffing to accommodate speed variations, the available capacity will be

$$= 4800 \times 0{\cdot}97$$
$$= 4656 \text{ bps}$$

We have

2 terminals with 600 bps data stream = 1200 bps.

5 terminals with 300 bps data stream = 1500 bps.

n terminals with 150 bps data stream = $n \times 150$ bps.

If $n = 13$ total capacity of channel used is

$$1200 + 1500 + 13 \times 150 = 4650 \text{ bps.}$$

Which is nearly equal to available capacity. Hence 13 terminals with 150 bps can be accommodated.

Example 6. *A four channel FDM system is to be implemented using the following design strategy. Channel 1 will be retained directly at baseband. A guard band equal to 20 percent of the bandwidth of channel 1 will be maintained between the upper edge of channel 1 and the lower edge of channel 2. Similarly, a guard band equal to 20 percent of the bandwidth at channel 2 will be maintained between the upper edge of channel 2 and the lower edge of channel 3. And so on.*

(a) Draw a spectral diagram for the composite baseband spectrum, label the frequencies where the channels begin and end, and compute the total transmission bandwidth for four data channels each having a 4-KHz baseband bandwidth.

(b) Repeat (a) for four data channels having the following baseband bandwidths.
Channel 1–2 KHz; Channel 2–4 KHz; Channel 3–8 KHz; Channel 4–16 KHz

Solution. (*a*)

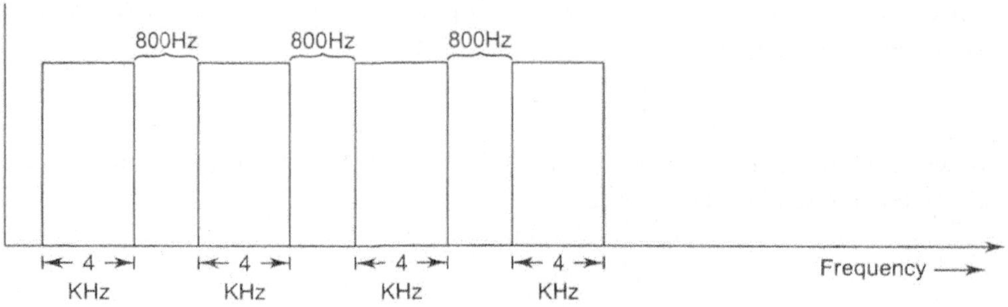

Fig. 5.10 *Spectral diagram*

Guard band $= \cdot 2 \times 4$ KHz $=$ 800 Hz

Total bandwidth $= 4 + \cdot 8 + 4 + \cdot 8 + 4 + \cdot 8 + 4$
$= 18.4$ KHz

(*b*)

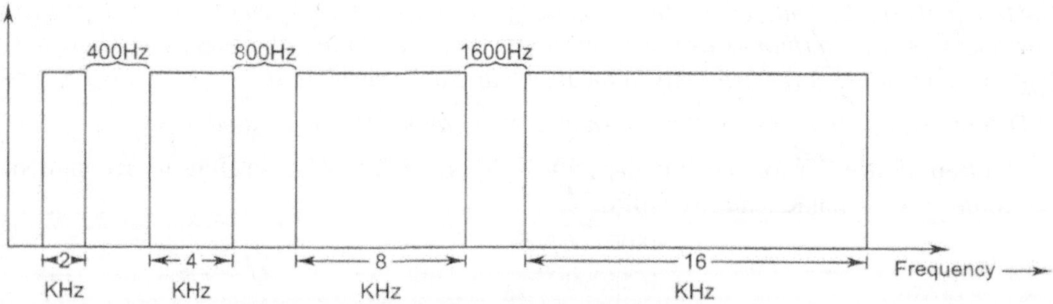

Fig. 5.11 *Spectral diagram*

Guard Band

GB(1) $= \cdot 2 \times 2$ KHz $=$ 400 Hz

GB(2) $= \cdot 2 \times 4$ KHz $=$ 800 Hz

GB(3) $= \cdot 2 \times 8$ KHz $=$ 1600 Hz

Total bandwidth $= 2 + \cdot 4 + 4 + \cdot 8 + 8 + 1\cdot 6 + 16 = 32.8$ KHz.

Example 7. *Assume that two 600 bps terminals, five 300 bps terminals and a number of 150 bps terminals are to be time multiplexed in character interleaved format over a 4800 bps digital line. The terminals send 10 bits/character and one synchronization character is inserted for every 99 data characters. All the terminals are asynchronous and 3% of the line capacity is allocated for pulse stuffing to accommodate variations in the terminal clock rates. Determine the number of 150 bps terminals that can be accommodated. Also sketch a possible framing pattern for the multiplexer.*

Solution. Given 1 character = 10 bits.

Total line capacity = 4800 bps = 180 characters.

Given data for terminals are :

(*i*) 2 terminals of 600 bps.

Therefore $600 \times 2 = 1200$ bps = 120 characters

(*ii*) 5 terminals of 300 bps = 5×300 bps = 1500 bps = 150 characters

(*iii*) Let there be n number of terminals of 150 bps

Therefore = $150\,n$ bps = $15\,n$ character

Total characters are = $270 + 15\,n$

No of synchronous character $= \dfrac{480}{99} = 5$ characters

Characters for stuffing $= \dfrac{310}{100} \times 4800 = 144$ bits = 14.4 characters.

Therefore $(270 + 15n) \times 10 + 5 \times 10 + 14.4 \times 10 \le 4800;\ n \le 13.3;\ n = 13.$

REVIEW QUESTIONS

1. Draw and explain TDM.
2. What is digital multiplexing ? Explain T1 carrier system.
3. Write short notes :
   (*a*) M12 MUX
   (*b*) Signaling technique of M12 MUX.
4. What is TDM? Comparision between TDM and FDM.
5. Define the synchronous TDM and Asynchronous TDM.

6

DIGITAL MODULATION TECHNIQUES

6.1 THE GRAM-SCHMIDT PROCEDURE

A function can be represented by a set of infinite orthogonal set of functions. But now we can construct an orthogonal set which can completely represent a signal but for now it has only finite number of functions. And Gram-Schmidt procedure describes us to do so.

Let us call the time functions which are to be expanded : $S_1(t), S_2(t), \ldots, S_N(t)$ and the orthogonal functions $U_1(t), U_2(t), \ldots, U_N(t)$. Then we require, given the function S (t), to find the function U (t) and to evaluate the coefficient S_{ij} in the expressions :

$$\left.\begin{aligned} S_1(t) &= S_{11} U_1(t) + S_{12} U_2(t) + \ldots + S_{1N} U_N(t) \\ S_2(t) &= S_{21} U_1(t) + S_{22} U_2(t) + \ldots + S_{2N} U_N(t) \\ &\vdots \\ S_N(t) &= S_{N1} U_1(t) + S_{N2} U_2(t) + \ldots + S_{NN} U_N(t) \end{aligned}\right\} \qquad (6.1)$$

or can be represented by expression :

$$S_i(t) = \sum_{j=1}^{N} S_{ij} u_j(t) \qquad (6.2)$$

$$i = 1, 2, \ldots, N.$$

The orthogonality of function U (t) over the interval T is expressed by

$$\int_T U_j(t) \, . \, U_k(t) \, dt = \begin{Bmatrix} 1 & ; & \text{if } j = k \\ 0 & ; & \text{if } j \neq k \end{Bmatrix} \qquad (6.3)$$

Now the Gram-Schmidt method procedure proceeds as follow :

Step 1. Now set for equation—, all the coefficient zero except S_{11} ; we have then

$S_1(t) = S_{11} U_1(t)$ So we can find that Squaring both sides and integrating for

$$T\int_T S_1^2(t)\,dt = \int_T S_{11}^2 U_1^2(t)\,dt$$

$$\Rightarrow \qquad S_{11} = \left[\int_T S_1^2(t)\,dt\right]^{\frac{1}{2}} \tag{6.4}$$

Now $U_1(t) \cong \dfrac{S_1(t)}{S_{11}(t)}$ can be determined. (6.5)

Step 2. Now set for equation — all coefficient zero except the first two S_{21} and S_{22}. We then have

$$S_2(t) = S_{21} U_1(t) + S_{22} U_2(t)$$

Multiplying both sides by $U_1(t)$ and integrating for interval T

$$\int_T S_2(t)\,U_1(t)\,.\,dt = \int_T S_{21} U_1(t)\,.\,U_1(t)\,dt + \int_T S_{22} U_1(t)\,.\,U_2(t)\,.\,dt \tag{6.6}$$

So,
$$S_{21} = \int_T S_2(t)\,U_1(t)\,.\,dt \tag{6.7}$$

Now that S_{21} is known we can evaluate S_{22}. We can write

$$S_2(t) - S_{21} U_1(t) = S_{22} U_2(t) \tag{6.8}$$

Squaring and integrating, we have

$$\int_T [S_2(t) - S_{21} U_1(t)]^2\,.\,dt = S_{22}^2 \int_T U_2^2(t)\,.\,dt \tag{6.9}$$

So that we have

$$S_{22} = \left\{\int_T [S_2(t) - S_{21} U_1(t)]^2\,dt\right\}^{\frac{1}{2}} \tag{6.10}$$

Finally, since now both S_{21} and S_{22} are known, we can again use equation to find $U_2(t)$ with the result

$$U_2(t) = \frac{1}{S_{22}}[S_2(t) - S_{21} U_1(t)] \tag{6.11}$$

$$= \frac{1}{S_{22}}\left[S_2(t) - \frac{S_{21} S_1(t)}{S_{11}}\right] \tag{6.12}$$

Step 3. Continuing the pattern as above we can write [setting $S_{33}(t) = 0$]

$$S_{33} = \left\{\int_T [S_3(t) - S_{31} U_1(t) - S_{32} U_2(t)]^2\,.\,dt\right\}^{\frac{1}{2}} \tag{6.13}$$

and finally we have,

$$U_3(t) = \frac{1}{S_{33}} [S_3(t) - S_{31} U_1(t) - S_{32} U_2(t)] \quad (6.14)$$

Step 4. Thus we continue the procedure until we have used all N equations and shall finally have N orthogonal function $U_1(t)$, $U_2(t)$, ..., $U_N(t)$.

In the above discussion, we have assumed that N function S (*t*) are linearly independent, that is that no one of the S (*t*) can be represented as a linear sum of the other S (*t*).

But in such a case where some expression of S (*t*) are such types

as

$$S_3(t) = C_1 S_1(t) + C_2 S_2(t) \quad (6.15)$$

where C_1 and C_2 are constant.

It means that we have total N functions out of them only M terms are linearly independent ($M \leq N$).

We can use this description of signal practically to draw the spectrum of a complicated signal.

6.2 DIGITAL MODULATION TECHNIQUES

When it is required to transmit digital data over a channel, it is necessary to modulate the incoming data on a carrier wave that is generally a high frequency sinusoidal signal.

Recall that modulation is defined as process by which some characteristics of a carrier is varied in accordance with a modulated wave.

As we have considered in analog communication system ; we varied amplitude, phase or frequency of carrier according to instantous value of message signal and called Amplitude Modulation (AM), Phase Modulation (PM) and Frequency Modulation (FM) respectively.

But in case of digital communication, the message signal (also called modulating wave) consists of digital data or an M-ary encoded version. For carrier again we use sinusoidal wave. So if we change amplitude, frequency, phase of carrier than it is called Amplitude Shift Keying (ASK), Frequency Shift Keying (FSK) and Phase Shift Keying (PSK) respectively. We have drawn the waveform of ASK, FSK and PSK respectively.

We can see in Fig. 6.1 that due to digital digits '1' and '0', we see the switching in Amplitude, Frequency and phase of carrier wave so it is called switching modulation.

Now we will study these modulation techniques one by one in detail.

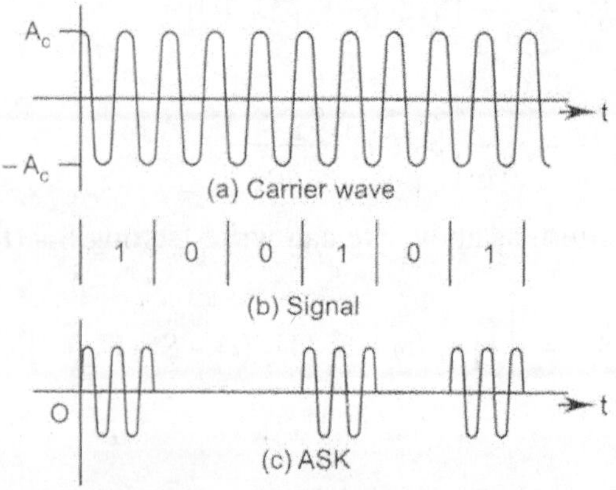

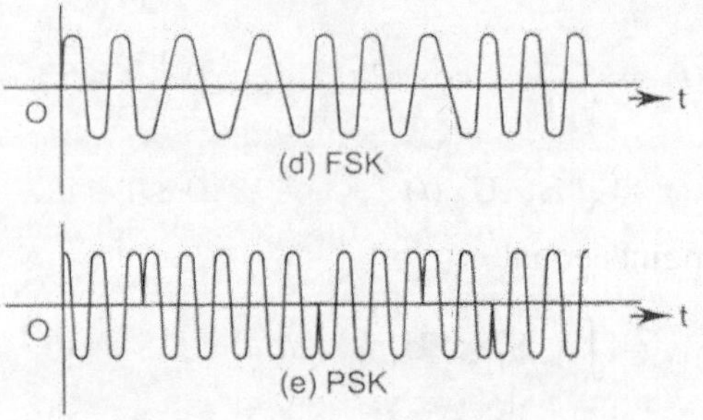

Fig. 6.1 *Digital modulation*

6.3 AMPLITUDE-SHIFT KEYING (ASK)

In ASK system, binary symbol '1' is represented by transmitting a sinusoidal carrier wave of fixed amplitude A_c and fixed frequency f_c for a bit duration T_b. Seconds whereas binary symbol 0 is represented by switching off the carrier for T_b seconds. This signal can be generated simply by turning the carrier on and off for prescribed period, thats why it is called On-Off Keying (OOK).

Let the sinusoidal carrier be represented by

$$S_c(t) = A_c \cos 2\pi f_c t$$

Then ASK signal can be represented by— (6.16)

$$S(t) = \begin{cases} A_c \cos 2\pi f_c t & ; \text{ Symbol 1} \\ 0 & ; \text{ Symbol 0} \end{cases} \quad (6.17)$$

In more general form

$$S(t) = m(t) \,.\, A_c \cos 2\pi f_c t \qquad 0 \le t \le T_b \quad (6.18)$$

where $m(t)$ = '0' or 1.

This signal has power $P = \dfrac{A_c^2}{2}$ (6.19)

So $A_c = \sqrt{2P}$, Now general equation comes

$$\left.\begin{aligned} S(t) &= \sqrt{2P} \cos 2\pi f_c t & 0 \le t \le T_b \\ &= \sqrt{PT}\sqrt{\frac{2}{T_b}} \,.\, \cos 2\pi f_c t & 0 \le t \le T_b \\ &= \sqrt{E_b}\sqrt{\frac{2}{T_b}} \cos 2\pi f_c t & 0 \le t \le T_b \end{aligned}\right\} \quad (6.20)$$

where E_b = PT is the energy contained in a bit duration

Now if we take

NOTE

$$\left\{\begin{aligned} S_1(t) &= S_{11} U_1(t) \\ S_1(t) U_1(t) &= S_{11} U_1^2(t) \\ \int_T S_1(t) U_1(t)\, dt &= S_{11} \end{aligned}\right\} \quad (6.21)$$

$$U_1(t) = \sqrt{\frac{2}{T_b}} \cdot \cos(2\pi f_c t) \qquad 0 \le t \le T_b \tag{6.22}$$

So,
$$S_1(t) = \sqrt{E_b}\, U_1(t) \qquad 0 \le t \le T_b \tag{6.23}$$

The co-ordinate message points equal

$$S_{11} = \int_0^{T_b} S_1(t)\, U_1(t) \cdot dt \tag{6.24}$$

$$= +\sqrt{E_b}\,. \tag{6.25}$$

The signal constellation diagram has been drawn in Fig. 6.2 (*b*).

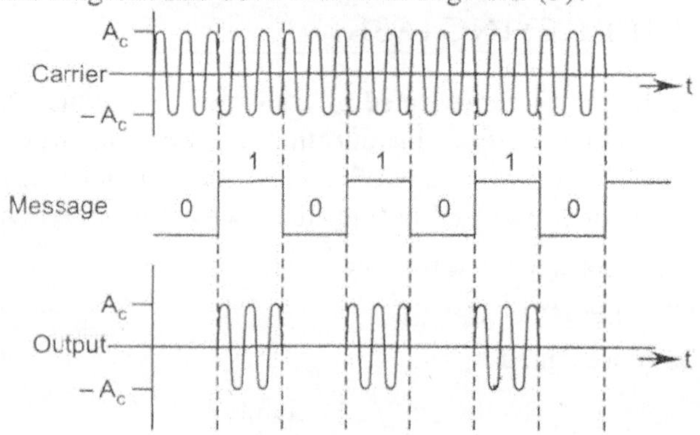

Fig. 6.2 *(a) ASK signal*

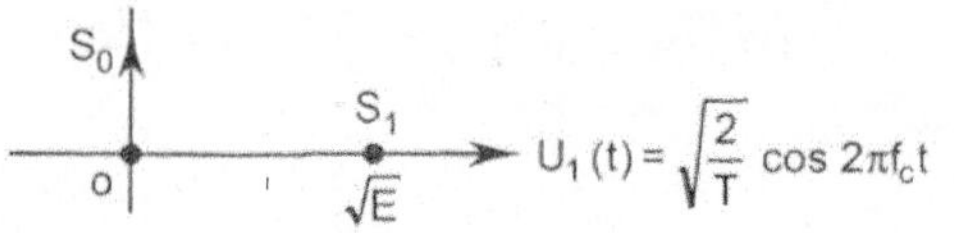

Fig. 6.2 *(b) Constellation diagram of ASK*

6.3.1 Generation of ASK Signal

ASK signal can be generated by applying the incoming binary data (should be in the unipolar formate) and the sinusoidal carrier to the two inputs of the **product modulator (Balanced Modulator)**. As we can understand that ASK signal is nothing but basically is allowing the carrier signal if we have input message bit '1' and not allowing carrier signal if message bit '0'.

Now to get the power density spectrum of the signal can be found by taking the fourier transform of ASK.

$$S(t) = M(t)\, A_c \cdot \cos 2\pi f_c t \tag{6.26}$$

that causes

$$\boxed{S(f) = \frac{A_c}{2} M(f - f_c) + \frac{A}{2} M(f + f_c).} \tag{6.27}$$

Thus, the spectrum of signal (message) is shifted by the modulated (carrier) frequency. The complete spectrum has been drawn. Thus, we can see that theoritically the bandwidth requirement of ASK is infinite but practically.

"For ASK, the bandwidth is defined as the bandwidth of an ideal bandpass filter centered at f_c whose output contains 95% of the total average power content of ASK signal". Here we can say that bandwidth of ASK is $\frac{3}{T_b}$ Hz.

6.3.2 Detection of ASK

To perform demodulation at the receiver, we have the choice of **coherent** 'or' **non-coherent detection.**

In ideal form of coherent detection, exact replicas of the possible arriving signals are available at the receiver. This means that the receiver has exact knowledge of the carrier wave. In coherent detection we cross relate the received signal with each one of replicas of exact signal and then making decision based on comparison with preselected thresolds.

The demodulation of ASK wave is done with the help of coherent detector as shown in figure. As discussed above the incoming ASK signal is applied to one input of product modulator. The other input of product modulator is supplied with a sinusoidal carrier generated with the help of local oscillator.

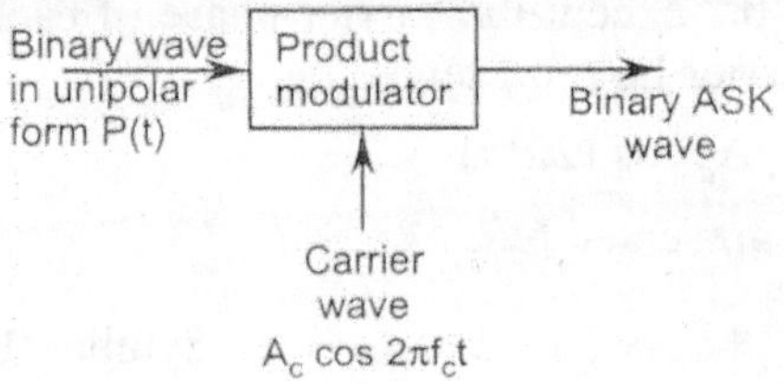

Fig. 6.3 *(a) ASK modulation*

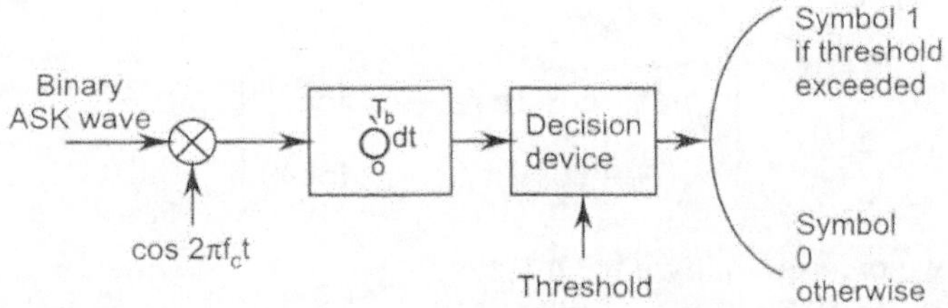

Fig. 6.3 *(b) ASK coherent demodulation*

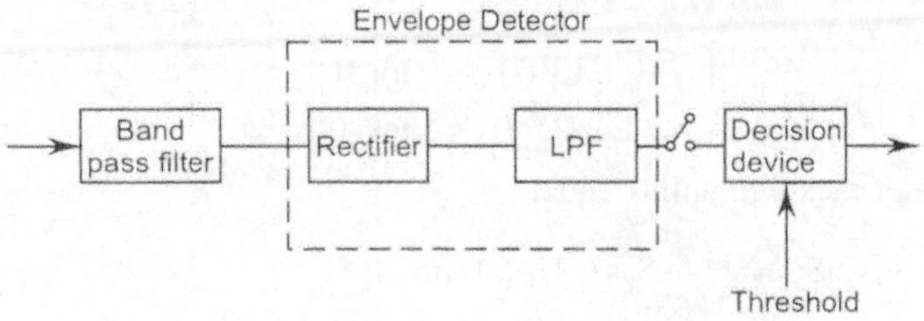

Fig. 6.3 *(c) ASK non-coherent demodulation*

The integrator operates on the output of the product modulator for successive bit intervals and essentially performs a low-pass filtering action. The output of the integrator goes to the input of a decision making device.

The decision making device compares the output of integrator with a preset threshold. It makes a decision in favour of symbol '1' if the threshold is exceed and in favour of symbol '0' otherwise. The coherent detection involves the use of linear operation. In the method discussed above it has been assumed that the local carrier is perfect synchronization with the carrier used in the transmitter. This means the frequency and phase of the locally generated carrier is same as those of the carrier used in the transmitter.

In Non-coherent detection, on the other hand, knowledge of the phase and frequency of carrier wave is not required. So that error caused due to phase and frequency synchronization is reduced. Thus, the complexicity of the receiver is thereby reduced but error performance is not as better as in the case of coherent detection.

ASK can be demodulated using Non-coherent detector using envelope detector. Basically if we get envelope, means bit '1' has been transmitted. If we get no envelope means no transmission ('0') bit.

6.4 PHASE SHIFT KEYING (PSK)

In a PSK system, a sinusoidal carrier wave of fixed amplitude and fixed frequency f_c is used to represent both symbols '1' and '0', except the carrier phase of each symbol differs by a phase of 180°. Let us the unmodulated carrier be represented by :

$$S_c(t) = A_c \cos(2\pi f_c t)$$

Then the PSK signal can be represented by :

$$S(t) = \begin{cases} A_c \cos 2\pi f_c t & ; \text{ Symbol '1'} \\ A_c \cos(2\pi f_c t + \pi) & ; \text{ Symbol '0'} \end{cases} \tag{6.28}$$

again for $E_b = PT$ we write

$$S(t) = \begin{cases} \sqrt{\frac{2E_b}{T_b}} \cdot \cos 2\pi f_c t & ; \text{ for '1'} \\ -\sqrt{\frac{2E_b}{T_b}} \cos 2\pi f_c t & ; \text{ for '0'} \end{cases} \tag{6.29}$$

For both equations we can see that we have :

$$U_1(t) = \sqrt{\frac{2}{T_b}} \cdot \cos 2\pi f_c t \tag{6.30}$$

So
$$S(t) = \begin{cases} \sqrt{E_b} \cdot U_1(t) & ; \text{ for '1'} \\ -\sqrt{E_b}\, U_1(t) & ; \text{ for '0'} \end{cases} \tag{6.31}$$

So the co-ordinates of message points equal

$$S_{11} = \int_0^{T_b} S_1(t)\, U_1(t) \, .dt$$

$$= +\sqrt{E_b} \tag{6.32}$$

$$S_{21} = \int_0^{T_b} S_2(t)\, U_1(t) \, .dt$$

$$= -\sqrt{E_b} \, . \tag{6.33}$$

The signal constellation diagram of message signal has been drawn here see Fig. 6.4 and Fig. 6.5.

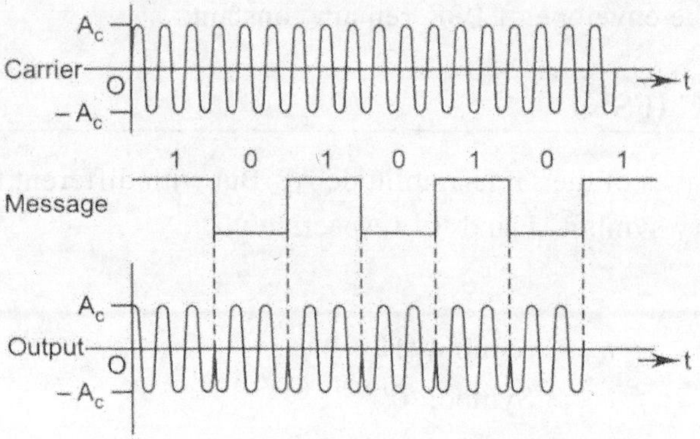

Fig. 6.4 *(a) PSK signal*

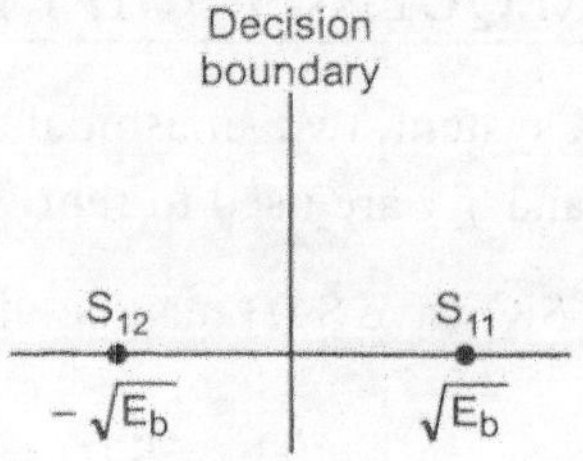

Fig. 6.4 *(b) Constellation diagram of PSK*

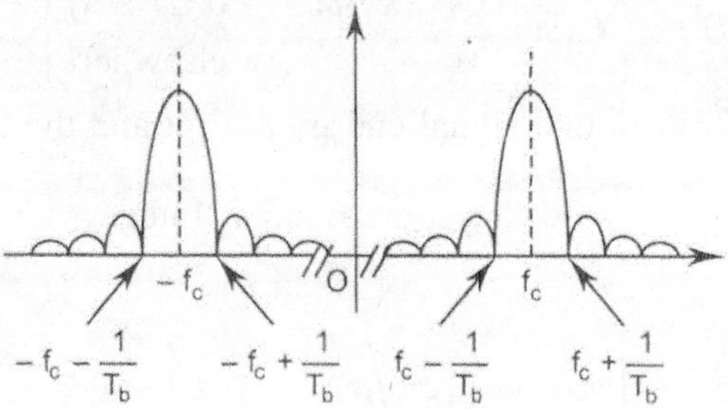

Fig. 6.4 *(c) PSD of PSK*

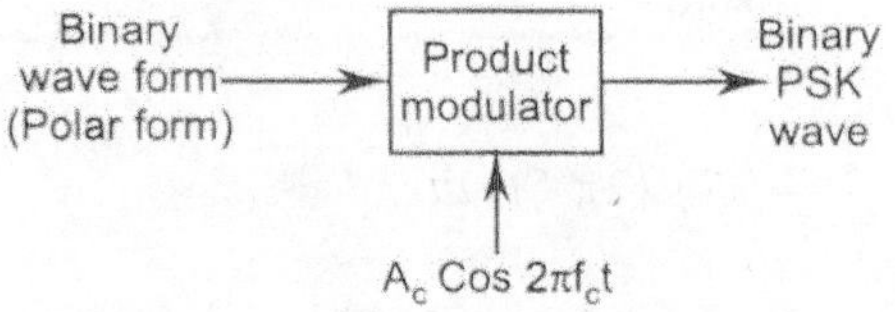

Fig. 6.5 *(a) PSK modulator*

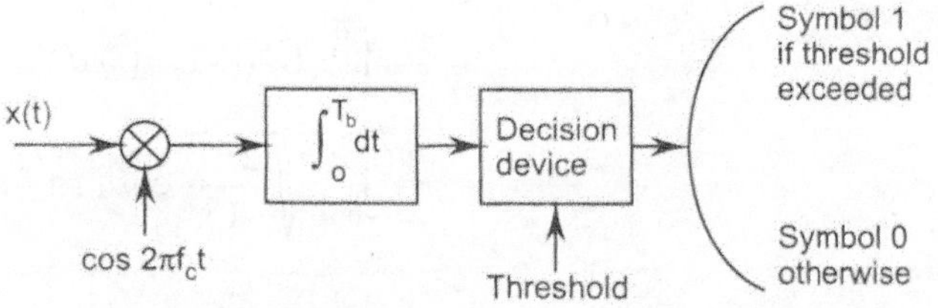

Fig. 6.5 *(b) Demodulator*

6.4.1 Generation of PSK Signal

PSK signal can be generated by using the same scheme as used in the generation of ASK. The only difference is that the incoming binary data should be in the polar form.

The PSD of PSK has been drawn. The shape of PSD of PSK and ASK signals are similar. The only difference is that the PSK does not have an impulse at the carrier frequency. The bandwidth requirement of ASK and PSK is same. But the most significant difference between these two is that ASK in a linear modulation where as PSK is a non-linear modulation scheme.

6.4.2 Detection of PSK

The coherent detection of PSK is similar to that of ASK. The only difference between coherent detection of PSK and ASK lies in the choice of the threshold.

The detection of PSK cannot be performed by using Non-coherent method because whether symbol '0' or symbol '1' is transmitted, the envelope of PSK remain constant.

6.5 FREQUENCY SHIFT KEYING (FSK)

In a FSK system, two sinusoidal carrier waves of the same amplitude A_c. But with different frequencies f_{c_1} and f_{c_2} are used to represent binary symbol '1'and '0' respectively.

The FSK wave S (*t*) may be viewed as

$$S(t) = \begin{Bmatrix} A_c \cos 2\pi f_{c_1} t & ; \text{ Symbol '1'} \\ A_c \cos 2\pi f_{c_2} t & ; \text{ Symbol '0'} \end{Bmatrix} \tag{6.34}$$

More generally it can be represented by

$$S_i(t) = \begin{Bmatrix} \sqrt{\frac{2E_b}{T_b}} \cos(2\pi f_i t) & ; \ 0 \le t \le T_b \\ 0 & ; \ \text{elsewhere} \end{Bmatrix} \tag{6.35}$$

where $i = 1, 2$, and E_b = is transmitted signal energy per pit and the transmitted frequency equals

$$f_i = \frac{n_c + i}{T_b} \quad \text{for some fixed } n_c \text{ integer.} \tag{6.36}$$

For the function :

$$U_1(t) = \begin{Bmatrix} \sqrt{\frac{2}{T_b}} \cos(2\pi f_i t) & ; \ 0 \le t \le T_b \\ 0 & ; \ \text{elsewhere} \end{Bmatrix} \tag{6.37}$$

where $i = 1, 2$. Correspondingly, the coefficient S_{ij} for $i = 1, 2$ and $j = 1, 2$ is defined by

$$S_{ij} = \int_0^{T_b} S_i(t)\, U_j(t)\, dt \tag{6.38}$$

$$= \int_0^{T_b} \sqrt{\frac{2E_b}{T_b}} \cos(2\pi f_i t) \cdot \sqrt{\frac{2}{T_b}} \cos(2\pi f_i t)\, dt$$

$$= \begin{Bmatrix} \sqrt{E_b} & ; \ i = j \\ 0 & ; \ i \ne j \end{Bmatrix} \tag{6.39}$$

Now we see that for FSK system has signal space that is two dimensional with two message points. The signal constellation has been drawn here (Fig. 6.6 and Fig. 6.7).

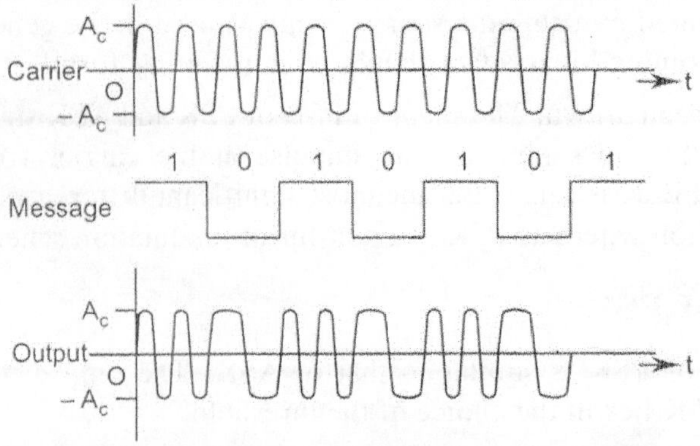

Fig. 6.6 *(a) FSK signal*

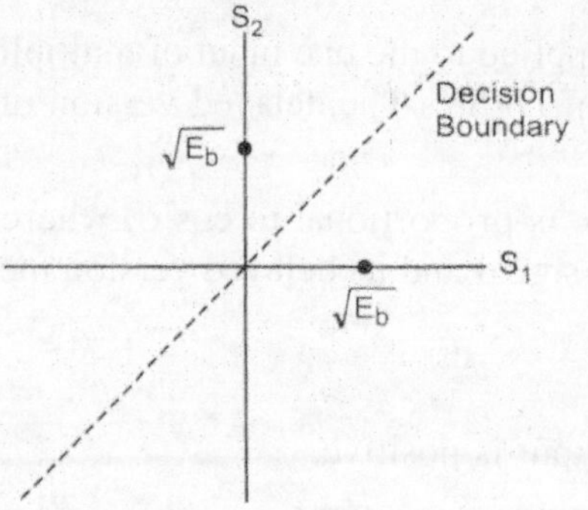

Fig. 6.6 *(b) Constellation diagram of FSK*

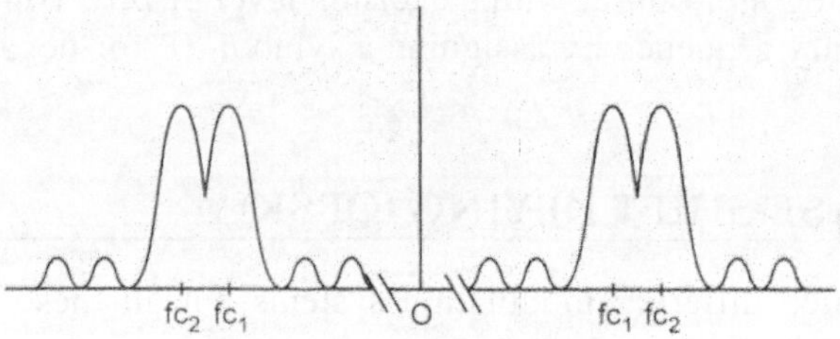

Fig. 6.6 *(c) PSD of FSK*

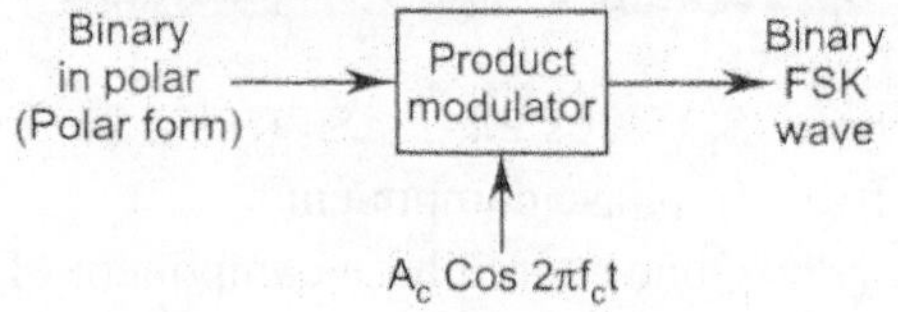

Fig. 6.7 *(a) FSK modulator*

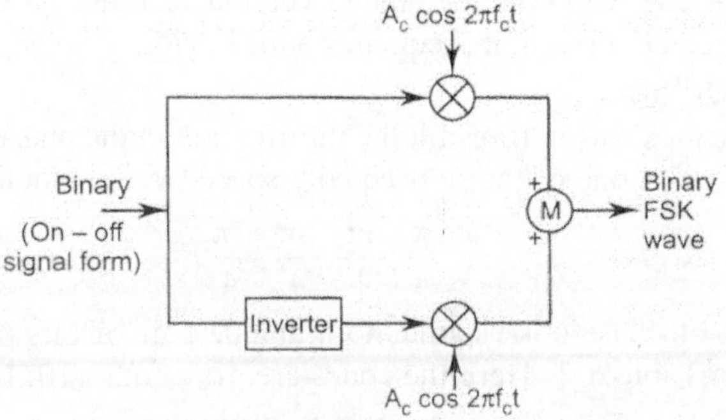

Fig. 6.7 *(b) FSK modulator*

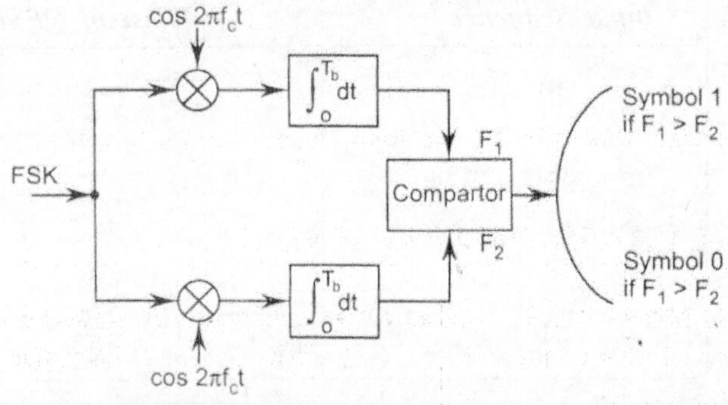

Fig. 6.7 *(c) FSK coherent demodulation*

Here the DPSK is directly applied to the one input of multiplier and the second input of multiplier is fed by the delayed version of DPSK. The delayed version of the received DPSK signal has been shown in 4th row of table.

The output of the difference is proportional to cos ϕ, where ϕ = difference between the carrier phase angle of received DPSK signal and its delayed version measured in the same bit interval. Now see the output of integrator.

When

$\phi = 0$, the integrator output is positive

$\phi = \pi$, the integrator output is negative.

By comparing the integrator output with a decision level of zero volt, the decision making device can reconstruct the binary sequence by assigning a symbol '0' for negative output and a symbol '1' for positive output.

6.7 QUADRIPHASE-SHIFT KEYING (QPSK)

Till now we have studied different modulation systems but all these techniques are not so much efficient in bandwidth. Now we will study two bandwidth conversing modulation schemes for transmission of binary data. Both are the example of quadrature carrier multiplexing system, which produces the output as :

$$S(t) = S_I(t) \cos 2\pi f_c t - S_Q(t) \sin 2\pi f_c t \tag{6.41}$$

where $S_I(t)$ = In phase component

$S_Q(t)$ = Quadrature phase component of modulated wave.

We first will study

Quadriphase shift keying which is a extended version of PSK. Next we will consider minimum shift keying that is a special version of frequency shift keying.

So let us start with QPSK.

In the PSK; the basic idea was to transmit the information in the phase change of carrier. In QPSK, the phase of carrier takes on one of the four equally spaced values such as

$$\frac{\pi}{4}, \frac{3\pi}{4}, \frac{5\pi}{4}, \frac{7\pi}{4}.$$

Each possible value of phase corresponds to a unique pair of bits called a *dibits*. Here we have complete description in Table 6.2. Here the codes are gray codes 10, 00, 01 and 11.

Table 6.2 DPSK

Input Sequence	*Phase of QPSK*
10	$\frac{\pi}{4}$
00	$\frac{3\pi}{4}$
01	$\frac{5\pi}{4}$
11	$\frac{7\pi}{4}$

6.7.1 QPSK Modulator

Fig. 6.9. The QPSK signal can be generally represented by :

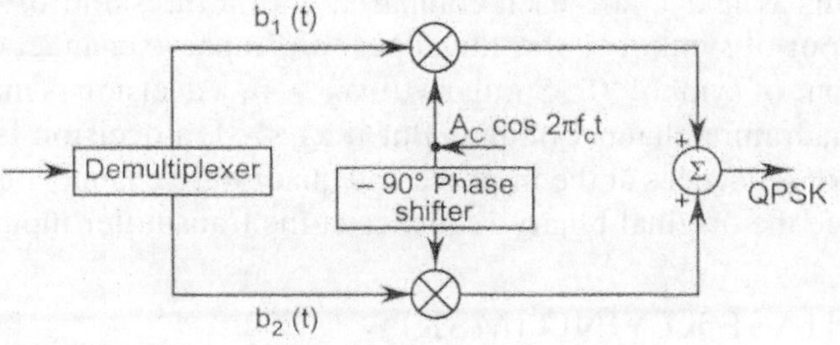

Fig. 6.9 *(a) QPSK generation*

$$S(t) = A_c \cos(2\pi f_c t + \phi(t)) \tag{6.42}$$

where
$$\phi(t) = \frac{\pi}{4}, \frac{3\pi}{4}, \frac{5\pi}{4}, \text{ or } \frac{7\pi}{4}. \tag{6.43}$$

Now equation can be expressed as

$$S(t) = A_c \cos\phi(t) . \cos 2\pi f_c t - A_c \sin\phi(t) . \sin 2\pi f_c t. \tag{6.44}$$

Here,

$A_c \cos\phi(t)$ = In phase component

$A_c \sin\phi(t)$ = Quadrature phase component.

On the basis of the above discussion the block diagram of QPSK generation has been drawn in Fig 6.9 *(a)*.

The input binary sequence is represented in polar form. This binary sequence is divided into odd and even numbered bits using a demultiplexer.

Now, even numbered bits (represents the in-phase component of Input) is multiplied by cos $2\pi f_c t$. Odd numbered bits (represents the out of phase component of input) is multiplied. Then these two streams are added to get QPSK. Note that the symbol duration, T, of a QPSK wave is twice as long as the bit duration, T_b, of the input binary wave. That is, for a given bit rate $\frac{1}{T_b}$, a QPSK wave requires half of the transmission bandwidth, a QPSK wave carriers twice as many bits of information as the corresponding binary PSK wave. This shows the basic advantage of QPSK.

6.7.2 QPSK Demodulator

The QPSK signal can be detected by using a pair of correlators (Multiplier-followed by integrator) in parallel. The circuit diagram has shown in Fig. 6.9 *(b)*.

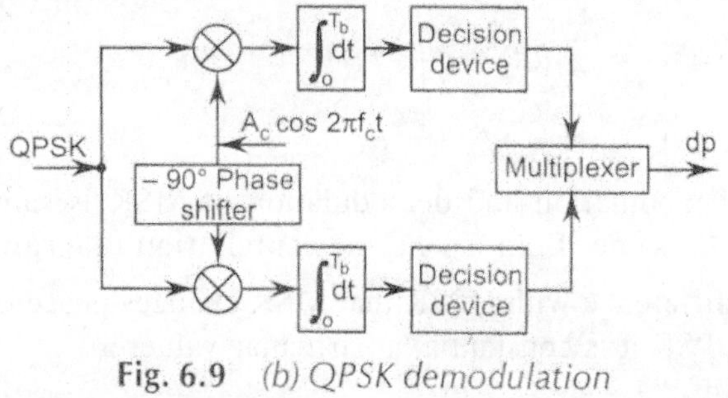

Fig. 6.9 *(b) QPSK demodulation*

The correlator in the upper path computes the cosine of the carrier wave where as the lower path computes the sine of the carrier wave.

The correlator outputs x_1 and x_2 are each compared with a threshold of zero volt. If $x_1 > 0$, a decision is made in favour of symbol '1' for the upper 'or' in phase channel output. But if $x_1 < 0$, a decision is made in favour of symbol '0'. Similarly, if $x_2 > 0$, a decision is made in favour of symbol '1' for the lower or quadrature channel output. But if $x_2 < 0$, a decision is favour of symbol '0'. Finally, these two binary sequences at the in-phase and quadrature channel outputs are combined in a multiplexer to reproduce the original binary sequence at the transmitter input.

6.8 MINIMUM PHASE KEYING (MSK)

The basic concept of FSK was changing in frequency. But we know that by $\theta = \dfrac{dw}{dt}$.

Due to change in frequency, we also have the change in phase. But FSK, maintain the phase continuity at the transition point so it is called Continuous Phase Frequency Shift Keying (CPFSK).

MSK is the special case of CPFSK, where the change in carrier frequency from symbol '0' to symbol '1' or symbol '1' to symbol '0' is taken place at one half bit rate of incoming data.

If Δf denotes the change in carrier frequency due to transition from one symbol to the other and T_b denotes the bit duration, the MSK wave will satisfy the condition.

$$\Delta f = \frac{1}{2T_b} \tag{6.45}$$

Now f_{c_1} and f_{c_2} (carriers frequencies for symbol '1' and symbol '0' respectively) can be represented in terms of Δf by

$$\left.\begin{aligned} f_{c_1} &= f_c + \frac{\Delta f}{2} \quad \text{where } f_c = \frac{f_{c_1} + f_{c_2}}{2} \\ f_{c_2} &= f_c - \frac{\Delta f}{2} \\ \text{So,} \quad \Delta f &= f_{c_1} - f_{c_2}. \end{aligned}\right\} \tag{6.46}$$

The MSK signal can be represented completely as shown in table. It is clear that MSK change state at every $t = T_b$; at $t = 2T_b$ the total phase shift is zero or 180°.

6.8.1 MSK Modulation and Demodulation

The MSK signal can be represented by the general equation

$$S(t) = A_c \cos(2\pi f_c t + \phi(t)) \tag{6.47}$$

where,

$$\phi(t) = \pm 2\pi \cdot \frac{\Delta f}{2} \cdot t \tag{6.48}$$

$$\phi(t) = \pm \pi \Delta f t. \tag{6.49}$$

The Block diagram of modulation and demodulation of MSK is same as that of QPSK as the general equation of both are same. Both have same **centellation diagrams**.

The MSK has a basic difference with QPSK that MSK changes phase twice for complete duration of symbol where as for QPSK it is constant at a particular value.

SOLVED EXAMPLES

Example 1. *Find the minimum bandwidth for an ASK signal transmitting at 2000 bps. The transmission mode is half duplex.*

Solution. In ASK, the baud rate and bit rate are the same. The baud rate is therefore 2000. An ASK signal rate is the same to its baud rate.

Therefore, the minimum bandwidth is 2000 Hz.

Example 2. *Given a bandwidth of 5000 Hz for an ASK signal, what are the baud rate and bit rate?*

Solution. In ASK the baud rate is the same as the bandwidth, which means the baud rate is 5000. But because the baud rate and the bit rate are also the same for ASK, the bit rate is 500 bps.

Example 3. *Find the minimum bandwidth for an FSK signal transmitting at 2000 bps. Transmission is in half duplex mode and the carriers must be separated by 3000 Hz.*

Solution. For FSK, if f_c and f_{c_0} are the carrier frequencies, then

$$\text{Bandwidth} = \text{Baud Rate} + \left(F_{c_1} - F_{c_0}\right)$$

However, the baud rate here is the same as the bit rats

Thus, $\text{B.W.} = 2000 + 3000 = 5000 \text{ Hz}$

Example 4. *What is the bandwidth efficiency for FSK, ASK, ASK, PSK and QPSK for a bit error rate of 10^{-7} on a channel with an SNR of 12 dB ?*

Solution. We have, $\dfrac{E_b}{N_o} = 12 \text{ dB} - \left(\dfrac{R}{B_r}\right)_{\text{dB}}$

for FSK and ASK, $\dfrac{E_b}{N_o} = 14.2 \text{ dB}$

$$\left(\frac{R}{B_r}\right)_{\text{dB}} = -2.2 \text{ dB or } \frac{R}{B_T} = 0.6$$

for PSK, $\dfrac{E_b}{N_o} = 11.2 \text{ dB}; \left(\dfrac{R}{B_T}\right)_{dB} = 0.8 \text{ dB or } \dfrac{R}{B_T} = 1.2$

The result for QPSK must take into account that the baud rate $D = \dfrac{R}{2}$ Thus, $\dfrac{R}{B_T} = 2.4$

So, ASK and FSK exhibit the same bandwidth efficiency, PSK is better, and even greater improvement can be achieved with multilevel signaling.

REVIEW QUESTIONS

1. What is modulation ? What are the advantages of modulation ?
2. Explain binary ASK. What are the various techniques to generate ASK ? Explain coherent and non-coherent demodulation technique of ASK.
3. Explain PSK. What are modulation and demodulation technique of PSK. Why coherent demodulation techniques cannot be applied for PSK detection ?
4. Explain binary FSK. What are modulation and demodulation techniques for BFSK ?
5. What is DPSK ? What are the differences between DPSK and PSK ? Explain modulation and demodulation of DPSK.
6. Explain QPSK. Is QPSK superior to PSK ? If yes why ? Explain modulation and demodulation of QPSK.
7. Explain merits of MSK.
8. Compare different digital modulation techniques.
9. Draw the block diagram of M-Ary FSK receiver. Compare M-Ary FSK with M-ary PSK
10. How is baud rate related to transmission bandwidth in QAM ?
11. What is the major factor that makes PSK superior to ASK ?
12. What are the four methods that convert a digital signal to an analog signal ?
13. The QAM demodulator corresponding to the QAM modulator. Show that this arrangement does recover the two signals $d1(t)$ and $d2(t)$, which can be combined to recover the original input.
14. An NRZ-L signal is passed through a filter with $r = 0.5$ and then modulated onto a carrier. The data rate is 2400 bps. Evaluate the bandwidth for ASK and FSK. For FSK assume that the two frequencies used are 50 kHz and 55 kHz.
15. Assume that a telephone line channel is equalized to allow bandpass data transmission over a frequency range of 600 to 3000 Hz. The available bandwidth is 2400 Hz. For $r = 1$, evaluate the required bandwidth for 2400 bps QPSK and 4800-bps, eight-level multilevel signaling. Is the bandwidth adequate ?
16. Explain the generation and demodulation BPSK signal. How DPSK is an improvement over BPSK ?
17. Explain the generation and demodulation BPSK signal. How DPSK is an improvement over BPSK?
18. Compare the following digital modulation techniques on the basis of bandwidth requirement and S/N ratio:

    (*i*) ASK (*ii*) PSK (*iii*) FSK

7

DETECTION OF DIGITAL SIGNALS

7.1 OPTIMUM RECEIVER FOR DATA COMMUNICATION

Till now we have discussed transmission. But for transmission through channel baseband transmission is not effective so we employed *digital modulation techniques*.

Now at receiver side signal is past through a filter so that at the sampling time (at time only signal is received) the signal voltage might be emphasized in comparison to Noise Voltage.

7.2 OPTIMUM FILTER

As we are considering the digital communication, the received waveform is a binary waveform. There are two bits let one binary digit (say '1') is represented by a signal waveform $S_1(t)$ while the other bit is represented by waveform $S_2(t)$ [Say $S_1(t)$ for '1' and $S_2(t)$ for '0'].

Now as shown in Fig. 7.1, the input [$S_1(t)$ and $S_2(t)$] is corrupted by addition of Noise $\eta(t)$. The Noise is Gaussian and has a spectural density $G(f)$ [In the most of the practical case; Gaussian Noise is white, so that $G(f) = \frac{\eta}{2}$].

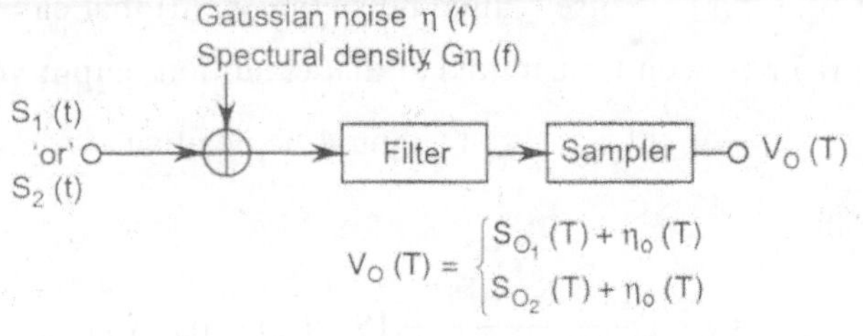

Fig. 7.1 *Receiver for binary data*

Now to recover the signal, this (signal + noise) is filtered and sampled at the rate of sampling (that was used).

So now we will concentrate upon the filters used at Receiver side. We will also try to find the optimum conditions for the designing of such filter.

At the transmitter (side). The output sample is either

or
$$V_{0_1}(T) = S_{0_1}(T) + \eta_0(T)$$
$$V_{0_2}(T) = S_{0_2}(T) + \eta_0(T) \quad (7.1)$$

where

$S_{0_1}(T)$ = is the output in the absence of noise ;

= This is output when we have transmitted signal $S_1(t)$.

= S_0 at the presence of noise ; the output $S_{0_2}(T)$ should be very less in comparison to $S_{0_1}(T)$ or

$$S_{0_1}(T) > S_{0_2}(T) \quad (7.2)$$

So that detected output is $S_1(T)$ same case is for $S_{0_2}(T)$ that is the output of $S_2(T)$ signal. So to detect $S_2(T)$, when ($S_{0_2}(T)$ has been detected) is $S_{0_2}(T) > S_{0_1}(T)$; (7.3)

So we have to make comparison between $S_{0_2}(T)$ and $S_{0_1}(T)$. And for that purpose, we have a decision boundary or threshold voltage

$$\boxed{V_0(T) = \frac{S_{0_1}(T) + S_{0_2}(T)}{2}} \quad (7.4)$$

Thus,

(*i*) If the transmitted signal is $S_2(T)$; then the difference

$V_0(T) - V_{0_2}(T)$ should be positive means if the received value of output signal is below than threshold, we will detect $S_2(t)$ [$S_2(t)$ is this time 0 binary]

(*ii*) If the transmitted signal is $S_1(T)$, then the difference

$V_{0_1}(T) - V_0(T)$ should be positive means if the received value of output signal is above than threshold we will detect $S_1(t)$ [$S_1(t)$ is this time binary '1'].

Now we will find the max limit of Noise (called noise threshold) that cannot cause any error.

If signal $S_2(t)$ (means '0') has been transmitted at that condition output voltage

$$= V_0(T) - V_{0_2}(T) \text{ should be positive}$$

Substitude values of both

$$= \frac{S_{0_1}(T) + S_{0_2}(T)}{2} - \left(S_{0_2}(T) + \eta_0(T)\right) \quad (7.5)$$

$$= \frac{S_{0_1}(T) + S_{0_2}(T) - 2S_{0_2}(T)}{2} - \eta_0(T)$$

$$= \frac{S_{0_1}(T) - S_{0_2}(T)}{2} - \eta_0(T) \tag{7.6}$$

should be positive so we will getting a error output. If

$$\boxed{\eta_0(T) \geq \frac{S_{0_1}(T) - S_{0_2}(T)}{2}} \tag{7.7}$$

It can also detect $S_1(t)$ at this time. If the value of $\boxed{S_{0_1}(T) > S_{0_2}(T)}$ because now it may come now positive.

So, if the value of $S_{0_1}(T) > S_{0_2}(T)$ and noise is positive and

$$\boxed{\eta_0(T) \geq \frac{S_{0_1}(T) - S_{0_2}(T)}{2}} \quad \text{(By equation (7.7))}$$

we are transmitting $S_2(T)$ but receiving $S_1(T)$.

The probability of error for Gaussian noise is given by :

$$P_e = \int_{\frac{S_{0_1}(T) - S_{0_2}(T)}{2}}^{\infty} \frac{e^{-n_0^2(T)/2\sigma^2}}{\sqrt{2\pi\sigma_0^2}} . \phi\, \eta_0(T) \tag{7.8}$$

Substitute, $$x = \frac{\eta_0(T)}{\sigma_0\sqrt{2}} \tag{7.9}$$

$$P_e = \frac{1}{2} \cdot \frac{2}{\sqrt{\pi}} \int_{\frac{S_0(T) - S_{0_2}(T)}{2\sqrt{2}\sigma_0}}^{\infty} e^{-x^2} . dx \tag{7.10}$$

The integration cannot be evaluated analytically, but can be determined using numerical methods. The form of equation is similar to the complementry error function which is given by :

$$erfc(U) = \frac{2}{\sqrt{\pi}} \int_{\mu}^{\infty} \exp(-z^2) . dz \tag{7.11}$$

So, $$\boxed{P_e = \frac{1}{2} . erfc\left[\frac{S_{0_1}(T) - S_{0_2}(T)}{2\sqrt{2}\sigma_0}\right]}$$

$$= \frac{1}{2}\, erfc\left[\frac{P_0(T)}{2\sqrt{2}\sigma_0}\right] \tag{7.12}$$

The complementry error function is a monotonically decreasing function. Thus, we can say that P_e will decrease as the value of difference $\left[S_{0_1}(T) - S_{0_2}(T)\right]$ becomes larger and the rms noise voltage σ_0 becomes smaller.

So the filter used by receiver is optimum filter if it maximise the ratio

$$\boxed{\gamma = \frac{S_{0_1}(T) - S_{0_2}(T)}{\sigma_0}} \tag{7.13}$$

But for the mathematical convenience, we shall actually maximize γ^2 rather than γ, now we will proceed to calculate a general expression for the transfer function of a *optimum filter*. So we can say that optimum filter operates on the principle on the maximization of the ratio γ; there a thing should be noted that γ ratio is nothing but it is signal to noise ratio of the optimum filter; so optimum filter is working on the principle of maximization of signal to noise ratio.

7.3 CALCULATION OF OPTIMUM—FILTER TRANSFER FUNCTION

The fundamental requirement we make of a binary encoded data receiver is that it distinguishes the voltages $S_1(t) + n(t)$ and $S_2(t) + n(t)$.

We have seen that the ability of receiver to do so depends on how large a particular receiver can make γ. It is important that y is proportional not to $S_1(t)$ nor to $S_2(t)$, but rather to the difference between them.

So we can easily say that the signal which is to be compare with noise is the difference signal

$$P(t) \equiv S_1(t) - S_2(t) \tag{7.14}$$

Thus, for the purpose of calculating the minimum error probability, we shall assume that the input signal to the optimum filter is P (*t*). The corresponding output signal of the filter is then,

$$P_0(t) \equiv S_{0_1}(t) - S_{0_2}(t) \tag{7.15}$$

We shall let $P(f)$ and $P_0(f)$ be the fourier transform, respectively of $P(t)$ and $P_0(t)$.

If $H(f)$ is the transfer function of filter ;

$$P_0(f) = H(f) \,.\, P(f) \tag{7.16}$$

So,

$$P_0(T) = \int_{-\infty}^{\infty} P_0(f) \,.\, e^{j2\pi fT} \,.\, df$$

$$= \int_{-\infty}^{\infty} H(f) \,.\, P(f) \,.\, e^{j2\pi fT} \,.\, df \tag{7.17}$$

The input noise of optimum filter is $n(t)$. The output noise is $n_0(t)$. If we assume the spectural density of input noise $G_n(f)$ and power spectural density of output noise is $G_{n_0}(f)$; then, they are related by the equation

$$G_{n_0}(f) = |H(f)|^2 \, G_n(f) \tag{7.18}$$

So the output noise power 'or' the noise variance is given by :

$$\sigma_0^2 = \int_{-\infty}^{\infty} G_{n_0}(f) \,.\, df$$

$$= \int_{-\infty}^{\infty} |H(f)|^2 \, G_n(f) \,.\, df \tag{7.19}$$

Now we can find the ratio γ^2 by substituting values from equation and equation is

$$\gamma^2 = \frac{P_0^2(T)}{\sigma_0^2} \tag{7.20}$$

$$\gamma^2 = \frac{\left| \int_{-\infty}^{\infty} H(f)\,P(f)\,.\,e^{j2\pi fT}\,df \right|^2}{\int_{-\infty}^{\infty} |H(f)|^2\,G_n(f)\,.\,df} \tag{7.21}$$

The Schwarz in equality states that given arbitrary complex function X (f) and Y (f) of a common variable (f), then,

$$\int_{-\infty}^{\infty} |X(f)\,.\,Y(f)\,.\,df|^2 \leq \left[\int_{-\infty}^{\infty} |X(f)|^2\,.\,df\right]\left[\int_{-\infty}^{\infty} |Y(f)|^2\,.\,df\right] \tag{7.22}$$

The equal sign applies when,

$$X(f) = K\,Y^*(f) \tag{7.23}$$

where K is an arbitrary constant and $Y^*(f)$ is the complex conjugate of Y (f).

We now apply the *Schwarz inequality* to equation, by making the following consideration :

$$X(f) = \sqrt{G_n(f)}\,H(f) \tag{7.24}$$

and

$$Y(f) = \frac{1}{\sqrt{G_n(f)}}\,.\,P(f)\,.\,e^{j2\pi fT} \tag{7.25}$$

Using these two relations and using Schwarz inequality equation; we may write for equation

$$\frac{P_0^2(T)}{\sigma_0^2} = \frac{\left| \int_{-\infty}^{\infty} X(f)\,Y(f)\,.\,df \right|^2}{\int_{-\infty}^{\infty} |X(f)|^2\,.\,df} \leq \int_{-\infty}^{\infty} |Y(f)|^2\,.\,df \tag{7.26}$$

Again put the value of Y (f), we have :

$$\frac{P_0^2(T)}{\sigma_0^2} \leq \int_{-\infty}^{\infty} |Y(f)|^2\,df = \int_{-\infty}^{\infty} \frac{|P(f)|^2}{G_n(f)}\,.\,df \tag{7.27}$$

The ratio $\frac{P_0^2(T)}{\sigma_0^2}$ will attain its maximum value when the equal sign in equation may be employed as in the case, when

$$X(f) = K\,Y^*(f) \qquad \text{(By equation (7.23))}$$

Substitute the value of X (f) and Y (f), and find the value of H (f).

$$\sqrt{G_n(f)}\,.\,H(f) = \frac{K}{\sqrt{G_n(f)}}\,.\,P^*(f)\,e^{-j2\pi fT} \tag{7.28}$$

or
$$H(f) = \frac{K}{G_n(f)} \cdot P^*(f) \cdot e^{-j2\pi fT} \tag{7.29}$$

That is the general expression for the transfer function of a optimum filter.

7.4 THE MATCHED FILTER

Till now we have calculated the transfer function of a optimum filter when noise is Gaussian. Matched filter is also a optimum filter that yields the maximum ratio $\frac{P_0^2(T)}{\sigma_0^2}$. But for white noise.

As we know that when input noise is white, the value of input noise power spectural density is given by

$$G_n(f) = \frac{\eta}{2} \tag{7.30}$$

So, now the expression for the transfer function of a matched filter becomes (by substituting the value of $G_n(f)$ in equation)

$$N(f) = K \cdot \frac{P^*(f)}{\eta/2} \cdot e^{-j2\pi Tf} \tag{7.31}$$

So, the impulse response of the filter (means response of the filter to the unit strength impulse applied at $t = 0$;) is

$$h(t) = f^{-1}[N(t)] \tag{7.32}$$

$$= \frac{2K}{\eta} \int_{-\infty}^{\infty} P^*(f) \cdot e^{-j2\pi fT} \cdot e^{j2\pi ft} \cdot df$$

$$h(t) = \frac{2K}{\eta} \int_{-\infty}^{\infty} P^*(f) \cdot e^{-j2\pi f(t-T)} \cdot df \tag{7.33}$$

A physically realizable filter will have an impulse response which is real 'or' not complex. Therefore

$$h(t) = h^*(t) \tag{7.34}$$

So equation can be modified by

$$h(t) = \frac{2K}{\eta} \int_{-\infty}^{\infty} P(f) \cdot e^{j2\pi f(T-t)} \cdot df$$

$$h(t) = \frac{2K}{\eta} P(T-t) \tag{7.35}$$

Finally, since $p(t) = S_1(t) - S_2(t)$ by equation, so we have

$$h(t) = \frac{2K}{\eta} [S_1(T-t) - S_2(T-t)] \tag{7.36}$$

Thus, we can say that the response of filter is nothing but the signal S (t) is related by axis $t = 0$; but delayed by a time amount T.

7.4.1 Probability of Error of the Matched Filter

As we have calculated for a optimum filter the ratio

$$\left|\frac{P_0^2(T)}{\sigma_0^2}\right|_{max} = \int_{-\infty}^{\infty} \frac{|P(f)|^2}{G_n(f)}.df \qquad \text{(By equation (7.27))}$$

and this is nothing but the signal to noise ratio of the optimum filter. Now if we write the expression for the matched filter;

$$G_n(f) = \frac{\eta}{2} \qquad \text{(By equation (7.30))}$$

So

$$(SNR)_{max} = \left[\frac{P_0^2(T)}{\sigma_0^2}\right]_{max} = \int_{-\infty}^{\infty} \frac{|P(f)|^2}{G_n(f)}.df$$

$$= \int_{-\infty}^{\infty} \frac{P|(f)|^2.df}{\eta/2}$$

$$= \frac{2}{\eta}\int_{-\infty}^{\infty} |P(f)|^2.df$$

$$\boxed{(SNR)_{max} = \frac{2}{\eta}\int_{-\infty}^{\infty} |P(f)|^2.df} \qquad (7.37)$$

But by *Parsevel's equations*.

$$\int_{-\infty}^{\infty} |P(f)|^2.df = \int_{-\infty}^{\infty} P^2(t).dt \qquad (7.38)$$

But signal persist only for time 0 to T where T is sampling period.

So

$$\int_{-\infty}^{\infty} |P(f)|^2.df = \int_{-\infty}^{\infty} P^2(t).dt = \int_0^T P^2(t).dt$$

But now $P(t) = S_1(t) - S_2(t)$ so.

$$(SNR)_{max} = \left[\frac{P_0^2(T)}{\sigma_0^2}\right]_{max} \qquad \text{(By equation (7.37))}$$

$$= \frac{2}{\eta}\int_{-\infty}^{\infty} |P(f)|^2.df$$

$$= \frac{2}{\eta}\int_0^T P^2(t).dt \qquad (7.39)$$

$$= \frac{2}{\eta}\int_0^T [S_1(t) - S_2(t)]^2.dt$$

$$= \frac{2}{\eta}\left\{\int_0^T S_1^2(t).dt + \int_0^T S_2^2(t).dt - 2\int_0^T S_1(t)S_2(t).dt\right\} \qquad (7.40)$$

But we know that

$$\int_0^T S_1^2(t)\,.\,dt = E_{S1} \text{ Energy of signal } S_1$$

$$\int_0^T S_2^2(t)\,.\,dt = E_{S2} \text{ Energy of signal } S_2.$$

and $$\int_0^T S_1(t)\,.\,S_2(t)\,.\,dt = E_{S12}$$

is the energy due to correlation between signal $S_1(t)$ and $S_2(t)$

So, $$(SNR)_{max} = \left[\frac{P_0^2(T)}{\sigma_0^2}\right]_{max} = \frac{2}{\eta}(E_{S1} + E_{S2} - 2E_{S12}) \tag{7.41}$$

But the optimum choice of $S_1(t)$ and $S_2(t)$ is that $S_1(t) = -S_2(t)$ (7.42)

So $$E_{S1}(t) = E_{S2}(t) = -E_{S12} = E_S \tag{7.43}$$

So,

$$[SNR]_{max} = \frac{2}{\eta}[E_S + E_S + 2E_S]$$

$$\boxed{[SNR]_{max} = \frac{8E_S}{\eta}} \tag{7.44}$$

So, now substitute in the equation of error probability expression of optimum filter

$$\left[\frac{P_0^2(T)}{\sigma_0^2}\right]_{max} = \left[\frac{8E_S}{\eta}\right] \quad \text{(By equation (7.37))}$$

So equation (7.12) becomes,

$$P_e = \frac{1}{2}\, erfc\left[\frac{P_0(T)}{2\sqrt{2}\sigma_0}\right] \quad \text{(By equation (7.12))}$$

$$= \frac{1}{2}\, erfc\left[\frac{P_0^2(T)}{8\sigma_0^2}\right]^{\frac{1}{2}} \tag{7.45}$$

$$(P_e)_{min} = \frac{1}{2}\, erfc\left\{\left[\frac{P_0^2(T)}{8\sigma_0^2}\right]_{max}\right\}^{\frac{1}{2}}$$

$$\boxed{(P_e)_{min} = \frac{1}{2}\, erfc\left(\frac{E_S}{\eta}\right)^{\frac{1}{2}}} \tag{7.46}$$

Thus, we can say that error probability depends upon only the energy of signal so if the signals of different but same energy have been applied to matched filter ; the error probability for them of a matched filter remained same.

7.4.2 Properties of Matched Filter

(*i*) Consider the expression for the matched filter :

$$[\text{SNR}]_{max} = \frac{8\text{E}_\text{S}}{\eta} \qquad \text{(By equation (7.44))}$$

As the factor 8 and η are constant. Thus, signal to noise ratio of a matched filter depends upon the energy of signal not shape. Thus, "the (SNR) of matched filter remains same for two signals of same energy but different shape".

(*ii*) Now consider the expression

$$h\,(t) = \frac{2\text{K}}{\eta}\,[\text{S}_1\,(\text{T} - t) - \text{S}_2\,(\text{T} - t)] \qquad (7.47)$$

So, if we apply the input signal to a optimum receiver

$$\text{P}(t) = \text{S}_1\,(t) - \text{S}_2\,(t) \qquad \text{(By equation (7.14))}$$

The output signal

$$\text{P}_0\,(f) = \text{H}\,(f)\,.\,\text{P}\,(f) \qquad \text{(By equation (7.16))}$$

or

$$\text{P}_0\,(t) = h\,(t) * \text{P}\,(t) \qquad (7.48)$$

$$= \frac{2\text{K}}{\eta}\,[\text{S}_1\,(\text{T} - t) - \text{S}_2\,(\text{T} - t)] * [\text{P}\,(t)]$$

$$\boxed{\text{P}_0\,(t) = \frac{2\text{K}}{n}\,[\text{S}_1\,(\text{T} - t) - \text{S}_2\,(\text{T} - t)] * [\text{S}_1\,(t) - \text{S}_2\,(t)]} \qquad (7.49)$$

So, we can say that the output of a matched filter is the *auto relation* of signal applied at it input: (*Auto-correlation — correlation* of signal with its own delayed function).

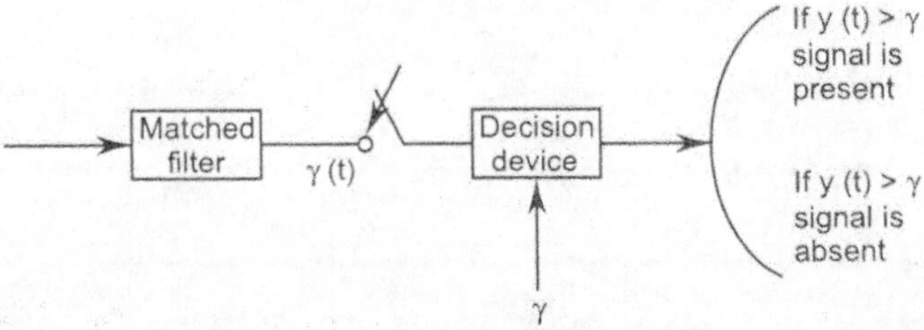

Fig. 7.2 *Matched filter receiver*

7.5 COHERENT RECEPTION : CORRELATION

We will discuss now an alternative type of receiving system (which really performs the same as the matched filter).

Again as shown in Fig. 7.3, the input is a binary data waveform $\text{S}_1\,(t)$ 'or' $\text{S}_2\,(t)$ corrupted by noise $n\,(t)$. The bit length is 'T'. Now the received signal plus noise

$$\text{V}_i\,(t) = \text{S}_i\,(t) + n\,(t) \qquad (7.50)$$

$$(\text{S}_i \text{ can be } \text{S}_1\,(t) \text{ 'or' } \text{S}_2\,(t))$$

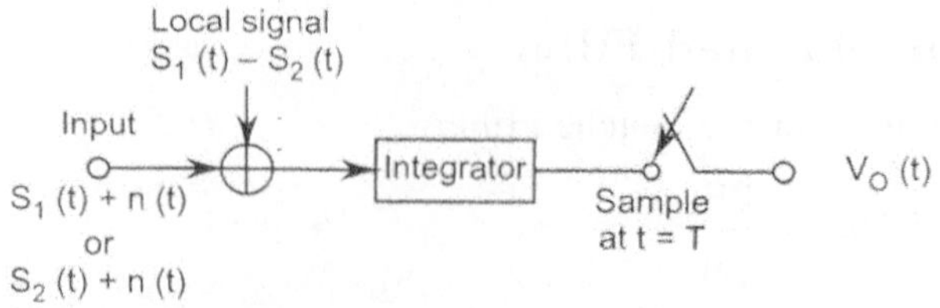

Fig. 7.3 *Correlation receiver*

Now, this received signal is multiplied by a locally generated signal.

$$[S_1(t) - S_2(t)] \tag{7.51}$$

Now, this signal is passed through a integrator. Now signal is sampled at the same rate as at the input sampling rate. Now the output consideration of this receiver :

Output signal

$$S_0(T) = \frac{1}{t}\int_0^T S_i(t)\left[S_1(t) - S_2(t)\right].dt \tag{7.52}$$

$$S_0(T) = \frac{1}{t}\int_0^T n(t)\left[S_1(t) - S_2(t)\right].dt \tag{7.53}$$

where t is the constant of integator.

Now, consider the matched filter;

If the transfer function of matched filter

$$h(t) = \frac{2K}{\eta}[S_1(T-t) - S_2(T-t)] \tag{7.54}$$

and applied signal is $V_i(t)$, the output of matched filter

$$V_0(f) = H(f)\,V_i(f) \tag{7.55}$$

or

$$V_0(t) = h(t) * V_i(f) \tag{7.56}$$

$$= \int_{-\infty}^{\infty} V_i(\lambda).h(t-\lambda).d\lambda \tag{7.57}$$

Substitute the values

$$V_0(t) = \frac{2K}{\eta}\int_0^T V_i(\lambda)\left[S_1(T-t+\lambda) - S_2(T-t+\lambda)\right].d\lambda \tag{7.58}$$

take at $t = T$

$$V_0(t) = \frac{2K}{\eta}\int_0^T V_i(\lambda)\left[S_1(\lambda) - S_2(\lambda)\right].d\lambda \tag{7.59}$$

But $V_i(\lambda) = S_i(\lambda) + \eta(\lambda)$

$$V_0(t) = \frac{2K}{\eta}\int_0^T \left[S_i(\lambda) + n(\lambda)\right]\left[S_1(\lambda) - S_2(\lambda)\right].d\lambda \tag{7.60}$$

$$= \frac{2K}{\eta}\int_0^T S_i(\lambda).\left[S_1(\lambda) - S_2(\lambda)\right].d\lambda + \frac{2K}{\lambda}\int_0^T n(\lambda)\left[S_1(\lambda) - S_2(\lambda)\right].d\lambda \tag{7.61}$$

But $V_0(t) = S_0(t) + n_0(t)$

$$S_0(t) + n_0(t) = \frac{2K}{\eta}\int_0^T S_i(\lambda)[S_1(\lambda) - S_2(\lambda)].d\lambda + \frac{2K}{\eta}\int_0^T n(\lambda).[S_1(\lambda) - S_2(\lambda)].d\lambda \quad (7.62)$$

Comparing both sides,

$$S_0(t) = \frac{2K}{\eta}\int_0^T S_i(\lambda)[S_1(\lambda) - S_2(\lambda)].d\lambda \quad (7.63)$$

and
$$n_0(t) = \frac{2K}{\eta}\int_0^T n(\lambda)[S_1(\lambda) - S_2(\lambda)].d\lambda \quad (7.64)$$

Thus calculated value of $S_0(t)$ and $n_0(t)$ for matched filter is identical to that of correlation filter. Thus, we can say that the performance of matched filter and correlation receiver is identical. So as a conclusion we can say :

'The matched filter and the correlation receiver are not two distinct, independent techniques; but they are simply the two way of defining a optimum filter'.

So matched filter works upon the principle of maximization of SNR ratio and correlation receiver works upon the principle of correlation of signal with $S_1(t) - S_2(t)$ means input signal itself (see equations (7.63) and (7.64)).

7.6 ERROR PROBABILITY IN THE COHERENT DETECTION OF BINARY ASK

In the case of Binary ASK ; we know that

$$\left.\begin{aligned} S_1(t) &= A\cos\omega_0 t \text{ symbol (1)} \\ S_2(t) &= 0 \qquad\quad \text{symbol (0)} \end{aligned}\right\} \quad (7.65)$$

Here we have drawn the receiver for the coherent detection of ASK (Fig. 7.4). Simply we can say that it is a correlation receiver.

Now first of we will calculate

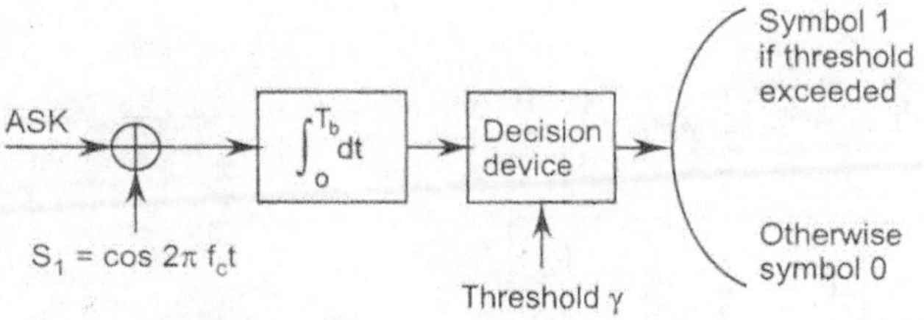

Fig. 7.4 *ASK coherent demodulation*

$$(SNR)_{max} = \left[\frac{P_0^2(T)}{\sigma_0^2}\right]_{max} = \frac{2}{\eta}\int_0^T P^2(t).dt \qquad \text{(By equation (7.39))}$$

$$= \frac{2}{\eta}\int_0^T [S_1(t) - S_2(t)]\,dt$$

$$= \frac{2}{\eta}\int_0^T S_1(t).dt$$

$$= \frac{2}{\eta} \int_0^T A \cos \omega_0 t \, . \, dt$$

$$= \frac{2}{\eta} \cdot \frac{A^2}{2} \cdot T$$

$$\left[\frac{P_0^2 (T)}{\sigma_0^2} \right]_{max} = \frac{2}{\eta} \times E_s \qquad (7.66)$$

where E_s = Energy of signal.

But in the case of ASK, the transmitted energy alternates between 0 (when the symbol '0' is sent) and the value $\frac{A^2 . T}{2}$ (when the symbol 1) transmitted. Therefore average signal energy per bit can be given by

$$E_{av} = \frac{E_{s_1} + E_{s_2}}{2} = \frac{A^2 T}{4} \qquad (7.67)$$

$$= \frac{E_s}{2} \qquad (7.68)$$

or

$$\boxed{E_s = 2E_{av}} \qquad (7.69)$$

Now the value of minimum error probability

$$P_e = \frac{1}{2} \, erfc \left[\frac{P_0^2 (T)}{8\sigma_0^2} \right]^{\frac{1}{2}} \qquad (7.70)$$

$$= \frac{1}{2} \, erfc \left[\frac{2}{\eta} \times E_s \times \frac{1}{8} \right]^{\frac{1}{2}}$$

$$= \frac{1}{2} \, erfc \left[\frac{1}{4} \cdot \frac{E_s}{n} \right]^{\frac{1}{2}}$$

$$\boxed{P_e = \frac{1}{2} \, erfc \sqrt{\frac{E_s}{4\eta}}} \qquad (7.71)$$

or

$$P_e = \frac{1}{2} \, erfc \sqrt{\frac{E_{av}}{2\eta}} \qquad (7.72)$$

7.7 ERROR PROBABILITY IN THE COHERENT DETECTION OF BINARY PHASE SHIFT KEYING

If we are using coherent detection of PSK ; Hence the input signal is

$$\left.\begin{aligned} S_1(t) &= A\cos\omega_c t \quad \text{for } \delta y.1 \\ S_2(t) &= -A\cos\omega_c t \quad \text{for } \delta y.0 \end{aligned}\right\} \qquad (7.73)$$

So to detect coherently, a locally generated signal $S_1(t) - S_2(t) = 2A\cos\omega_c t$ needs to be provided for the multiplier as shown in Fig. 7.2.

So as PSK; fullfil the requirement

$$S_1(t) = -S_2(t)$$

so equation derived for error probability is equally valid for PSK

$$\boxed{P_e = \frac{1}{2}\, erfc\left(\frac{E_S}{\eta}\right)^{\frac{1}{2}}} \qquad (7.74)$$

As we have used

$$\left.\begin{aligned} &S_1(t) = -S_2(t) = A\cos\omega_c t \\ &\text{So the value of} \quad E_s = \frac{A^2}{2}\,.\,T \end{aligned}\right\} \qquad (7.75)$$

(If a bit is extended for a time T)

So the error probability becomes,

$$P_e = \frac{1}{2}\, erfc\sqrt{\frac{A^2T}{2\eta}}$$

$$= \frac{1}{2}\, erfc\sqrt{\frac{E_s}{\eta}}. \qquad (7.76)$$

7.8 ERROR PROBABILITY IN THE COHERENT DETECTION OF BINARY FREQUENCY SHIFT KEYING

In frequency shift keying the received signal is either

$$\left.\begin{aligned} S_1(t) &= A\cos(\omega_c + \Omega)\,t \text{ for Symbol 1} \\ S_2(t) &= A\cos(\omega_c - \Omega)\,t \text{ for Symbol 0} \end{aligned}\right\} \qquad (7.77)$$

Now the locally generated signal to detect this signal is

$$S_1(t) - S_2(t) = A\cos(\omega_c + \Omega)\,t - A\cos(\omega_c - \Omega)\,t \qquad (7.78)$$

Now we cannot use the expression $P_e = \frac{1}{2}\, erfc\left(\frac{E_s}{\eta}\right)^{\frac{1}{2}}$ because it was a general expression derived for the condition $S_1(t) = -S_2(t)$. But this condition is not fullfilled ; so use the basic equation for the measurement of error probability.

$$\frac{P_0^2(t)}{\sigma_0^2} = \frac{2}{\eta}\left\{\int_0^T \left[S_1(t) - S_2(t)\right]^2 . dt\right\} \tag{7.79}$$

Now substitute the values of $S_1(t)$ and $S_2(t)$

$$\frac{P_0^2(t)}{\sigma_0^2} = \frac{2}{\eta}\int_0^T \left[A . \cos(\omega_c + \Omega)t - A\cos(\omega_c\ \Omega)t\right]^2 . dt \tag{7.80}$$

Now solving this integration, we will get result :

$$\left[\frac{P_0^2(T)}{\sigma_0^2}\right]_{max} = \frac{2A^2T}{\eta}\left[1 - \frac{\sin 2\Omega T}{2\Omega T} + \frac{1}{2} . \frac{\sin\left[2(\omega_c + \Omega)T\right]}{2(\omega_c + \Omega)T} - \frac{1}{2}\sin\frac{\left[2(\omega_c - \Omega)T\right]}{2(\omega_c - \Omega)T} - \frac{\sin 2\omega_c T}{2\omega_c T}\right] \tag{7.81}$$

Now consider the two practical situations :

(*a*) The off-set angular frequency Ω is very small in comparision with carrier angular frequency ω_0; then the last three terms becomes in the form $\frac{\sin 2\omega_c T}{2\omega_c T}$. This ratio approaches zero as $\omega_0 T$ increases.

(*b*) We may further assume that :

$\omega_c T >> 1$, so we may neglect these last three terms :

Thus finally we have

$$\left[\frac{P_0^2(T)}{\sigma_0^2}\right]_{max} = \frac{2A^2T}{\eta}\left(1 - \frac{\sin 2\Omega T}{2\Omega T}\right) \tag{7.82}$$

The quantity $\left[\frac{P_0^2(T)}{\sigma_0^2}\right]_{max}$ in the equation attains its largest value when Ω is selected so that $2\Omega T = \frac{3\pi}{2}$ for this value of Ω we find :

$$\left[\frac{P_0^2(T)}{\sigma_0^2}\right]_{max} = 2.42\,\frac{A^2T}{\eta}$$

$$= 4.84\,\frac{\left(\frac{A^2}{2}\right)T}{\eta} \tag{7.83}$$

So now the probability of error can be given by equation

$$Pe_{min} = \frac{1}{2}\, erfc\left\{\left[\frac{P_0^2(T)}{8\sigma_0^2}\right]_{max}\right\}^{\frac{1}{2}} \qquad \text{(By equation (7.45))}$$

$$= \frac{1}{2}\, erfc \left\{ \frac{1}{8} \times 4.84 \,.\, \frac{\left(\frac{A^2}{2}\right) T}{\eta} \right\}^{\frac{1}{2}} \tag{7.84}$$

$$(P_e)_{min} = \frac{1}{2}\, erfc \left\{ 0.6 \frac{\left(\frac{A^2}{2}\right) T}{\eta} \right\}^{\frac{1}{2}}$$

But $\frac{A^2}{2}\,.\,T = E_s$ (Energy of signal)

$$(P_e)_{min} = \frac{1}{2}\, erfc \left(0.6 \frac{E_s}{\eta} \right)^{\frac{1}{2}} \tag{7.85}$$

If two orthogonal signal are used for FSK, then $2\Omega T = m\pi$ (m is an integer)

So,

$$P_e = \frac{1}{2}\, erfc \left(\frac{E_s}{2\eta} \right)^{\frac{1}{2}} \tag{7.86}$$

If we compare the FSK and PSK coherent detection.

(*a*) Comparison of equation (7.76) and equation (7.86) suggests that PSK is better than the FSK on the basis on noise performance.

(*b*) The equal probability of error in each system can be achieved if the signal energy in the PSK signal is 0.6 times as large as the signal energy of FSK.

7.9 ERROR PROBABILITY OF COHERENT DETECTION OF QUADRATURE PHASE SHIFT KEYING

Recall the QPSK ; QPSK is a special case of PSK in which the carrier assumes any one of the four possible equispaced phase shifts *e.g.*, $\pm \frac{\pi}{4}, \pm \frac{3\pi}{4}$ in response to the one of the four possible digits (00, 01, 11, 10 using gray code).

Now here we have redrawn the arrangement of coherent detection of QPSK signal. We have assumed that channel noise is zero. Mean while Gaussian with power spectural density $\frac{\eta}{2}$, we have also assumed that all four gray code encoded digits 00, 10, 11, 01 occur with equal probability.

Here as we have discussed we can see from Fig. 6.9 of Chapter 6 that QPSK demodulation may be viewed as a quadrature multiplexed version of two coherent binary PSK receivers, one receiver operates with carrier $\omega_c t$ while other operates with quadrature carrier sin $\omega_c t$.

We may therefore equate the probability of error $P_e(I)$ for the inphase channel and probability of error $P_e(Q)$ for the quadrature channel as in the case of PSK.

The error probability can be written in similar form to that of coherent detection of PSK. But considering the effect of both channels. Thus, we may write

$$P_e(I) = P_e(Q) = \frac{1}{2} \cdot erfc\left(\sqrt{\frac{E}{2\eta}}\right) \tag{7.87}$$

where E = transmitted signal energy be symbol.

Simply because now each symbol has two bits so that we have substitute

$$E_s = \frac{E}{2} \text{ for each bit}$$

or

$$E = 2E_s.$$

$$\boxed{P_e(I) = P_e(Q) = \frac{1}{2} \cdot erf\sqrt{\frac{E_s}{\eta}}} \tag{7.88}$$

for each bit.

So we can say that bit error probability of QPSK and PSK remained same.

The probability that the QPSK receiver will correctly identify the transmitted data sequence is equal to the probabilities that both the (in phase) and quadrature correlators of the receiver yield the correct result. Let P_c denotes the average probability of correct reception. Then, we may write

$$P_c = (1 - P_e(I))(1 - P_e(Q)) \tag{7.89}$$

Substituting

$P_e(I) = P_e(Q)$ and assuming both the values are small compared to unity, we may write

$$P_c = 1 - 2P_e(I) \tag{7.90}$$

The average probability of symbol error in QPSK receiver is therefore,

$$P_e = 1 - P_c \tag{7.91}$$

$$= 2P_e(I)$$

$$= 2 \times \frac{1}{2} \, erfc\left(\sqrt{\frac{E}{2\eta}}\right)$$

$$= erfc\sqrt{\frac{E}{2\eta}} \tag{7.92}$$

or Average probability of bit error in QPSK

$$E_s = \frac{E}{2}$$

$$\boxed{P_e = erfc\sqrt{\frac{E_s}{\eta}}} \tag{7.93}$$

So QPSK provide very less efficient noise performance in comparison to PSK.

7.10 ERROR PROBABILITY OF MSK

Although QPSK and MSK have different methods of generation and detection. But the noise performance for both of them are identical. Thus, the expression for the error probability of QPSK is equally valid for MSK ;

$$\boxed{P_e = erfc\sqrt{\frac{E_s}{\eta}}} \qquad (7.94)$$

Error probability for Non-coherent Detection :

Till now we have calculated the noise probability of coherent detection technique of the all the digital modulation techniques ; as in the non-coherent detection (only possible for ASK, FSK, DPSK), we employ matched filter followed by *envelope detector*.

Due to the envelope detector used and its non-linear performance, the calculation becomes very complicated, so we have only information here about these noise performance not exact calculation.

(*a*) Non-coherent detection of FSK has error probability.

$$P_e = \frac{1}{2}\exp\left(-\frac{E_s}{2\eta}\right) \qquad (7.95)$$

(*b*) Non-coherent detection of ASK has error probability

$$P_e = \frac{1}{2}\exp\left(-\frac{E_{av}}{2\eta}\right) \qquad (7.96)$$

Here E_{av} = average signal energy per bit in ASK is same as signal energy (E_s) in the case of FSK.

(*c*) The probability of error associated with non-coherent detection of DPSK signal is

$$\boxed{P_e = \frac{1}{2}\exp\left(-\frac{E_b}{\eta}\right)} \qquad (7.97)$$

Since in a DPSK receiver decisions are made on the basis of the signal received in two successive bit intervals, the bit error tends to occur in pair.

7.11 COMPARISON OF NOISE PERFORMANCE OF DIGITAL SYSTEMS

Here we have drawn table 7.1 that shows the error probabilities of each modulation scheme. From Fig. 7.5, it should also be clear that the error rates of all system decreases monotonically with increase in $\frac{E_s}{\eta}$. For a given value of $\frac{E_s}{\eta}$, the coherent PSK produces smaller error rate in comparison to other modulation schemes. But it should be clear that the comparison of expression doesn't only fullfil the requirement due to different fundamentals of modulation system. As we have seen that for same bit pattern DPSK produces higher error than that of PSK but the expression for error is same for path.

Table 7.1 Comparison

	Digital Modulation Scheme	*Error Probability*
(*i*)	Coherent BASK	$\frac{1}{2}\, erfc \sqrt{\frac{E_s}{4\eta}}$
(*ii*)	Coherent BFSK	$\frac{1}{2}\, erfc \sqrt{\frac{E_s}{\eta}}$
(*iii*)	Coherent BPSK	$\frac{1}{2}\, erfc \sqrt{\frac{E_s}{2s}}$
(*iv*)	Coherent BFSK, Coherent BPSK	$erfc \sqrt{\frac{E_s}{\eta}}$
(*v*)	Non-coherent BFSK	$\frac{1}{2} \exp\left(-\frac{E_s}{2\eta}\right)$
(*vi*)	Non-coherent BASK	$\frac{1}{2} \exp\left(-\frac{E_{av}}{2\eta}\right)$
(*vii*)	DPSK	$\frac{1}{2} \exp\left(-\frac{E_b}{\eta}\right)$.

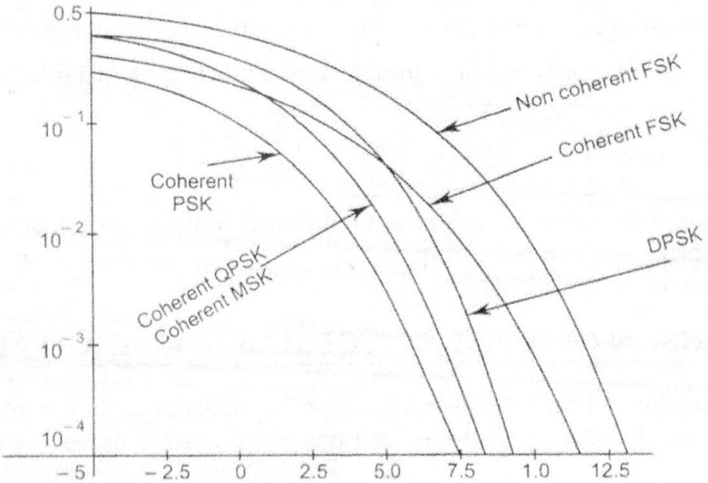

Fig. 7.5 *Comparison of different modulation techniques*

SOLVED EXAMPLES

Example 1. *Binary data is being transmitted over a telephone link using ASK. If the Bandwidth of telephone link is 4KHz and the maximum achievable signal to power ratio is 6 dB at the output. Bandwidth of ASK signal to be $4r_b$. (r_b = bit rate), calculate the error probability for ASK signal.*

Solution. Given that :

$$10\log_{10}\left(\frac{\text{Average-Signal Power}}{\text{Noise Power}}\right) = 6\ dB$$

$$\frac{\text{Average Signal Power}}{\text{Noise Power}} = 4$$

$$\frac{E_{av}}{N_0} = 4$$

$$\frac{E_{av}}{\eta B} = 4$$

$$\frac{E_{av}}{\eta \times 4000} = 4$$

$$\frac{E_{av}}{\eta} = 4 \times 4000$$

But we know that our discussion was for per bit ; so

$$\left(\frac{E_{av}}{\eta}\right) \text{ for one bit} = \frac{4 \times 4000}{r_b}$$

$$= \frac{4 \times 4000}{1000}$$

as

$$4r_b = B\Omega \text{ given}$$

$$r_b = \frac{B\Omega}{4}$$

$$= \frac{4 \times 10^3}{4} = 1000 \text{ Bits/second.}$$

No substitute in the expression

$$P_e = \frac{1}{2}\, erfc\left(\sqrt{\frac{E_{av}}{2\eta}}\right)$$

$$= \frac{1}{2}\, erfc\left(\sqrt{\frac{4 \times 4000}{2 \times 1000}}\right)$$

$$= \frac{1}{2}\, erfc\left(\sqrt{8}\right)$$

$$\boxed{P_e = 0.0001.}$$

Example 2. *A binary data is transmitted ones a link with bandwidth 10 MHz at a rate 6.8 Mb/s using ASK. The carrier amplitude is 2 mV. The mean power spectural density of noise is 10^{-15} Ω/Hz. Calculate the error probability for ASK detection.*

Solution. We know by the formula, error probability

$$P_e = \frac{1}{2}\, erfc\left(\sqrt{\frac{E_{av}}{2\eta}}\right)$$

But
$$E_{av} = \frac{A^2T}{4}$$

$$= \frac{(2 \times 10^{-3})^2 \times \left(\frac{1}{r_b}\right)}{4}$$

$$= \frac{4 \times 10^{-6} \times 1}{4 \times 6.8 \times 10^6}$$

given that,

$$\frac{\eta}{2} = 10^{-15}$$

$$\eta = 10^{-15} \times 2$$

So substituting the values

$$P_e = \frac{1}{2}\, erfc\left(\sqrt{\frac{4 \times 10^{-6}}{4 \times 6.8 \times 10^{+6} \times 4 \times 10^{-15}}}\right)$$

$$= \frac{1}{2}\, erfc\,(\sqrt{6.06})$$

$$\boxed{P_e \simeq 3 \times 10^{-5}.}$$

Example 3. *Binary data is transmitted using PSK at a bit rate of 1 Mb/s over a R.F. link having a bandwidth of 3 MHz. Assume the noise power spectural density at a coherent receiver input to be 10^{-10} M/Hz. If required probability of error is limited to 10^{-6}, find signal power per bit at receives.*

Solution. We know that for PSK, error probability

$$P_e = \frac{1}{2}\, erfc\left(\sqrt{\frac{E_s}{\eta}}\right)$$

But
$$\frac{\eta}{2} = 10^{-10}\ \text{W/Hz},\ T_b = \frac{1}{r_b}$$

$$T_b = 10^{-6}\ \text{sec.}$$

So,
$$P_e \leq 10^{-6}$$

$$\frac{1}{2}\, erfc\left(\sqrt{\frac{E_s}{\eta}}\right) \leq 10^{-6}$$

or
$$\sqrt{\frac{E_s}{\eta}} \leq 3.3$$

Therefore,
$$E_s \geq 10.89 \times 10^{-10}\ \text{watts.}$$

Now Average Power needed per bit.

So, Average power per bit

$$\geq \frac{10.89 \times 10^{-10}}{T_b}$$

$$\geq \frac{10.89 \times 10^{-10}}{10^{-6}}$$

$$\geq 10.89 \times 10^{-4} \text{ W/B.}$$

Example 4. *Binary data is to be Transmitted via. a telephone link that has a usable BW = 3000 MHz and a maximum achievable signal to noise ratio (Power ratio) of 6 dB at its ouput.*

(a) Determine the maximum signalling rate and P_e if a where nb ASK scheme is used for transmitting binary data through this channel.

(b) If the data rate is maintained at 525 bits/sec. Calculate the error probability.

Solution. (*a*) If it is assumed that an ASK signal requires a BW = 3 r_b Hz. Then the maximum signalling rate permissible is r_b = 1000 bit per sec.

The probability of error P_e can be calculated as follows.

$$\text{Average signal power} = \frac{A^2}{4}$$

$$\text{Noise power} = \eta \text{ BW} = 3000\,\eta$$

$$\frac{\text{average signal power}}{\text{Noise power}} = \text{antilog}\left(\frac{6}{10}\right) = 4 = \frac{A^2}{12000\,\eta}$$

From which $\dfrac{A^2}{\eta} = 48{,}000$

Hence $\dfrac{A^2}{4\eta} r_b = 12$ and $P_e = Q\left(\sqrt{12}\right) = 3 \times 10^{-4}$

(*b*) If the bit rate is reduced to 500 bit/sec then

$$\frac{A^2}{4\eta r_b} = \frac{64000}{4 \times 500} = 32$$ [if bit rate is reduced the error probability is also reduced.]

$$P_e = Q\sqrt{32} = \sqrt{5.56}\ Q.$$

Hence $\dfrac{A^2}{4\eta r_b} = 16$ and $P_e\ Q\sqrt{16} = 3 \times 10^{-4}$

Note that if the bit rate reduced the error probability is also reduced 100 fold.

Example 5. *Binary data is transmitted over an RF bandpass channel with a usable band width of 10 MHz at a rate of 5.0 megabits/sec using an ASK signalling method. The carrier amplitude at the receive antenna 2 milli volt and noise power spectral density at the receiver input is 10^{-16} watt/Hz.*

(a) Find the error probability of a coherent receiver.

(b) Find the error probability of a non coherent receiver.

Solution. (*a*) The bit error probability for the coherent demodulator is

$$P_e = Q\left(\sqrt{\frac{A^2T_b}{4n}}\right); \; A = 2\,mv \; T_b = \frac{10^{-6}}{5.0}$$

$$\frac{\eta}{2} = 10^{-16} \text{ watt/Hz. Hence } P_e = Q\sqrt{\frac{A^2T_b}{4n}}$$

$$\frac{A^2T_b}{4\eta} = \frac{A^2}{\eta} \times \frac{T_b}{4} = \frac{\dfrac{2 \times 10^{-3} \times 2 \times 10^{-3} \times 10^{-6}}{2 \times 10^{-16} \times 5.0 \times \frac{1}{2}}}{10^{-4}}$$

$$\frac{1}{10^{-4} \times 5 \times 2} = \frac{1}{10^{-3}} = 10^3 \left(P_e = Q\sqrt{10^3}\right)$$

(*b*) $A^2 >> \sigma^2$ actual calculation shows

$$A^2 = 2 \times 10^{-3} \times 2 \times 10^{-3} = 4 \times 10^{-6}$$

$$\sigma_0^2 = 2\,\eta\, r_b = 2 \times 2 \times 10^{-6} \times \frac{\dfrac{1}{10^{-6}}\left[r_b = \dfrac{1}{T_b}\right]}{5}$$

$$= \frac{4 \times 10^{-16} \times 5}{10^{-6}} = 20 \times 10^{-10}$$

$$P_{e_1} = Q\sqrt{\frac{A^2}{4\sigma_0^2}} \quad \text{and} \quad P_{e_0} = \exp\left(-\frac{A^2}{8\sigma_0^2}\right)$$

$$P_e = \frac{1}{2}\left[P_{e_0} + P_{e_1}\right] P_e$$

of non–coherent receiver is approx. 2000 time larger than that for coherent receivers.

Example 6. *Binary data is transmitted over an RF bandpass channel with a usable band width of 10 MHz at a rate of 5.0 megabits/sec using an ASK signalling method. The carrier amplitude at the receive antenna 2 milli volt and noise power spectral density at the receiver input is 10^{-16} watt/Hz.*

(a) Find the error probability of a coherent receiver.

(b) Find the error probability of a non coherent receiver.

Solution. (*a*) The bit error probability for the coherent demodulator is

$$P_e = Q\left(\sqrt{\frac{A^2T_b}{4n}}\right);\ A = 2\ mv;\ T_b = \frac{10^{-6}}{5.0}$$

$$\frac{\eta}{2} = 10^{-16}\ \text{watt/Hz. Hence } P_e = Q\sqrt{\frac{A^2T_b}{4n}}$$

$$\frac{A^2T_b}{4\eta} = \frac{A^2}{\eta}\times\frac{T_b}{4} = \frac{\dfrac{2\times10^{-3}\times2\times10^{-3}\times10^{-6}}{2\times10^{-16}\times5.0\times\frac{1}{2}}}{10^{-4}}$$

$$\frac{1}{10^{-4}\times5\times2} = \frac{1}{10^{-3}} = 10^3\ \left(P_e = Q\sqrt{10^3}\right)$$

(*b*) $A^2 >> \sigma^2$ actual calculation shows

$$A^2 = 2\times10^{-3}\times2\times10^{-3} = 4\times10^{-6}$$

$$\sigma_0^2 = 2\,\eta\,r_b = 2\times2\times10^{-6}\times\frac{\dfrac{1}{10^{-6}}\left[r_b = \dfrac{1}{T_b}\right]}{5}$$

$$= \frac{4\times10^{-16}\times5}{10^{-6}} = 20\times10^{-10}$$

$$P_{e_1} = Q\sqrt{\frac{A^2}{4\sigma_0^2}}\quad\text{and}\quad P_{e_0} = \exp\left(-\frac{A^2}{8\sigma_0^2}\right)$$

$$P_e = \frac{1}{2}\left[P_{e_0} + P_{e_1}\right]P_e$$

of non-coherent receiver is approx. 2000 time larger than that for coherent receivers.

REVIEW QUESTIONS

1. What is the concept of optimum receiver ? Explain the fundamental of optimum filter.
2. Calculate the transfer function for a optimum filter used for detection of digital signals. Assume channel noise as a Gaussian white noise.
3. What is matched filter ? What is the error probability for matched filter ? Explain the various properties of matched filter.
4. What is the difference between matched Filter and correlation filter ?
5. Find the error probability for

   (*a*) BASK (*b*) BPSK (*c*) BFSK (*d*) QPSK (*e*) MSK.

   Compare the results.

8

INFORMATION THEORY

Information theory is originally known as 'Mathematical Theory of Communication' and it deals with mathematical modeling and analysis of a communication system rather than with physical sources and physical channel.

By the information theory, we can consider the efficient way to communicate the data. It basically provides limits on :

(*i*) The minimum number of bits per symbol required to fully represent the source.

(*ii*) The maximum rate at which reliable communication can take over the channel.

8.1 CONCEPT OF INFORMATION

The communication system never be described in the deterministic sense, it can be considered of statistical nature. It means to describe a communication system completely we have to use its unpredictable 'or' uncertain behaviour.

It can be easily understand by example that each transmitter transmit the information randomly, we cannot predict which one message, transmitter is going to be transfer just next moment. But we know the probability of transmitting a particular message.

So to define a system completely, we need statistical study of system and statistical study of system is performed with the help of concept of probability.

Now, just an example of two messages :

(*a*) Dog bites a man.

(*b*) Man bites a Dog.

Sentence (*a*) has minimum information, but Sentence (*b*) has maximum information as this Sentence (*b*) has least probability to occur. So we can say there is a sort of inverse relationship between the probability of event and the amount of information associated with it.

Thus, we can say $$I(x_j) = f\left[\frac{1}{P(x_j)}\right] \tag{8.1}$$

where x_j is an event with a probability of $P(x_j)$ and the amount of information associated with it is $I(x_j)$.

Generally for shake of simplicity we define the logarithm measurement of information. So, we can say finally

$$I(x_j) = \log \frac{1}{P(x_j)}$$

$$\boxed{I(x_j) = -\log P(x_j)} \tag{8.2}$$

The advantage of logarithm presentation is that if we have probability of joint event $P(x_j, y_k)$ and if both are statistically independent

$$P(x_j, y_k) = P(x_j) \, . \, P(y_k) \tag{8.3}$$

So, $$I(x_j, y_k) = \log \frac{1}{P(x_j, y_k)}$$

$$= \log \frac{1}{P(x_j) \, . \, P(y_k)} = \log \frac{1}{P(x_j)} + \log \frac{1}{P(y_k)}$$

$$\boxed{I(x_j, y_k) = I(x_j) + I(y_k)} \tag{8.4}$$

Thus, we can say that now mathematical calculation is quite easy.

As we know that the base of logarithm can be different, so we may have different units of information all of them has been shown in Table 8.1.

Table 8.1 Conversion of information units

Unit	*Bits (Base 2)*	*Nats (Base e)*	*Decits (Base 10)*
Bit (Base 2)	–	1 Bit $= \frac{1}{\log_2 e}$ $= 0.6932$ nat	1 bit $= \frac{1}{\log_2 10}$ $= 0.3010$ decit
Nats (Base e)	1 nat $= \frac{1}{\log 2}$ $= 1.4426$ bits	–	1 nat $= \frac{1}{\log 10}$ $= 0.4342$ decit
Decit (Base 10)	1 decit $= \frac{1}{\log_{10} 2}$ $= 3.3219$ bits	1 decit $= \frac{1}{\log_{10} e}$ $= 2.3026$ nats	

8.1.1 Properties of Information Content

This definition exhibits the following important properties that are intuitively satisfying:

(*i*) $I(s_k) \geq 0 \quad \text{for} \quad 0 \leq p_k \leq 1$ (8.5)

That is to say, the occurrence of an event $S = s_k$ either provides some or no information, but never brings about a loss of information.

(*ii*) $I(s_k) > I(s_i) \quad \text{for} \quad p_k < P_i$ (8.6)

That is, the less probable an event is, the more information we gain when it occurs.

(*iii*) $I(s_k) = 0 \quad \text{for} \quad p_k = 1$ (8.7)

Obviously, if we are absolutely certain of the outcome of an event, even before it occurs, there is no information gained.

(*iv*) The amount of information $I(s_k)$ produced by the source during an arbitrary signaling interval depends on the symbols s_k emitted by the source at that time. Indeed, $I(s_k)$ is a discrete random variable that takes on the values $I(s_0)$, $I(s_1)$, ..., $I(s_{k-1})$ with probabilities $p_0, p_1, \ldots, p_{k-1}$, respectively. The mean of $I(s_k)$ over the source alphabet ℓ is given by

$$H(\ell) = E[I(s_k)]$$

$$= \sum_{k=0}^{k-1} P_k I(s_k) = \sum_{k=0}^{k-1} P_k \log_2\left(\frac{1}{P_k}\right) \tag{8.8}$$

The important quantity $H(\ell)$ is called the entropy of a discrete memoryless source with source alphabet ℓ. It is a measure of the average information content per source symbol.

(*v*) $I(s_k S_1) = I(s_k) + I(S_1)$ if S_k and S_l are statistically independent. (8.9)

The resulting unit of information is called the bit (a contraction of binary digit). We thus write

$$I(s_k) = \log_2\left(\frac{1}{p_k}\right) = -\log_2 p_k \quad \text{for} \quad k = 0, 1, \ldots, k-1 \tag{8.10}$$

When $p_k = 1/2$, we have $I(s_k) = 1$ bit, Hence, one bit is the amount of information that we gain when one of two possible and equally likely (*i.e.*, equiprobable) events occurs. Note that the information $I(s_k)$ is positive, since the logarithm of a number less than one, such as a probability, is negative.

8.1.2 Average Information, Entropy

Suppose we have M different and independent messages m_1, m_2 ... with probabilities of occurrence p_1, p_2 ... suppose further during a long period of transmission of sequence of L messages has been generated thus if L is very large that we expect that in L message sequence we transmit p_1L messages of m_1, $p_2 l$ messages of m_2, etc. The total in such a sequence will be

$$I_{\text{total}} = P_1 L \log_2 \frac{1}{p_1} + P_2 L \log_2 \frac{1}{p_2} \tag{8.11}$$

The average information per message internal, represented by symbol H will be

$$H = \frac{I_{\text{total}}}{L}$$

$$H = p_1 \log_2\left(\frac{1}{p_1}\right) + p_2 \log_2\left(\frac{1}{p_2}\right) + \ldots H = \sum_{k=1}^{M} P_k \log_2\left(\frac{1}{p_k}\right) \tag{8.12}$$

This average information is termed as entropy. For example we have two messages with probabilities P and (1-P). Then average information per message interval is

$$H = p\log_2\left(\frac{1}{p}\right) + (1-p)\log_2\left(\frac{1}{1-p}\right) \quad (8.13)$$

A plot of H as a function of P, as shown in Fig. 8.1 H = 0 at P = 0 and P = 1. The maximum value of H may be located by setting the value of $\frac{dH}{dp} = 0$

$$\because \quad H = P\log_2\frac{1}{p} + (1-p)\log_2\frac{1}{1-p} \quad (8.14)$$

Differentiating equation (8.13) with respect to P but before this as we know that

$$\log_2 x = \frac{\log_e x}{\log_2 2}$$

After applying this equation (8.14) look like

$$H = p\frac{\log_e\left(\frac{1}{p}\right)}{\log_e 2} + (1-p)\frac{\log_e\left(\frac{1}{1-p}\right)}{\log_e 2}$$

$$H = \frac{1}{\log_e 2}\left[p\log_e\frac{1}{p} + (1-p)\log_e\frac{1}{1-p}\right] \quad (8.15)$$

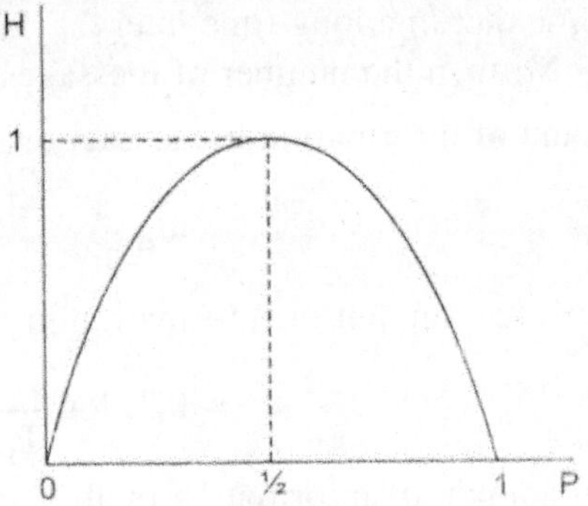

Fig. 8.1

Now differentiating equation (8.15)

$$H = \frac{-1}{\log_e 2}\left[p \times \log_e p + (1-p)\log_e(1-p)\right]$$

$$\frac{dH}{dp} = \frac{-1}{\log_e 2}\left[p \times \frac{1}{p} + \log_e p + (1-p) \times \frac{-1}{(1-p)} + \log_e(1-p)(-1)\right]$$

$$= \frac{-1}{\log_e(2)}\left[1 + \log_e p - 1 - \log_e(1-p)\right] = \frac{-1}{\log_e(2)}\left[\log_e p - \log_e(1-p)\right]$$

$$= -\log_2 p + \log_2(1-p) = \log_2\left(\frac{1-p}{p}\right) \quad (8.16)$$

To find out the maximum value of p, equate $\frac{dH}{dp}$ to zero

$$\therefore \quad \log_2\left(\frac{1-p}{p}\right) = 0 \quad \text{or} \quad \left(\frac{1-p}{p}\right) = 2^0 \Rightarrow \left(\frac{1-p}{p}\right) = 1$$

$$\Rightarrow \quad 1 - p = p \quad \text{is} \quad 2p = 1 \Rightarrow p = \frac{1}{2}$$

Hence maximum value of H is

$$H_{max} = \frac{1}{2}\log_2 2 + \frac{1}{2}\log_2 2$$

$$H_{max} = \log_2 2 = 1 \text{ bit/message}$$

when there are M messages, it may likewise be proved that H becomes maximum when all messages are equally likely. In this case each message has probability $p = \frac{1}{M}$.

8.2 ENTROPY

In the communication system, we don't have only a single message, but have a number of messages. So to calculate the total information instead of calculating information due to individual messages and adding them, we calculate the average information of the system knows as entropy of the source.

Let there be M different messages $m_1, m_2, \ldots, m_m$ with their respective probability $P_1, P_2, \ldots P_M$. Let us assume that in a long time interval, L messages have been generated. Let L be very large so that L >> M; then the number of messages $m_1 = P_1L$.

The amount of information in message m_1

$$= \log \frac{1}{P_1} \tag{8.17}$$

Thus, the total amount of information in m_1

$$= P_1L \log \frac{1}{P_1} \tag{8.18}$$

The total amount of information in all L messages will be

$$I_t = P_1L \log \frac{1}{P_1} + P_2L \log \frac{1}{P_2} + \ldots + P_ML \log \frac{1}{P_M} \tag{8.19}$$

So, the average information will be

$$H \equiv \frac{I_t}{L} = P_1 \log \frac{1}{P_1} + P_2 \log \frac{1}{P_2} + \ldots + P_M \log \frac{1}{P_M}$$

$$\boxed{H = \sum_{K=1}^{M} P_K \log \frac{1}{P_K}} \tag{8.20}$$

Thus the unit of entropy will be information/message (according to units of these both).

8.3 RATE OF INFORMATION

If a message source generates message at the rate of r messages per second ; then rate of information

$$\boxed{R = rH \text{ bits/second}} \tag{8.21}$$

8.4 SELF ENTROPY

I (x) is called self information and simply H (x) is called self entropy.

8.5 JOINT ENTROPY AND CONDITIONAL ENTROPY

Till now we have consider self information (also called marginal information) that was one dimensional calculation ; we have considered only transmitter. Simply we can consider only receiver also that is also a marginal information).

But now we will proceed for study of transmitter and receiver simultaneously. That is called two dimensional study.

Suppose, we have totally m messages

$$[X] = [x_1, x_2, x_3, \ldots, x_m] \tag{8.22}$$

And now at receiver, we receive totally n messages

$$[Z] = [y_1, y_2, y_3, \ldots, y_n] \tag{8.23}$$

Then,

$P(x_j)$ = called marginal pı bability of x messages.

$P(y_k)$ = called marginal probability of y messages.

Thus, marginal entropy

$$H(x) = -\sum_{j=1}^{m} P(x_j) \log P(x_j) \tag{8.24}$$

where $P(x_j)$ may be defined as

$$P(x_j) = \sum_{k=1}^{n} P(x_j, y_k) \tag{8.25}$$

So,

$$\boxed{H(x) = -\sum_{j=1}^{m}\sum_{k=1}^{n} P(x_j)\, P(x_j, y_k)} \tag{8.26}$$

In same manner, marginal entropy for y.

$$\boxed{H(y) = -\sum_{k=1}^{n}\sum_{j=1}^{m} P(y_k)\, P(x_j, y_k)} \tag{8.27}$$

We know that,

$P(x_j, y_k)$ = joint probability of event x and y

So joint entropy of x and y

$$\boxed{H(X, Y) = -\sum_{j=1}^{m}\sum_{k=1}^{n} P(x_j, y_k) \log P(x_j, y_k)} \tag{8.28}$$

But we also know that P (X/Y) and P (Y/X) are called conditional probability that will be clear by their definitions.

P (X/Y) = Probability of X when Y has been received.

P (Y/X) = Probability of Y when X has been transmitted.

For conditional entropy; we use the relation

$$\boxed{H\ (Y/X) = -\sum_{j=1}^{m}\sum_{k=1}^{n} P\left(x_j, y_k\right) \log\left(y_k/x_j\right)} \tag{8.29}$$

Similarly

$$\boxed{H\ (X/Y) = -\sum_{j=1}^{m}\sum_{k=1}^{n} P\left(x_j, y_k\right) \log\left(x_j/y_k\right)} \tag{8.30}$$

8.6 RELATION BETWEEN THE DIFFERENT ENTROPIES

We may prove the relations between joint, conditional and marginal entropies and given by,

$$\left.\begin{aligned} H\ (X, Y) &= H\ (X/Y) + H\ (Y) \\ \text{or} \quad H\ (X, Y) &= H\ (Y/X) + H\ (X) \end{aligned}\right\} \tag{8.31}$$

For example start with

$$H\ (X, Y) = -\sum_{j=1}^{m}\sum_{k=1}^{n} P\left(x_j, y_k\right) \log P\left(x_j, y_k\right) \tag{8.32}$$

But by Bay's formula,

$$\frac{P\ (x, y)}{P\ (x)} = P\ (Y/x) \tag{8.33}$$

So,

$$= -\sum_{j=1}^{m}\sum_{k=1}^{n} P\left(x_j, y_k\right) \log\left[P\left(x_j\right) \times P\left(y_k/x_j\right)\right]^{-1} \tag{8.34}$$

$$= -\sum_{j=1}^{m}\sum_{k=1}^{n} P\left(x_j, y_k\right) . \log P\left(x_j\right) - \sum_{j=1}^{m}\sum_{k=1}^{n} P\left(x_j, y_k\right) \log P\left(y_k/x_j\right)$$

But

$$\sum_{j=1}^{m}\sum_{k=1}^{n} P\left(x_j, y_k\right) \log P\left(x_j\right)$$

$$= \sum_{j=1}^{m} P\left(x_j\right) \log P\left(x_j\right)$$

So equation Simplifies

$$\boxed{H\ (X, Y) = H\ (X) + H\ (Y/X)} \tag{8.35}$$

Thus all relations can be easily proved.

8.7 MUTUAL INFORMATION

Now consider

$$P(x_j) = \text{Probability of transmitting } x_j,$$

$$P(x_j/y_k) = \text{Probability of transmitting } x_j, \text{ when } y_k \text{ has been received.}$$

Thus P (x_j) shows the probability 'or' uncertainty of x; when we have not received any thing called prior. Uncertainty and P (x_j/y_k) called final uncertainty of 'x' when we have received that at receiver side and the difference of these uncertainties called *mutual information*.

$$I(x_j, y_k) = \text{Initial uncertainty} - \text{Final uncertainty} \tag{8.36}$$

$$= -\log P(x_j) - \left[-\log P\left(x_j/y_k\right)\right]$$

$$= -\log P(x_j) + \log P(x_j/y_k)$$

$$\boxed{I(x_j, y_k) = \log \frac{P\left(x_j/y_k\right)}{P\left(x_j\right)}} \tag{8.37}$$

$$= \log \left[\frac{P\left(x_j, y_k\right)}{P\left(y_k\right) . P\left(x_j\right)}\right]$$

$$= \log \frac{P\left(y_k/x_j\right)}{P\left(y_k\right)}$$

$$\boxed{I(x_j, y_k) = I(y_k, x_j)} \tag{8.38}$$

Entropy corresponding to mutual information is given by

$$I(x : y) = \sum_{j=1}^{m}\sum_{k=1}^{n} P\left(x_j', y_k\right) I\left(x_j; y_k\right) \tag{8.39}$$

$$= \sum_{j=1}^{m}\sum_{k=1}^{n} P\left(x_j, y_k\right) \frac{P\left(x_j/y_k\right)}{P\left(x_j\right)}$$

After showing

$$\boxed{I(x : y) = H(X) - H(X/Y)} \tag{8.40}$$

$$\left.\begin{aligned} &= H(X) + H(Y) - H(X, Y) \\ &= H(Z) - H(Y/X) \end{aligned}\right\} \tag{8.41}$$

8.8 CHANNEL CAPACITY

The mutual information calculated above also shows the average information per symbol transmitted in the system. And it will be practical as the Shannon has also showed that capacity of channel can be said as the max practical rate of information. So capacity of channel is given by :

$$C = \max I(X : Y)$$

$$= \max \left[H(X) - H(X/Y)\right] \tag{8.42}$$

8.9 THE TRANSMISSION EFFICIENCY AND REDUNDANCY

Transmission efficiency may be defined as the ratio of actual transmission rate to the max transmission rate.

$$\eta = \frac{I(x ; y)}{\text{Max } I(x ; y)} \tag{8.43}$$

The redundancy of the channel is defined as :

$$\boxed{R = 1 - \eta} \tag{8.44}$$

8.9.1 Formulation for Efficiency

We know that rate of information

$$R = rH \text{ bits/sec} \tag{8.45}$$

If all symbols convery the same amount of information then,

$$H = \log_2 M \tag{8.46}$$

So, we may write

$$R = r \log_2 M \tag{8.47}$$

Now consider we are using a encoder that converts the incoming symbols to codewords consisting of binary digits produced at same fixed rate. Then

$$rH = r_b \log_2 2 \tag{8.48}$$

$$rH = r_b \tag{8.49}$$

If the symbols have different probabilities, then

$$r_H \leq r_B \tag{8.50}$$

or

$$\frac{r_b}{r} \geq H \tag{8.51}$$

Here $\left(\frac{r_b}{r}\right)$ represents a very important parameter and called average code length denoted by $\overline{N}$; thus

$\overline{N} = \frac{r_b}{r}$. So equation becomes

$$\boxed{\overline{N} \geq H} \tag{8.52}$$

For a optimum source coding

$$\boxed{\overline{N} = H} \tag{8.53}$$

Thus average code length is equal to the average information of channel. But practically

$$\overline{N} \geq H$$

$$\boxed{\frac{H}{\overline{N}} \leq 1} \tag{8.54}$$

That is also called efficiency of coding technique.

8.10 KRAFT INEQUALITY

A necessary and sufficient condition for a binary code to be uniquely decipherable, the code length must be such that,

$$K = \sum_{i=1}^{M} 2^{-N_i} \leq 1 \tag{8.55}$$

The above condition is known as *Kraft inequality*.

Now simplest coding is that we generate a fixed length code in which all codes have the same length given by :

$$N_i = \overline{N} \text{ So}$$

$$K = M \cdot 2^{-\overline{N}} \leq 1 \tag{8.56}$$

This means that for decipherable (equally identified) codes in the case of fixed length coding, we need

$$\overline{N} \geq \log_2 M \tag{8.57}$$

So, the resulting efficiency can be calculated as :

$$\boxed{\frac{H}{\overline{N}} \leq \frac{H}{\log_2 M}} \tag{8.58}$$

Here the result of this discussion is that if $H < \log_2 M$ and we need higher efficiency, we have to reduce the average code length $\overline{N}$. That is why we use *variable code length code*.

Example 1. *An event has total six outcomes with probability* $P_1 = \frac{1}{2}$, $P_2 = \frac{1}{4}$, $P_3 = \frac{1}{8}$, $P_4 = \frac{1}{16}$, $P_5 = \frac{1}{32}$ *and* $P_6 = \frac{1}{32}$. *Find the entropy of the system. Also find the rate of information if these are 18 outcomes per second.*

Solution. By the formula of entropy $= \sum_{i=1}^{6} P_i \log \frac{1}{P_i}$

$$H = P_1 \log \frac{1}{P_1} + P_2 \log \frac{1}{P_2} + P_3 \log \frac{1}{P_3} + P_4 \log \frac{1}{P_4} + P_5 \log \frac{1}{P_5} + P_6 \log \frac{1}{P_6}$$

$$= \frac{1}{2}\log 2 + \frac{1}{4}\log 4 + \frac{1}{8}\log 8 + \frac{1}{16}\log 16 + \frac{1}{32}\log 32 + \frac{1}{32}\log 32$$

$$\boxed{H = \frac{31}{16} \text{ bits/message}}$$

But now $r = 18$ outcomes/second

The rate of Information.

$$R = rH$$

$$= 18 \times \frac{31}{16}$$

$$\boxed{R = 34.875 \text{ bits/sec.}}$$

Example 2. *A discrete source emits one of five symbol once every microsecond with probabilities* $\frac{1}{2}, \frac{1}{4}, \frac{1}{8}, \frac{1}{16}, \frac{1}{32}$ *respectively. Determine the entropy and information rate.*

Solution. By the formula $= \sum_{i=1}^{5} P_i \log \frac{1}{P_i}$

$$= \frac{1}{2}\log_2 2 + \frac{1}{4}\log_2 4 + \frac{1}{8}\log_2 8 + \frac{1}{16}\log_2 16 + \frac{1}{32}\log_2 32$$

$$= \frac{1}{2} + \frac{2}{4} + \frac{3}{8} + \frac{4}{16} + \frac{5}{32}$$

$$= \frac{16}{32} + \frac{16}{32} + \frac{12}{32} + \frac{8}{32} + \frac{5}{32}$$

$$= \frac{57}{32} \text{ bits/symbol.}$$

$$\text{Symbol rate} = \frac{1}{10^{-6}} = 10^6 \text{ symbols/second.}$$

So information rate $= rH$

$$= 10^6 \times \frac{57}{32}$$

$$= 1781.25 \times 10^3 \text{ bits/sec.}$$

Example 3. *Find all the entropies of given channels shown in Fig. 8.2.*

Solution. Given that $P(X_1) = 0.2$, $P(X_2) = 0.5$, $P(X_3) = 0.3$

For the given graph

$$P(Y_1/X_1) = 0.8, \; P(Y_2/X_1) = 0.2$$

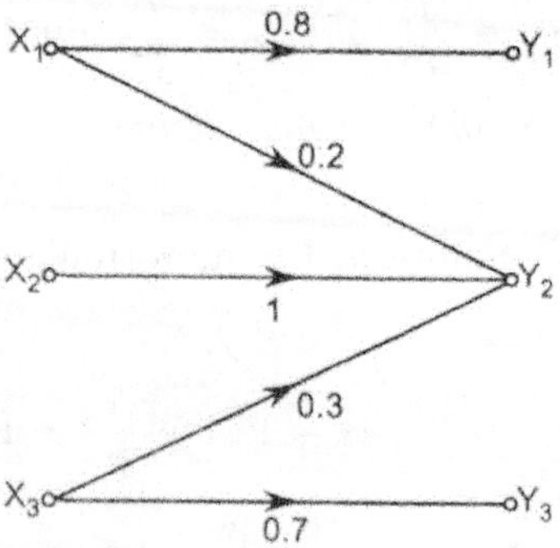

Fig. 8.2

$$P(Y_2/X_2) = 1,\ P(Y_2/X_3) = 0.3,\ P(Y_3/X_3) = 0.7$$

So, we can find the matrix

$$P(Y/X) = \begin{matrix} & Y_1 & Y_2 & Y_3 \\ X_1 & 0.8 & 0.2 & 0 \\ X_2 & 0 & 1 & 0 \\ X_3 & 0 & 0.3 & 0.7 \end{matrix}$$

$$P(X, Y) = \begin{bmatrix} 0.8 \times 0.2 & 0.2 \times 0.2 & 0 \times 0.2 \\ 0 \times 0.5 & 1 \times 0.5 & 0 \times 0.5 \\ 0 \times 0.3 & 0.3 \times 0.3 & 0.7 \times 0.3 \end{bmatrix}$$

$$P(X, Y) = \begin{bmatrix} 0.16 & 0.04 & 0 \\ 0 & 0.5 & 0 \\ 0 & 0.09 & 0.21 \end{bmatrix}$$

So, we can calculate

$$P(Y_1) = 0.16,\ P(Y_2) = 0.04 + 0.5 + 0.09 = 0.63$$
$$P(Y_3) = 0.21$$

We can calculate,

$$P(X/Y) = \frac{P(X, Y)}{P(Y)}$$

$$= \begin{bmatrix} \frac{0.16}{0.16} & \frac{0.4}{0.63} & \frac{0}{0.21} \\ \frac{0}{0.16} & \frac{0.5}{0.63} & \frac{0}{0.21} \\ \frac{0}{0.16} & \frac{0.09}{0.63} & \frac{0.21}{0.21} \end{bmatrix} = \begin{bmatrix} 1 & 0.06 & 0 \\ 0 & 0.80 & 0 \\ 0 & 0.15 & 1 \end{bmatrix}$$

So, calculate :

$$(a)\ H(X) = P(X_1) \log \frac{1}{P(X_1)} + P(X_2) \log \frac{1}{P(X_2)} + P(X_3) \log \frac{1}{P(X_3)}$$

$$= 0.2 \log \frac{1}{0.2} + 0.5 \log \frac{1}{0.5} + 0.3 \log \frac{1}{0.3}$$

$$= 0.3218 + 0.3466 + 0.3612 \text{ nat}$$

$$= 1.0296 \text{ nat/message}$$

$$= 1.0296 \times 1.4426 \text{ bits/message}$$

$$\boxed{H(X) = 1.4853 \text{ bits/message}}$$

$$(b)\quad H(Y/X) = 0.8 \log \frac{1}{0.8} + 0.2 \log \frac{1}{0.2} + 1 \log \frac{1}{1} + 0.3 \log \frac{1}{0.3} + 0.7 \log \frac{1}{0.7}$$

$$= 0.1785 + 0.3218 + 0 + 0.3612 + 0.2496$$

$$= 1.1111 \text{ nat/message}$$

$$= 1.1111 \times 1.4426$$

$$\boxed{H(Y/X) = 1.6029 \text{ bits/message}}$$

(*c*) $$H(X, Y) = 0.16 \log \frac{1}{0.16} + 0.04 \log \frac{1}{0.04} + 0.5 \log \frac{1}{0.5} + 0.09 \log \frac{1}{0.09} + 0.21 \log \frac{1}{0.21}$$

$$\boxed{= 1.8942 \text{ bits/message.}}$$

(*d*) $$H(Y) = 0.16 \log \frac{1}{0.16} + 0.63 \log \frac{1}{0.63} + 0.21 \log \frac{1}{0.21}$$

$$\boxed{H(Y) = 1.3157 \text{ bits/message.}}$$

(*e*) $$H(X/Y) = 1 \log \frac{1}{1} + 0.06 \log \frac{1}{0.06} + 0.80 \log \frac{1}{0.80} + 0.15 \log \frac{1}{0.15} + 1 \log \frac{1}{1}$$

$$= 0.5364 \times 1.4426$$

$$\boxed{H(X/Y) = 0.7738 \text{ bits/message.}}$$

Example 4. *A transmitter has an alphabet of four letters [x_1, x_2, x_3, x_4] and the receiver has an alphabet of three letters. The joint probability matrix,*

$$P(X, Y) = \begin{bmatrix} 0.3 & 0.05 & 0 \\ 0 & 0.25 & 0 \\ 0 & 0.15 & 0.05 \\ 0 & 0.05 & 0.15 \end{bmatrix}$$

Calculate all entropies.

Solution. Now to calculate P (X), add the rows of joint matrix ;

$$P(x_1) = 0.3 + 0.05 + 0 = 0.35$$

$$P(x_2) = 0 + 0.25 + 0 = 0.25$$

$$P(x_3) = 0 + 0.15 + 0.05 = 0.20$$

$$P(x_4) = 0 + 0.05 + 0.15 = 0.20$$

To calculate P (Y_2), add the coloumb

$$P(Y_1) = 0.3,\ P(Y_2) = 0.5,\ P(Y_3) = 0.2$$

So we can calculate

(*a*) $$H(X) = 0.35 \log \frac{1}{0.35} + 0.25 \log \frac{1}{0.25} + 0.20 \log \frac{1}{0.20} + 0.20 \log \frac{1}{0.20}$$

$$\boxed{H(X) = 1.96 \text{ bits/message.}}$$

(*b*) $$H(Y) = 0.3 \log \frac{1}{0.3} + 0.5 \log \frac{1}{0.5} + 0.2 \log \frac{1}{0.2}$$

$$\boxed{H(Y) = 1.49 \text{ bits/message.}}$$

(*c*) $$H(X, Y) = 0.3 \log \frac{1}{0.3} + 0.05 \log \frac{1}{0.05} + 0.25 \log \frac{1}{0.25} + 0.15 \log \frac{1}{0.15} +$$

$$0.05 \log \frac{1}{0.05} + 0.05 \log \frac{1}{0.05} + 0.15 \log \frac{1}{0.15}$$

$$\boxed{H(X, Y) = 2.49 \text{ bits/message}}$$

(*d*) $$H(X/Y) = H(X, Y) - H(Y)$$

$$= 2.49 - 1.49$$

$$\boxed{H(X/Y) = 1.00 \text{ bits/message.}}$$

(*e*) $$H(Y/X) = H(X, Y) - H(X)$$

$$= 2.49 - 1.96$$

$$\boxed{H(Y/X) = 0.53 \text{ bits/message.}}$$

Example 5. *A card is drawn from a deck of playing cards. (i) You are informed that the card you draw is a spade. How much information did you receive (in bits)? (ii) How much information do you receive if you are told that the card that you drew is an ace? (iii) How much information do you receive if you are told that the card you drew is an ace of spades? (iv)Is the information content of the message "ace of spades" the sum of the information contents of the messages "spade" and "ace"?*

Solution. (*i*) P (card you drawn is spade)

$$= \frac{13_{C_1}}{52_{C_1}} = \frac{1}{4}$$

$$\therefore \quad I = \log_2 1/P = \log_2 (4) \qquad \therefore \quad I = 2 \text{ bits}$$

(*ii*) P (card you drawn is an ace)

$$= \frac{4_{C_1}}{52_{C_1}} = \frac{1}{13} \qquad \therefore \quad I = \log_2 (13)$$

$$= \frac{\log_e (13)}{\log_e (2)} = 3.697 \qquad \therefore \quad I = 3.697 \text{ bits}$$

(*iii*) P(card is ace of spades)

$$= \frac{13_{C_1}}{52_{C_1}} = \frac{1}{4}, \; I = \log_2 4$$

(*iv*)

$$I_{total} = I_{ace} + I_{spade}$$

$$I_{total} = 3.697 + 2 = 5.697.$$

Example 6. *A Black & white TV picture consists of 525 line of picture information. Assume that each line consist of 525 picture element & each element have 256 brightness level. Pictures are repeated at the rate of 30/sec. Calculate average rate of information content by TV set to viewer.*

Solution. Given TV picture consists of 525 lines of information. Each line consist of 525 picture elements

∴ 525 lines consists of 525 × 525 picture element. Now rate of 525 × 525 elements are

$$r = 525 \times 525 \times 30/\text{sec}$$

$$= 8268750 \text{ elements/ sec}$$

Now

$$H = +\sum_{i=1}^{m} \frac{1}{256} \log_2 256$$

Now

$$H = +\sum_{i=1}^{256} \frac{1}{256} \log_2 256$$

{Because here information are transmitted *i.e.*, element have 256 brightners levels *i.e.*, total no. of message = 256. Each levels are equiprobable ∴ 256 P = 1, $p = \frac{1}{256}$ [∵ sum of probability of all messages are equal to 1]}

$$H = \log_2 256$$

$$H = 8 \text{ bits/element}$$

Now information Rate

$$R = rH$$

$$R = 8268750 \times 8$$

$$R = 66.15 \times 10^6 \text{ bit/sec.}$$

$$R = 6.552 \times 10^6 \text{ bit/sec.}$$

Example 7. *An analog signal bandlimited to 10KHz is quantized into 8 levels of a pcm system with probability of 1/4, 1/5, 1/5, 1/10, 1/10, 1/20, 1/20 and 1/20 respectively. Find the entropy and rate of information.*

Solution. According to the sampling theorem, the signal must be sampled at a frequency given as

$$f_s = 10 \times 2 \text{ KHz}$$

$$= 20 \text{ KHz}$$

Considering each of the eight quantized levels as a message, the entropy of the source may written as

$$H = \frac{1}{4}\log_2(4) + \frac{1}{5}\log_2(5) + \frac{1}{5}\log_2(5) + \frac{1}{10}\log_2(10) + \frac{1}{10}\log_2(10)$$

$$+ \frac{1}{20}\log_2(20) + \frac{1}{20}\log_2(20) + \frac{1}{20}\log_2(20)$$

$$H = \frac{1}{4}\log_2(4) + \frac{2}{5}\log_2(5) + \frac{3}{20}\log_2(20) + \frac{2}{10}\log_2(10)$$

$$H = 2.74 \text{ bits/msg.}$$

The sampling frequency is 20 KHz. Then the rate at which message is produced, will be given as r = 20,000 message/sec.

the information rate is

$$R = rH$$
$$= 20{,}000 \times 2.74$$
$$R = 54{,}800 \text{ bits/sec.}$$

Example 8. *An analog signal is bandlimited to f_m Hz and the sampled at Nyquist rate. The samples are quantized into 4 levels. Each level represent one message. Thus there are 4 messages. The probabilities of occurrence of these 4 level (Message) are $P_1 = P_4 = 1/8$ and $P_2 = P_3 = 3/8$. Obtain information rate of the source.*

Solution. We are given four message with probabilities $P_1 = P_4 = 1/8$ and $P_2 = P_3 = 3/8$. Average information is expressed as

$$H = P_1 \log_2\left(\frac{1}{P_1}\right) + P_2 \log_2\left(\frac{1}{P_2}\right) + P_3 \log_2\left(\frac{1}{P_3}\right) + P_4 \log_2\left(\frac{1}{P_4}\right)$$

or $$H = \frac{1}{8}\log_2(8) + \frac{3}{8}\log_2\left(\frac{8}{3}\right) + \frac{3}{8}\log_2\left(\frac{8}{3}\right) + \frac{1}{8}\log_2(8)$$

or $$H = 1.8 \text{ bits/message}$$

It is given that the signal is sampled at Nyquist rate. Nyquist rate for f_m Hz bandlimited signal is N.P. Nyquist rate $= 2f_m$ samples/sec.

Every sample generates one message signal message per second, $r = 2f_m$ messages/sec Information rate $R = rH$

$$R = 2f_m \text{ message/sec} \times 1.8 \text{ bits/message}$$
$$R = 3.6\, f_m \text{ bits/sec}$$

Example 9. *Consider a Telegraph Source having two symbols, dot and dash. The dot duration is 0.2S. The dash duration is 3 times the dot duration. The probability of the dot's occurring is twice that of the dash and the time between symbols is 0.2 sec. calculate the information rate of the telegraph source.*

Solution. $$p(\text{dot}) = 2p(\text{dash})$$

$$p(\text{dot}) + p(\text{dash}) = 3p(\text{dash}) = 1$$

$$p(\text{dash}) = \frac{1}{3} \text{ and } p(\text{dot}) = \frac{2}{3}$$

$$H(X) = -p(\text{dot}) \log_2 p(\text{dot}) - p(\text{dash}) \log_2 p(\text{dash})$$
$$= 0.667\ (0.585) + 0.0333\ (0.585) = 0.92 \text{ b/symbol}$$
$$t \text{ dot} = 0.2\text{s},\ t \text{ dash} = 0.6\text{s},\ t \text{ space} = 0.2\text{s}$$

Thus, the average time per symbol is

$$T_s = p\ (\text{dot})\ t \text{ dot} + p\ (\text{dash})\ t \text{ dash} + t \text{ space}$$
$$= 0.5333 \text{ sec/symbol.}$$

and the average symbol rate is

$$r = \frac{1}{T_s} = 1.875 \text{ symbols/s}$$

Thus, the average information rate of the Telegraph source is

$$R = r\,H\,(x) = 1.875\ (0.92) = 1.725 \text{ b/sec.}$$

Example 10. *A high resolution black and white TV picture consists of about 2×10^6 picture elements and 16 different brightness levels. Pictures are repeated at the rate of 32 per second. All picture elements are assumed to be independent and all level have equal likelihood of occurrence. Calculate the average rate of information conveyed by this TV picture source.*

Solution.
$$H(X) = -\sum_{i=1}^{16} \frac{1}{16} \log_2 \frac{1}{16} = 4 \text{ b/element}$$
$$r = 2 \times (10^6)\ (32) = 64 \times 10^6 \text{ elements/sec}$$
$$R = r\,H(X) = 64\ (10^6)\ (4) = 256\ (10^6) \text{ b/sec}$$
$$= 256 \text{ Mb/sec.}$$

Example 11. *A binary data source has $P_o = \frac{3}{8}$, $P_1 = \frac{5}{8}$ and inter symbol interference extending over groups of two successive symbols such that $P(1/0) = \frac{3}{4}$ and $P(0/1) = \frac{1}{16}$. Calculate the conditional entropy he and compare with H_{max}.*

Solution.
$$P\,(1/1) = 1 - P\,(0/1) = \frac{1}{16} = \frac{15}{16}$$
$$P\,(0/0) = 1 - P\,(1/0) = 1 - \frac{3}{4} = \frac{1}{4}$$

$$\text{Entropy } H = \sum_{v} \sum_{j} P\,(i)\,P\left(\frac{j}{i}\right) \log \left(\frac{1}{P(j/i)}\right)$$

$$= p\,(0)\,p\,(1/0) \log \frac{1}{p\,(1/0)} + p\,(1)\,P\,(0/1) \log \frac{1}{p\,(0/1)}$$

$$+ p\,(0)\,p\left(\frac{0}{0}\right) \log \frac{1}{p\left(\frac{0}{0}\right)} + p\,(1)\,p\left(\frac{1}{1}\right) \log \left(\frac{1}{p\left(\frac{1}{2}\right)}\right)$$

$$= \frac{3}{5}\cdot\frac{3}{4}\log\frac{4}{3}+\frac{5}{8}\cdot\frac{1}{16}\log\frac{16}{1}+\frac{3}{8}\cdot\frac{1}{4}\log\frac{4}{1}+\frac{5}{8}\cdot\frac{15}{16}\log\frac{16}{15}$$

$$= \frac{3}{8}\left[\frac{3}{4}\log\frac{4}{3}+\frac{1}{4}\log 4\right]+\frac{5}{8}\left[\frac{5}{16}\log\frac{16}{15}+\frac{1}{16}\log 16\right]$$

$$= 0.51 \text{ bits/symbols}$$

$$H_{max} = \log_2 2 = 1 \text{ bits/symbols}$$

Example 12. *Consider a DMS with the alphabet (S_0, S_1, S_2) with probabilites $P_0 = \frac{1}{4}$, $P_1 = \frac{1}{4}$, $P_2 = \frac{1}{2}$. Find out the entropy of orignal source and second order extension entropy?*

Solution. Given $P_0 = \frac{1}{2}, P_1 = \frac{1}{4}, P_2 = \frac{1}{2}$

∴ Entropy of orignal source H (α)

$$H(\alpha) = \sum_{i=0}^{2} P_i \log_2 \frac{1}{P_i}$$

$$= \frac{1}{4}\log_2(4)+\frac{1}{2}\log_2(2)+\frac{1}{2}\log_2(2) = \frac{3}{2} \text{ bits}$$

Now consider second order extension source. Source alphabet $\alpha = (S_0, S_1, S_2)$ consists of three symbols. [k = number of symbols = 3]

Source alphabet of second order extension source consists of nine symbols

$$k^2 = 3^2 = 9$$

These symbols are

symbols $\quad \sigma_0\ \sigma_1\ \sigma_2\ \sigma_3\ \sigma_4\ \sigma_5\ \sigma_6\ \sigma_7\ \sigma_8$

$$\alpha^2 = (s_0s_1, s_0s_0, s_0s_2, s_1s_0, s_1s_1, s_1s_2, s_2s_0, s_2s_1, s_2s_2)$$

Probabilites of these in blocks that consists of 2 symbols are.

$$P(\sigma_i) = \left(\frac{1}{16},\frac{1}{16},\frac{1}{8},\frac{1}{16},\frac{1}{16},\frac{1}{8},\frac{1}{8},\frac{1}{8},\frac{1}{4}\right)$$

Entropy of second order extension:

$$H(\alpha^2) = \sum_{i=0}^{8} P(\sigma_i)\log_2\left(\frac{1}{p}\right)(\sigma_i)$$

$$H(\alpha^2) = \frac{1}{16}\log_2(16)+\frac{1}{16}\log_2(16)+\frac{1}{8}\log_2(8)+\frac{1}{16}\log_2(16)+\frac{1}{16}\log_2(16)$$

$$+\frac{1}{8}\log_2(8)+\frac{1}{8}\log_2(8)+\frac{1}{8}\log_2(8)+\frac{1}{4}\log_2(4)$$

$$= 3 \text{ bits}$$

cross-check the ans. by formula

$$H(\alpha^2) = 2H(\alpha)$$

$$3 = 2\times\frac{3}{2}$$

Exmaple 13. *Consider a DMS with source alphabet* $\alpha = (s_0, s_1, s_2, s_3)$ *with prob.* $\frac{1}{4}, \frac{1}{4}, \frac{1}{4}, \frac{1}{4}$, *find out second & third order extension entropy?*

Solution. $\alpha^2 = s_0s_0\ s_0s_1\ s_0s_2\ s_0s_3\ s_1s_0\ s_1s_1\ s_1s_2\ s_1s_3$

[Block of 2 symbol] $s_2s_0\ s_2s_1\ s_2s_2\ s_2s_3\ s_3s_0\ s_3s_1\ s_3s_2\ s_3s_3$

total number of symbols $= 4^2 = 16$

probabilities of each of 16 blocks of 2 symbols are 1/16.

$$H(\alpha^2) = \sum_{i=1}^{16} P_i \log_2 \frac{1}{P_i}$$

$$H(\alpha^2) = \sum_{i=1}^{16} \frac{1}{16}\log_2(16) = \log_2(16) = 4\text{bits/symbol}$$

now third order extension

α^3 (Block of 3 symbols) $= (s_0\, s_1\, s_2 \,.\, s_0 s_1 s_3 \,\ldots\ldots)$

No. of blocks $= 4^3 = 64$

$$H(\alpha^3) = \sum_{i=1}^{64} \frac{1}{64}\log_2(64) = 6 \text{ bits/symbol}$$

$$H(\alpha) = \text{first order extension}$$

$$\sum_{i=1}^{4} \frac{1}{4}\log_2(4) = 2\text{bits/symbol}$$

we can cross-check by $H(\alpha^n) = nH(\alpha)$

8.11 SOURCE CODING THEOREM

The last stage of digital system is encoding. So we can say that in communication system ; for a efficient and error free transmission of data ; coding techniques are very important. Basically we have two things in our mind:

(*a*) Codes generated should be in binary form.

(*b*) The code should be decipherable means it should be uniquely identified so that at the receiver side ; it should be decoded easily and without any error at the receiver side.

8.12 PREFIX CODING

As we have seen that codes generated by encoder should be unique for minimum error during decoding. But due to this; every source have limitation on the number of codes generated. For the purpose that we have codes that are uniquely identified can be solved by using *'prefix coding'*.

'A prefix code is defined as a code in which no code word is the prefix of any other code-word'.

This can be understand by the following example.

	Code I	*Code II (Prefix code)*
S_0	0	0
S_1	1	1 0
S_2	0 0 S_0 is the prefix	1 1 0
S_3	1 1 S_1 is the prefix	1 1 1

It can be easily see that these code can be easily decoded by decoder as this follows the **Kraft inequality**.

It is also found that there are some codes that can follow the Kraft inequality but not prefix code, so they can be decoded without any error.

For example,

Code III
0
0 1
0 1 1
0 1 1 1

Now we will study different type of coding techniques of efficient transmission.

8.13 SHANNON-FANO CODING

For variable length coding ; if we apply a practical concept that the generally used codes (means codes with high-probability) should be coded in the minimum length and rarely used codes should be coded in the long length so that don't effect the efficiency to much.

Shannan-Fano coding generates efficient codes in which the word length increases as the probability of symbol decreases.

In this method; first of all arrange message according to the descending order of probability. Now draw a line that divides the symbol into two groups such that the group probabilities are as nearly as equal as possible.

Then assign the digit 0 to each symbol in the group above the line and digit 1 to each symbol in the group below the line. For all the subsequent steps, subdivide each group into subgroups and again assign digits following the previous rule. Whenever a group contains one symbol no further subdivision is possible and code word for that symbol is completed. When all the groups have been reduced to one symbol, the code words for each symbol is assigned.

8.14 HUFFMAN CODING

The Huffman coding (compact coding) is a optimum coding in the sense that no other uniquely decodable set of code-words has a smaller average code-word length for a given source.

The Huffmann encoding algorithm proceeds as follows :

1. The source symbols are arranged in the descending order. The two source symbols of lowest probability are assigned *a* 0 and *a* 1.
2. These two source symbols are regarded as being combined into a new source symbol with probability equal to the sum of the two original probabilities. The probability of the new symbol is placed in the list in accordance with its value.
3. The procedure is repeated until we are left with a final list of source. Symbol of only two for which a '0' and a '1'' are assigned. The code for each symbol is found by working backward and tracing the sequence of 0s and 1s assigned to that symbol as well as its successor.

8.15 CONTINUOUS CHANNEL

The information theory concept of discrete channels can be extended for the continuous channels.

In the case of discrete symbols ; the definitions of different entropies were based on the concept of different averages. But in case of continuous channels, it will be defined on the basis of continuous distributions.

If P (x) is the probability density function associated with signal $x\ (t)$, then the entropy of the source is :

$$H(x) = -\int_{-\infty}^{\infty} P(x) \log P(x)\, dx \tag{8.59}$$

$$H(y) = -\int_{-\infty}^{\infty} P(y) \log P(y)\, .dy \tag{8.60}$$

$$H(x, y) = -\int_{-\infty}^{\infty}\int_{-\infty}^{\infty} P(x, y) \log P(x, y)\, dx, dy \tag{8.61}$$

$$H(x/y) = -\int_{-\infty}^{\infty}\int_{-\infty}^{\infty} P(x, y) \log P(x/y)\, dx\, .\, dy \tag{8.62}$$

8.15.1 Capacity of Gaussian Channel : Shannon Hartely Theorem

The noise characteristics of channels practically observed are assumed Gaussian. If the channel is not gaussian, then this concept can be extended to Non-Gaussian channel also.

For a Gaussian channel, $P(x) = \frac{1}{\sqrt{2\pi\sigma^2}} e^{-\frac{x^2}{2\sigma^2}}$ (8.63)

for zero mean value,

Hence

$$H(x) = -\int_{-\infty}^{\infty} P(x) \log P(x) \, . dx$$

But $$-\log P(x) = \log \sqrt{2\pi\sigma^2} + \log e^{\frac{x^2}{2\sigma^2}} \quad (8.64)$$

So, $$H(x) = \int_{-\infty}^{\infty} P(x) \log \sqrt{2\pi\sigma^2} + \int_{-\infty}^{\infty} P(x) \log e^{\frac{x^2}{2\sigma^2}}$$

After calculating, we get

$$H(x) = \log \sqrt{2\pi e\sigma^2} \text{ bits/message} \quad (8.65)$$

Now if the signal is bandlimited to BHz. Then it may be uniquely specified by taking 2B samples per second according to sampling theorem. Hence rate of information is

$$R(x) = 2B\,H(x) \quad (8.66)$$

$$= 2B \log \sqrt{2\pi e\sigma^2}$$

$$= B \log (2\pi e\sigma^2) \quad (8.67)$$

Now for Gaussian noise $\sigma^2 = N$.

$$R(x) = B \log (2\pi eN)$$

$$R(n) = B \log (2\pi eN) \quad (8.68)$$

Now for receiver side, we get signal with noise

$$R(y) = B \log (2\pi e\,(S + N)) \quad (8.69)$$

So, we can calculate; rate of signal by

$$R = R(y) - R(n) \quad (8.70)$$

$$= B \log (2\pi e\,(S + N) - B \log 2\pi eN)$$

$$= B \log \left(\frac{S+N}{N}\right)$$

$$\boxed{R = B \log \left(1 + \frac{S}{N}\right)} \quad (8.71)$$

Channel capacity can be calculated by

$$C = \max\,[R]$$

$$\boxed{C = B \log \left(1 + \frac{S}{N}\right)} \quad (8.72)$$

So channel capacity of a white-bandlimited gaussian channel is

$$C = B \log\left(1 + \frac{S}{N}\right) \text{ Bits/second}$$

But for a gaussian white noise

$$N = \eta B \tag{8.73}$$

$$\boxed{C = B \log\left(1 + \frac{S}{\eta B}\right)} \tag{8.74}$$

where C = Channel capacity

B = Bandwidth

8.15.2 Capacity of a Channel of an Infinite Bandwidth

For a noiseless channel $\left(\frac{S}{N} = \infty\right)$ has an infinite capacity. On the other hand, for a $B \to \infty$, we have

$$C = \lim_{B \to \infty} B \log\left(1 + \frac{S}{\eta B}\right) \tag{8.75}$$

$$= \lim_{B \to \infty} \frac{S}{\eta} \times \frac{\eta B}{S} \log\left(1 + \frac{S}{\eta B}\right)$$

But,

$$\lim_{X \to \infty} \left\{X \log_2\left(1 + \frac{1}{X}\right)\right\} = \log_2 e$$

$$= 1.44$$

$$\boxed{C = 1.44 \frac{S}{\eta}} \tag{8.76}$$

This also shows the trade off $\frac{S}{N}$ ratio to bandwidth; if we increase $\frac{S}{N}$ ratio, bandwidth B is decreased.

Example 14. *For the given message sequence with their probabilities, Apply shannon-fano coding, calculate the code efficiency*

$$[x] = [x_1 \quad x_2 \quad x_3 \quad x_4 \quad x_5 \quad x_6 \quad x_7 \quad x_8]$$

$$[P] = \left[\frac{1}{4} \quad \frac{1}{8} \quad \frac{1}{16} \quad \frac{1}{16} \quad \frac{1}{16} \quad \frac{1}{4} \quad \frac{1}{16} \quad \frac{1}{8}\right].$$

Solution. Arrange the probabilities in descending order.

Message	*Probability*	*Encoding*				*Code length (No.)*
x_1	$\frac{1}{4}$	0	0			2
x_6	$\frac{1}{4}$	0	1			2
x_2	$\frac{1}{8}$	1	0	0		3
x_8	$\frac{1}{8}$	1	0	1		3
x_3	$\frac{1}{16}$	1	1	0	0	4
x_4	$\frac{1}{16}$	1	1	0	1	4
x_5	$\frac{1}{16}$	1	1	1	0	4
x_7	$\frac{1}{16}$	1	1	1	1	4

Now length / message, $\bar{L} = \sum_{i=1}^{8} P_i N_i$.

$$= \frac{1}{4}\times 2 + \frac{1}{4}\times 2 + \frac{1}{8}\times 3 + \frac{1}{8}\times 3 + \frac{1}{16}\times 4 + \frac{1}{16}\times 4 + \frac{1}{16}\times 4 + \frac{1}{16}\times 4$$

$$\boxed{\bar{L} = 2.75 \text{ letters/message.}}$$

Now

$$H(X) = \sum_{i=1}^{8} P_i \log \frac{1}{P_i}$$

$$= \frac{1}{4}\log 4 + \frac{1}{8}\log 8 + \frac{1}{16}\log 16 + \frac{1}{16}\log 16 + \frac{1}{16}\log 16 + \frac{1}{4}\log 4 + \frac{1}{16}\log 16 + \frac{1}{8}\log 8$$

$$\boxed{H(X) = 2.75 \text{ bits/message.}}$$

So code efficiency, $\eta = \frac{H(x)}{\bar{L}} \times 100$

$$= \frac{2.75}{2.75} \times 100$$

$$\boxed{\eta = 100\%.}$$

Example 15. *Apply the Shannon-Fano coding and find the code efficiency*

$$[x] = [x_1 \quad x_2 \quad x_3 \quad x_4 \quad x_5 \quad x_6 \quad x_7]$$

$$[P] = [0.4 \quad 0.2 \quad 0.12 \quad 0.08 \quad 0.08 \quad 0.08 \quad 0.04]$$

Solution. Arrange the probabilities in the descending order.

Message	*Probability*	*Encoding*				*Length*	*Letters/Message*
x_1	0.4	0	0			2	$2 \times 0.4 = 0.8$
x_2	0.2	0	1			2	$2 \times 0.2 = 0.4$
x_3	0.12	1	0	0		3	$3 \times 0.12 = 0.36$
x_4	0.08	1	0	1		3	$3 \times 0.08 = 0.24$
x_5	0.08	1	1	0		3	$3 \times 0.08 = 0.24$
x_6	0.08	1	1	1	0	4	$4 \times 0.08 = 0.32$
x_7	0.04	1	1	1	1	4	$4 \times 0.04 = 0.32$

$$\bar{L} = \sum_{i=1}^{7} P_i N_i = 2.52 \text{ letter/message.}$$

$$\text{Entropy} = 0.4 \log \frac{1}{0.4} + 0.2 \log \frac{1}{0.2} + 0.12 \log \frac{1}{0.12} + 0.08 \log \frac{1}{0.08}$$

$$+ 0.08 \log \frac{1}{0.08} + 0.08 \log \frac{1}{0.08} + 0.04 \log \frac{1}{0.04}$$

$$H(x) = 2.42 \text{ bits/message.}$$

So code efficiency

$$= \frac{H(x)}{\bar{L}}$$

$$= \frac{2.42}{2.52} \times 100$$

$$= 96.03 \%.$$

Example 16. *We have 5 symbols for a discrete source*

S_0	S_1	S_2	S_3	S_4
0.4	0.2	0.2	0.1	0.1

Obtain Huffmann coding, average code word length, entropy of the given system.

Solution. (*a*) Entropy of the system is given by :

$$H(x) = \sum_{i=0}^{4} P_i \log \frac{1}{P_i}$$

$$H(x) = 0.4 \log \frac{1}{0.4} + 0.2 \log \frac{1}{0.2} + 0.2 \log \frac{1}{0.2} + 0.1 \log \frac{1}{0.1} + 0.1 \log \frac{1}{0.1}$$

$$\boxed{H(x) = 2.12193 \text{ bits/Message.}}$$

(*b*) Huffman coding can be performed as follow :

Symbol	*Step I*		*Step II*		*Step III*		*Step IV*	
S_0	0.4	0 0	0.4	0 0	0.4	1	0.6	0
S_1	0.2	10	0.2	01	0.4	0 0	0.4	1
S_2	0.2	11	0.2	10	0.2	01		
S_3	0.1	010	0.2	11				
S_4	0.1	011						
		↑ **Final codes**						

Thus we have finally

Symbols	*Code*	*Code length (N)*
S_0	0 0	2
S_1	1 0	2
S_2	1 1	2
S_3	0 1 0	3
S_4	0 1 1	3

(*c*) Average code length:

$$I = \sum_{i=0}^{4} N_i P_i$$

$$= 2 \times 0.4 + 2 \times 0.2 + 2 \times 0.2 + 3 \times 0.1 + 3 \times 0.1 = 2.2 \text{ letter/message.}$$

Example 17. *A message source generates ten messages with probabilities 0.1, 0.13, 0.01, 0.04, 0.08, 0.29, 0.06, 0.22, 0.05 and 0.02. The rate of message generation is 300 message/sec. Find the entropy of source and information rate. Obtain the Huffmann codes for message and calculate the average number of bits/message. What is code redundancy ?*

Solution. (*a*) Entropy of source can be calculated :

$$H = \sum_{i=1}^{10} P_i \log_2 \frac{1}{P_i} = -\sum_{i=1}^{10} P_i \log_2 P_i$$

$$= -[0.1 \log_2 0.1 + 0.13 \log_2 0.13 + 0.01 \log_2 0.01 + 0.04 \log_2 0.04 + 0.08 \log_2 0.08 + 0.29 \log_2 0.29 + 0.06 \log_2 0.06 + 0.22 \log_2 0.22 + 0.5 \log_2 0.5 + 0.02 \log_2 0.02$$

$$= 0.31 + 0.32 + 0.05 + 0.11 + 0.22 + 0.05 + 0.21 + 0.41 + 0.15 + 0.10$$

$$\boxed{H = 2.38 \text{ bits/message.}}$$

(*b*) Information rate $R = rH$

$$= 300 \times 2.38$$

$$= 714 \text{ bits/sec.}$$

(*c*) Huffmann coding : Place all the probabilities in descending order. Now applying the same method as previous.

	Step I	Step II	Step III	Step IV	Step V	Step VI	Step VII	Step VIII	Step IX
S_6	0.29	0.29	0.29	0.29	0.29	0.29	0.29	0.43	0.57
S_8	0.22	0.22	0.22	0.22	0.22	0.22	0.28	0.29	0.43
S_2	0.13	0.13	0.13	0.13	0.15	0.21	0.22	0.28	
S_1	0.10	0.10	0.10	0.11	0.13	0.15	0.21		
S_5	0.08	0.08	0.08	0.10	0.11	0.13			
S_7	0.06	0.06	0.07	0.08	0.10				
S_9	0.05	0.05	0.06	0.07					
S_4	0.04	0.04	0.05						
S_{10}	0.02	0.03							
S_3	0.01								

We have symbols with codes :

Message	S_6	S_8	S_2	S_1	S_5	S_7	S_9	S_4	S_{10}
Code	1	0 0	0 0 1	0 1 1	0 0 0 0	0 1 0 0	0 1 0 1	0 0 0 1 1 0	0 0 0 1 1 0.

J	S_3	
...	0 0 0 1	1 1

(*d*) Average code length:

$$\bar{L} = \sum_{i=0}^{9} P_i N_i$$

$$= 0.29 \times 1 + 0.22 \times 2 + 0.13 \times 3 + 0.1 \times 3 + 0.08 \times 4 + 0.06 \times 4 + 0.05 \times 4 + 0.05 \times 4 + 0.04 \times 5 + 0.02 \times 6 + 0.01 \times 6.$$

$= 2.43$ letters/message.

(*e*) Code efficiency:

$$\eta = \frac{H}{\bar{L}} = \frac{2.38}{2.43} \times 100$$

$= 97.94\%$.

So Redundency $= 1 - \eta$

$= 1 - 0.9794$

$= 2.06\%$.

Example 18. *For 7 message signals have the following probabilities*

x_1	x_2	x_3	x_4	x_5	x_6	x_7
0.4	0.2	0.12	0.08	0.08	0.08	0.04

find the Huffman coding for M = 3. Calculate code efficiency of system.

Solution. Here we have M = 3 that means that we have now to combine '3' probabilities. Let the encoding alphabet be – 1, 0, 1 otherwise no change in process.

Message	*Codeword*	*Probability*	*First step*	*Second step*
x_1	1	0.4	0.4	0.4 1
x_2	– 1	0.2	0.2	0.4 0
x_3	0 0	0.12	0.2 1	0.2 – 1
x_4	0 – 1	0.08	0.12 0	
x_5	0 1 1	0.08 1	0.08 – 1	
x_6	0 1 0	0.08 0		
x_7	0 1 – 1	0.04 – 1		

$$I = \sum_{i=1}^{7} P_i N_i$$

$$= (0.4 \times 1) + (0.2 \times 1) + (0.12 \times 2) + (0.08 \times 2) + (0.08 \times 3) + (0.08 \times 3) + (0.08 \times 3) + (0.04 \times 3)$$

$$= 1.5 \text{ letter/message.}$$

$$H(x) = \sum_{i=1}^{7} P(x) \log P(x)$$

$$= 2.42 \text{ bits/message (Calculated in example)}$$

$$\eta = \frac{H(x)}{L \log m} = \frac{2.42}{1.5 \log 3} = 99.5\%.$$

Example 19. *We have two message signals*

$$m_0 = 0.4$$

$$m_1 = 0.2$$

Now find the Shannon-Fano coding for second order system.

Solution. Second order system means a system that will transmit two message at a time. Thus, now we have messages for IInd order

$$M_0 = m_0 \,.\, m_1 = 0.4 \times 0.4 = 0.16$$

$$M_1 = m_0 \,.\, m_1 = 0.4 \times 0.2 = 0.08$$

$$M_2 = m_1 \,.\, m_0 = 0.2 \times 0.4 = 0.08$$

$$M_3 = m_1 \,.\, m_1 = 0.2 \times 0.2 = 0.04.$$

Now arranging them in descending order (Now method is same as general coding)

Message	*Probability*	*Encoding*	
M_0	0.16	0	0
M_1	0.08	0	1
M_2	0.08	1	0
M_3	0.04	1	1

Here one thing should be clear that when we are dividing the probabilities in two group having equal addition result ; If we are not getting equal part then it is conventional that the upper half should be greater than lower half. That's why we have taken

$$0.16 + 0.08 = 0.24 \text{ and}$$

$$0.08 + 0.08 = 0.16$$

and don't take

0.16 and

$$0.8 + 0.8 + 0.4 = 0.20$$

(Generally it is conventional not hard and fast rule).

Example 20. *Calculate the capacity of a low pass channel with a usable bandwidth of 3 KHz and* $\frac{S}{N}$ *= 100 at channel output. Assume the channel noise to be white Gaussian.*

Solution. The capacity C is given by :

$$C = B \log_2 \left(1 + \frac{S}{N}\right)$$

$$= 3000 \log_2 (1 + 100)$$

$$C \simeq 20 \text{ Kb/s}.$$

Example 21. *The terminal of a computer used to enter alphanumeric data is connected to the computer through a voice grade telephone line having a usable bandwidth of 3 KHz and a output of 10 dB. Determine :*

(a) Capacity of channel

(b) The max rate of transmission without error.

Assume that terminal has 128 characters.

Solution. (*a*) Capacity is given by :

$$C = B \log_2 \left(1 + \frac{S}{N}\right)$$

$$= 3000 \log_2 (1 + 10)$$

$$= 10378 \text{ bits/second}.$$

(*b*) Average information:

$$H = \log_2 (128)$$

$$= 7 \text{ bits/message}$$

Therefore, rate

$$R = r_S \times H$$

for error free transmission

$$r_S H < C$$

$$7 r_S < 10378$$

$$\boxed{r_S < 1482 \text{ character/second.}}$$

Example 22. *A DMS has 4 symbols x_1, x_2, x_3, x_4 with the probability $\frac{1}{2}, \frac{1}{4}, \frac{1}{8}, \frac{1}{8}$. Construct shannon's fano code & determine entropy & no. of bits/message?*

Solution. Step 1 : Write down messages in decreasing probability.

Step 2 : Divide these messages into the two group, so that sum of prob's of both groups are same.

Step 3 : Assign to one group bit 0 & to another bit 1.

Setp 4 : Repeat process until no. further partition is possible.

TABLE

Message	*Probability*		*Step I*	*Step II*	*Step III*	*No. of bits*
m_0	$\frac{1}{2}$	Σ0.5	0			1
m_1	$\frac{1}{4}$		1	0		2
m_2	$\frac{1}{8}$	Σ0.5	1	1	0	3
m_3	$\frac{1}{8}$		1	1	1	3

Entropy
$$H = \sum_{i=0}^{3} P_i \log_2 \frac{1}{P_i} = \frac{1}{2}\log_2(2) + \frac{1}{4}\log_2(4) + \frac{1}{8}\log_2(8) + \frac{1}{8}\log_2(8)$$
$$= 1.75 \text{ bits/symbol}$$

Number of
$$N = \sum_{i=0}^{3} P_i\, n_i$$

bits/message
$$= 1\left(\frac{1}{2}\right) + 2\left(\frac{1}{4}\right) + 3\left(\frac{1}{8}\right) + 3\left(\frac{1}{8}\right) = \frac{1}{2} + \frac{1}{2} + \frac{3}{8} + \frac{3}{8} = \frac{14}{8} = \frac{7}{4} = 1.75$$

$$\eta = \frac{H}{N} = 1$$

If the 4 messages are directly coded into Binary digit, then codes would be {00, 01, 10, 11}, each symbol requiring 2 digit and giving $\Sigma P_i \eta_i = 2$ bits/symbol $> H(x)$

so
$$\eta = \frac{H(x)}{N} = \frac{1.75}{2} = 0.875$$

and it is less than to 1.

Example 23. *For message probabilities $P(m_1) = \frac{1}{2}$, $P(m_2) = \frac{1}{8}$, $P(m_3) = \frac{1}{8}$, $P(m_4) = \frac{1}{16}$, $P(m_5) = \frac{1}{16}$, $P(m_6) = \frac{1}{16}$, $P(m_7) = \frac{1}{32}$, $P(m_8) = \frac{1}{32}$, construct shannon-fano code.*

Solution.

Message	*Probability*	*Step I*	*Step II*	*Step III*	*Step IV*	*Step V*	*No. of bits of message*	*Code*
m_1	$\frac{1}{2}$	0					1	0
m_2	$\frac{1}{8}$	1	0	0			3	100
m_3	$\frac{1}{8}$	1	0	1			3	101
m_4	$\frac{1}{16}$	1	1	0	0		4	1100
m_5	$\frac{1}{16}$	1	1	0	1		4	1101
m_6	$\frac{1}{16}$	1	1	1	0		4	1110
m_7	$\frac{1}{32}$	1	1	1	1	0	5	11110
m_8	$\frac{1}{32}$	1	1	1	1	1	5	11111

The average information per message interval can be obtained as

$$H = \sum_{i=1}^{8} P_i \log \frac{1}{P_i}$$

$$= \frac{1}{2}\log_2 2 + \frac{1}{8}\log_2 8 + \frac{1}{8}\log_2 8 + \frac{1}{16}\log_2 16 + \frac{1}{16}\log_2 16$$

$$= \frac{1}{2} + \frac{6}{8} + \frac{12}{16} + \frac{10}{32}$$

$$= \frac{1}{2} + \frac{3}{4} + \frac{3}{4} + \frac{5}{16} = \frac{37}{16} = 2\frac{5}{16} = 2.3125 \text{ bits/message} \quad \ldots(i)$$

Further, the average number of bits per message can be obtained from the table No. $\overline{M}$ = Average bits/message

$$= \frac{1}{2}\times 1 + \frac{1}{8}\times 3 + \frac{1}{8}\times 3 + \frac{1}{16}\times 4 + \frac{1}{16}\times 4 + \frac{1}{32}\times 5 + \frac{1}{32}\times 5$$

$$= \frac{1}{2} + \frac{3}{4} + \frac{3}{4} + \frac{5}{16} = 2\frac{5}{16} \quad \ldots(ii)$$

From equation (*i*) and (*ii*) we find that each bit on the average carries $2\frac{5}{16}\Big/2\frac{5}{16} = 1$ bit of information. Thus the efficiency $\frac{H}{M}$ is look which is the maximum. If we head not used this efficient coding, we would have required 3 bits per message which is about 30 percent lower than $2\frac{5}{16}$ calculated above. The saving in average number of bits per second has the primary

1. It allows more time per bit that reduces bandwidth and improve noise immunity
2. If the bit rate is kept fixed the bits not used for information transmission may be utilized for ever detection error correction or both.

Example 24. *Apply Shannon–Fano coding procedure for the following message ensemble*

$$[x] = [x_1 \quad x_2 \quad x_3 \quad x_4 \quad x_5 \quad x_6 \quad x_7]$$

$$[P] = [0.4 \quad 0.2 \quad 0.12 \quad 0.08 \quad 0.08 \quad 0.08 \quad 0.04]$$

Take $M = 2$

Solution. For the first partitioning, there are two ways

(*i*) $[x_1] = [x_1, x_2]$, $[x_2] = [x_3\, x_4\, x_5\, x_6\, x_7]$

(*ii*) $[x_1] = [x_1]\, [x_2] = [x_2\, x_3\, x_4\, x_5\, x_6\, x_7]$

(*i*)

Message	*Probability*	*Enclosed message*	*Length*
x_1	0.4	00	2
x_2	0.2	01	2
x_3	0.12	100	3
x_4	0.08	101	3
x_5	0.08	110	3
x_6	0.08	1110	4
x_7	0.04	1111	4

$$\bar{L} = \sum_{k=1}^{7} p_k n_k = (0.4 \times 2) + (0.2 \times 2) + (0.12 \times 3) + (0.08 \times 3)$$

$$+ (0.08 \times 3) + (0.08 \times 4) + (0.04 \times 4)$$

$$= 2.52 \text{ letter/message}$$

(*ii*)

Message		*Probability*	*Enclosed message*	*Length*
$x_1 \longrightarrow$	x_1	0.4	0	1
$x_{21} \longrightarrow$	x_2	0.2	100	3
	x_3	0.12	101	3
$x_2 \longrightarrow$	x_4	0.08	1100	4
	x_5	0.08	1101	4
$x_{22} \longrightarrow$	x_6	0.08	1110	4
	x_7	0.04	1111	4

$$\bar{L} = \sum_{k=1}^{7} P_k n_k = (0.4 \times 1) + (0.2 \times 3) + (0.12 \times 3) + (0.08 \times 4)$$

$$+ (0.08 \times 4) + (0.08 \times 4) + (0.04 \times 4)$$

$$= 2.48 \text{ letter/message}$$

Thus, it can be seen that the second method is better as it gives a lower value for $\bar{L}$

$$H(x) = -\sum_{k=1}^{7} P_k \log r_k = [0.4\log 0.4 + 0.2\log 0.2 + 0.12\log 0.12$$

$$+ 0.08\log 0.08 + 0.08\log 0.08 + 0.05\log 0.05 + 0.04\log 0.4]$$

$$= 2.42 \text{ bits/message}$$

Hence, effieicency in the second method is

$$\eta = \frac{H(x)}{\bar{L} \log m} = \frac{2.42}{2.48 \log 2} = 97.6\,\%$$

Example 25. *For the following messages construct huffman's code*

$P(m_1) = 0.30$ $\quad$ $P(m_2) = 0.25$ $\quad$ $P(m_3) = 0.15$

$P(m_4) = 0.12$ $\quad$ $P(m_5) = 0.10$ $\quad$ $P(m_6) = 0.08$

Solution.

Message	*Probability*	*Code*	S_1	S_2	S_3	S_4
m_1	0.30	00	0.30 (00)	0.30 (00)	0.43 (1)	0.57 (0)
m_2	0.25	10	0.25 (00)	0.27 (01)	0.30 (00)	0.43 (1)
m_3	0.15	010	0.18 (11)	0.25 (10)	0.27 (01)	
m_4	0.12	011	0.15 (010)	0.18 (11)		
m_5	0.10 (110)	110	0.12 (011)			
m_6	0.08 (111)	111				

$$L = \sum_{i=1}^{6} P_i L_i = (0.3 \times 2) + (0.25 \times 2) + (0.15 \times 3) + (0.12 \times 3)$$

$$+ (0.1 \times 3) + (0.08 \times 3)$$

$$= 2.45 \text{ binary digits}$$

The entropy $H(n)$ of the source is given by

$$H(n) = \sum_{i=1}^{n} P_i \log \frac{1}{P_L} = 2.418 = 2.42$$

$$= 0.3 \log \frac{1}{0.3} + 0.25 \log_2 \frac{1}{0.25} + 0.15 \log \frac{1}{0.15} + 0.12 \log \frac{1}{0.12}$$

$$+ 0.10 \log \frac{1}{0.1} + 0.08 \log \frac{1}{0.08}$$

$$= \frac{H(n)}{L \log_2 M} = \frac{2.42}{2.45 \log_2 2} = \frac{2.42}{2.45} = 0.988$$

redundancy $= 1 - \eta = 1 - 0.988 = 0.011$

Example 26. *A Gaussian channel has a bandwidth of 4kHz and a two sideband noise power spectral density η /2 of 10-14 watts/Hz. The signal power at the receiver has to be maintained at a level less than or equal to 1/10 of a milliwatt. Calculate the capacity of this channel.*

Solution. $B = 4\text{kHz} = 4000$ Hz.

$$N = \eta B = 2 \times 10^{-14} \times 4000 = 8 \times 10^{-11} \text{ watts}$$

Signal power $S = (1/10) \times 10^{-3} = 10^{-4}$ watt.

$$\text{S/N Ratio} = \left(10^{-4}\right)/8 \times 10^{-11} = 1.25 \times 10^{6}$$

$$C = 4000 \log_2 \left(1 + 1.25 \times 10^{6}\right)$$

$$= 4000 \log_2 \left(\log_2 1.25 + \log_2 10^{6}\right)$$

$$= 81000 \text{ bp sec} \approx 81013.99 \text{ bp sec.}$$

Example 27. *An Analog signal has a 4kHz bandwidth. The signal is sampled at 2.5 times the Nyquist rate and each sample is quantized into one of 256 equally likely levels. Assume that the successive sample are statistically independent.*

(a) What is the information rate of this source ?

(b) can the output of this source be transmitted without errors over a Gaussian channel with a bandwidth of 50 KHz and S/N ratio of 23 dB?

(c) What will be the bandwidth requirement of an analog channel for transmitting the output of the source without error if the S/N ratio is 10dB?

Solution. *(a)* $H(x) = \log_2 (256) = 8$ bit/sample

$$r_s = \left(\frac{1}{\tau}\right) = 2.5\ (2\text{ dB}) = 5 \times 4000 = 20\text{kHz}$$

Therefore information Rate R = 8 × 20 kHz = 160 kbpsec. Since R = 160 kbpsec is less than C = 382 kbpsec output of this source can be transmitted without errors.

(*b*)
$$C = B\log_2\left(1+\frac{S}{N}\right) = 50\times 10^3 \log_2 (1+199.52)$$

$$= 50\times 10^3 \log_2 (200.52)$$

$$= 50\times 10^3 \times 7.647 = 382.38\,k\ bps$$

(*c*)
$$C = B \log_2(1+10)$$

$$160\times 10^3 = B \log_2 (11)$$

$$B = \frac{160\times 10^3}{3.459} = 46.25 \text{ KHz}$$

$$= 3.5\times 10^3 (3.45)$$

$$= 12.10\times 10^3$$

Here $R < C$

We can believe on it.

Example 28. *A transmission channel has a bandwidth of 4KHz and signal to noise power ratio is 31.*

(*a*) *How much should the bandwidth be in order to have the same channel capacity if S/N ratio is reduce to 15 ?*

(*b*) *What will be the signal power required of the bandwidth is reduced to 3KHz for the source channel capacity.* [Raj. Univ. 2003]

Solution. (*a*) Channel capacity where B = bandwidth of interest

$$C = B\log_2\left(1+\frac{S}{N}\right) \text{ bits/sec}$$

$$= 4\times 10^3 \log_2 (1+31) = 20\times 10^3 \text{ bits/sec.}$$

$$= 20 \text{ K bits/sec}$$

when S/N ratio is reduced from 31 to 15

$$= 20\times 10^3 = B \log_2 (1+15) = B \log_2 16$$

$$= 20\times 10^3 = B\times 4$$

$$B = 5 \text{ KHz}$$

(*b*) Bandwidth $B = 3$ KHz

$$20\times 10^3 = 3\times 10^3 \log_2\left(1+\frac{S}{N}\right)$$

$$\log_2\left(1+\frac{S}{N}\right) = \frac{20}{3}$$

$$\frac{S}{N} = 101.5$$

Thus, we find that when the bandwidth was 4KHz, $\frac{S}{N}$ power ratio was *31*. With decrease in channel bandwidth from *4KHz to 3KHz* the $\frac{S}{N}$ ratio has increased from 31 to 101.5 for obtaining the some channel capacity. Since with decrease in B.W, the noise power is also reduced to $\frac{3}{4}$ of the original noise power. Then the signal power must be increased by the

$$\text{factor} = \frac{3}{4} \times \frac{101.5}{31} = 2.45 = 2.45 \text{ times.}$$

Example 29. *For a signal the bandwidth is 3 KHz and* $\frac{S}{N}$ *ratio is 15.*

(*a*) *Calculate the channel capacity.*

(*b*) *If the bandwidth is increased to 4 KHz and the signal is transmitted over the same channel calculate the required S / N ratio, and the percentage change in the signal power.*

Solution. (*a*) $$C = B \log_2\left(1+\frac{S}{N}\right) \Rightarrow \left(\frac{S}{N}\right)_1 = 15 \text{ where B = bandwidth of interest}$$

$$= 3 \log_2 (1 + 15) = 3 \log_2 16 = 3 \times 4$$

$$= 12\ k \text{ bits /sec}$$

(*b*) Here $4 \log_2 \left(1+\frac{S}{N}\right) = 12$

$$\log_2\left(1+\frac{S}{N}\right) = 3$$

$$1+\frac{S}{N} = 8$$

$$\left(\frac{S}{N}\right)_2 = 7$$

Clearly the S/N ratio decrease from 15 to 7 and the noise power is proportional to the bandwidth in case each. That is $N = \eta B$

Therefore $$\frac{N_2}{N_1} = \frac{4}{3} = \frac{nB_1}{B_1} = \frac{B_2}{B_1} = \frac{4\text{KHz}}{3\text{KHz}} = \frac{4}{3}$$

$$\frac{\left(\frac{S}{N}\right)_2}{\left(\frac{S}{N}\right)_1} = \frac{S_2}{S_1}\frac{N_1}{N_2} = \frac{3}{4}\cdot\frac{S_2}{S_1} = \frac{7}{15}$$

$$\frac{S_2}{S_1} = \frac{4}{3}\cdot\frac{7}{15} = 20.622$$

$$\Delta S = 37.8\%$$

It means for a 37.8 percent increased in B. W the signal power reduces by 37.8%.

8.16 SPECIAL CHANNELS

8.16.1 Lossless Channel

A channel described by a channel matrix with only one non zero element in each column is called a lossless channel. An example of a lossless channel is shown in Fig. 8.3 and the corresponding channel matrix is shown in equation (8.77)

$$[P(Y/X)] = \begin{bmatrix} \frac{3}{4} & \frac{1}{4} & 0 & 0 & 0 \\ 0 & 0 & \frac{1}{3} & \frac{2}{3} & 0 \\ 0 & 0 & 0 & 0 & 1 \end{bmatrix} \qquad \ldots(8.77)$$

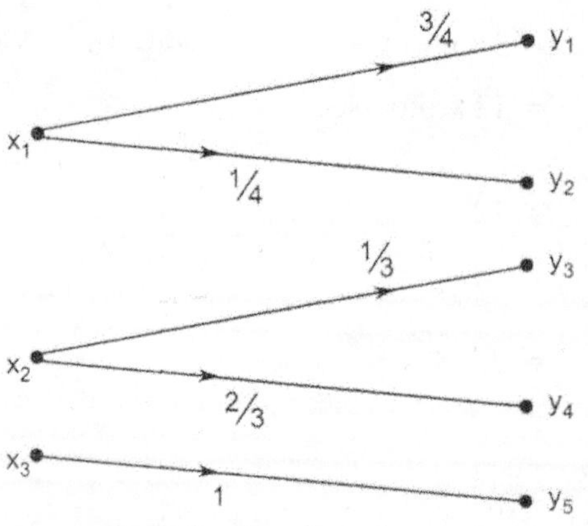

Fig. 8.3 *Lossless channel*

It can be shown that in the lossless channel no source information is lost in transmission.

8.16.2 Deterministic Channel

A channel described by a channel matrix with only one non-zero element in each row is called a *deterministic channel*. An example of a deterministic channel is shown in Fig. 8.4 and the corresponding channel matrix is shown in equation (8.78)

$$[P(Y/X)] = \begin{bmatrix} 1 & 0 & 0 \\ 1 & 0 & 0 \\ 0 & 1 & 0 \\ 0 & 1 & 0 \\ 0 & 0 & 1 \end{bmatrix} \quad \ldots(8.78)$$

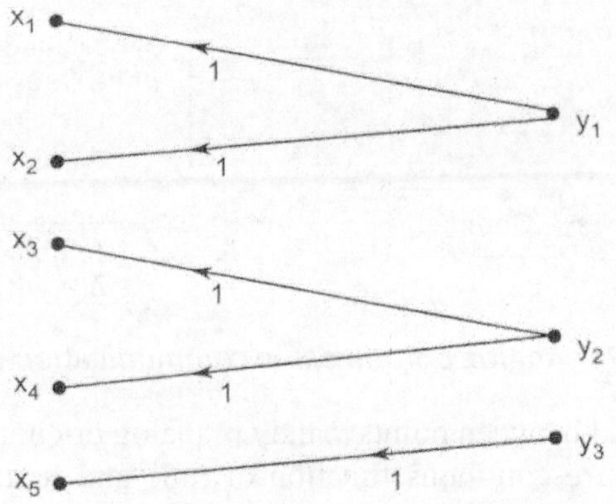

Fig. 8.4 *Deterministic channel*

8.16.3 Noiseless Channel

A channel is called noiseless if it is both lossless and deterministic. A noiseless channel is shown in Fig. 8.5. The channel matrix has only one element in each row and in each column, and this element is unity. Note that the input and output alphabets are of the same size; that is, $m = n$ for the noiseless channel.

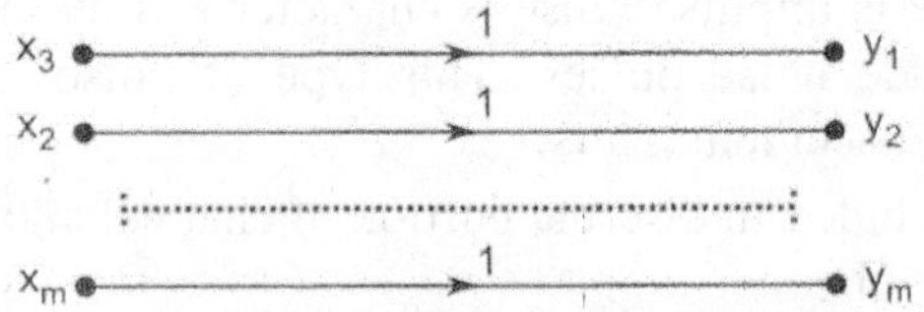

Fig. 8.5 *Noiseless channel*

8.16.4 Binary Symmetric Channel

The Binary Symmetric Channel (BSC) is defined by the channel diagram shown in Fig. 8.6 and channel matrix is given by

$$[P(Y/X)] = \begin{bmatrix} 1-p & p \\ p & 1-p \end{bmatrix} \quad \ldots(8.79)$$

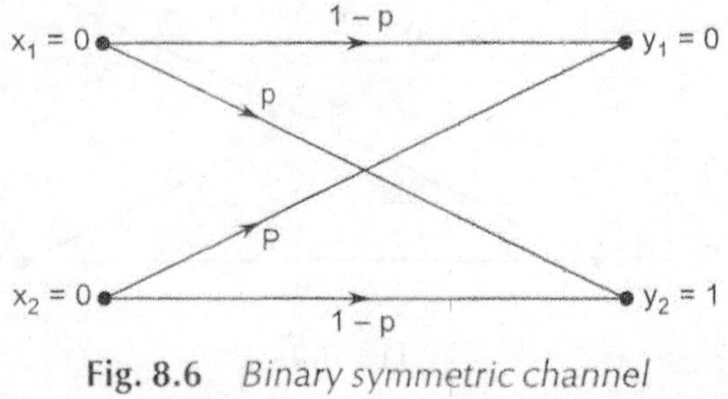

Fig. 8.6 *Binary symmetric channel*

The channel has two input ($x_1 = 0$, $x_2 = 1$) and two output ($y_1 = 0$, $y_2 = 1$). The channel is symmetric because the probability of receiving a 1 if a 0 is sent is the same as the probability of receiving a 0 if a 1 is sent. This common transition probability is denoted by p.

8.17 CONTINUOUS CHANNEL

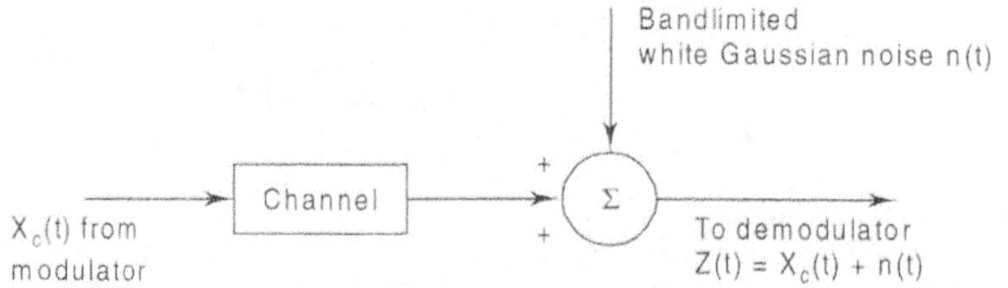

Fig. 8.7 *Analog portion of the communication channel*

The communication channel between points d and f is analog or continous in nature. In this portion of channel the input signals are continous function of time and function of channel is to produce wavefrom presented at its input A real channel accomplish this approximately.

(*i*) Channel modifies the waveform in a deterministic fashion, and this effect can be adequately modeled by treating the channel as linear system. The channel also modifies the input waveform in random fashion due to additive and multiplicative noise. Additive noise can be *Gaussian or* implusive in nature. *Gaussian noise* includes *thermal and shot noise* from equipment and radiation picked up by receiver antenna. We can reduce the Gaussian noise effects by modulation and demodulation techniques

(*ii*) A second type of noise is impulse noise is characterized by long quiet intervals followed by bursts of high amplitude noise pulses. This type of noise is due to lightning discharges, switching transients and accidental hits.

Impulse noise are often included in discrete portion of channel and only Gaussian noise is include in model of analog portion of channel.

Here in Fig. 8.7 the input to channel is random process $X_c(t)$ which consists of the collection of all waveforms generated by modulator. The bandwidth of $X_c(t)$ and the channel is assumed to be B Hz. The additive noise at the channel output is zero mean, bandlimited Gaussian white noise n(t). The capacity of this portion of channel is found by maximizing rate of information transmission w.r.t. distribution of $X_c(t)$.

Example 30. *Consider a binary channel as shown in Fig. 8.8.*

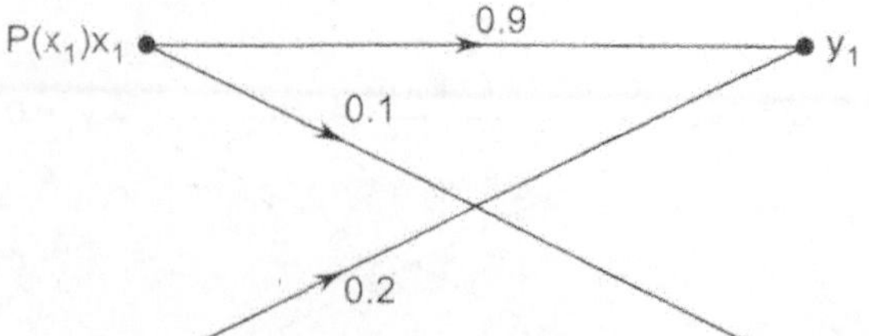

Fig. 8.8

(a) Find channel matrix of channel.

(b) Find $P(y_1)$ & $P(y_2)$ when $P(x_1) = P(x_2) = 0.5$.

Solution. (*a*)

$$P(y/x) = \begin{bmatrix} P(y_1|x_1) & P(y_2|x_1) \\ P(y_1|x_2) & P(y_2|x_2) \end{bmatrix}$$

$$P\,(y/x) = \begin{bmatrix} 0.9 & 0.1 \\ 0.2 & 0.8 \end{bmatrix}$$

(*b*) Find $P(y_1)$ & $P(y_2)$ when $P(x_1) = P(x_2) = 0.5$

$$[P(y)] = [P(x)]\,[P(y/x)]$$

$$[P(y)] = [0.5 \quad 0.5]\begin{bmatrix} 0.9 & 0.1 \\ 0.2 & 0.8 \end{bmatrix}$$

$$[P(y)] = [0.55 \quad 0.45], \quad \therefore \quad P(y_1) = 0.55$$

$$P(y_2) = 0.45$$

Example 31. *Two binary channels are connected in cascade as shown in Fig. 8.9.*

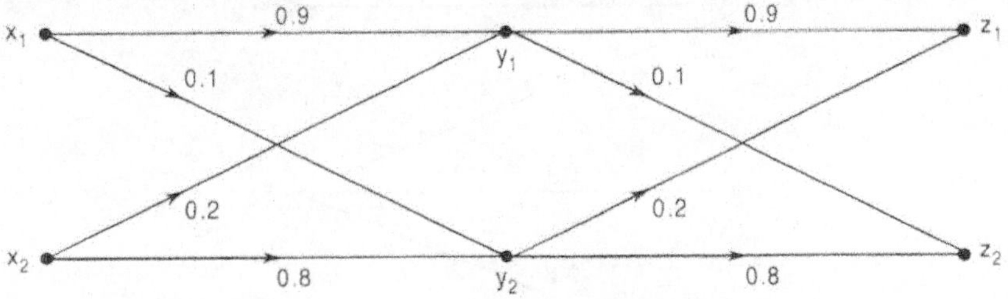

Fig. 8.9

(a) Find overall channel matrix & equivalent channel diagram.

(b) Find $P(z_1)$ & $P(z_2)$.

Solution.

$$[P(y)] = [P(x)]\,[P(y/x)] \qquad \ldots(8.80)$$

$$[P(z)] = [P(y)]\,[P(z/y)] \qquad \ldots(8.81)$$

from (8.80) and (8.81) $\quad = [P(x)]\,[P(y/x)]\,[P(z/y)] = [P(x)]\,[P(z/x)]$

$$p(z/x) = [P(y/x)]\,[P(z/y)]$$

$$= \begin{bmatrix} 0.9 & 0.1 \\ 0.2 & 0.8 \end{bmatrix}\begin{bmatrix} 0.9 & 0.1 \\ 0.2 & 0.8 \end{bmatrix} = \begin{bmatrix} 0.83 & 0.17 \\ 0.34 & 0.66 \end{bmatrix}$$

$\therefore$ Resultant's Matrix Diagram

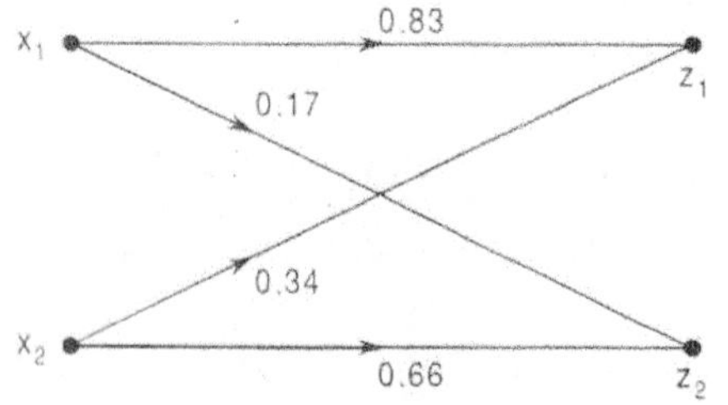

Fig. 8.10

(*b*) $[P(z)] = [P(x)]\,[P(z/x)]$

$$= \begin{bmatrix} 0.5 & 0.5 \end{bmatrix} \begin{bmatrix} 0.83 & 0.17 \\ 0.34 & 0.66 \end{bmatrix}$$

$$= \begin{bmatrix} 0.585 & 0.415 \end{bmatrix}$$

$$P(z_1) = 0.585,\ P(z_2) = 0.415$$

Example 32. *A discrete source transmits messages x_1, x_2 and x_3 with the probabilities 0.3, 0.4 and 0.3. The source is connected to the channel as shown in Fig. 8.11. Calculate all the entropies.*

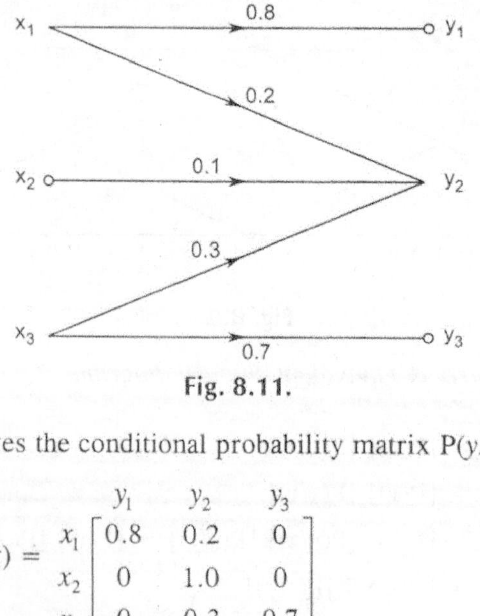

Fig. 8.11.

Solution. Figure. 8.11 gives the conditional probability matrix $P(y/x)$ as

$$p(y/x) = \begin{array}{c} \\ x_1 \\ x_2 \\ x_3 \end{array} \begin{array}{c} \begin{array}{ccc} y_1 & y_2 & y_3 \end{array} \\ \begin{bmatrix} 0.8 & 0.2 & 0 \\ 0 & 1.0 & 0 \\ 0 & 0.3 & 0.7 \end{bmatrix} \end{array}$$

$$P[x] = [0.3,\ 0.4,\ 0.3]$$

The joint probability matrix $P(x, y)$ can be obtained by multiplying the rows of $P(y/x)$ by $P(x_1)$, $P(x_2)$ and $P(x_3)$ *i.e.*, by 0.3, 0.4 and 0.3 respectively, giving

$$P(x, y) = \begin{bmatrix} 0.8 \times 0.3 & 0.2 \times 0.3 & 0 \\ 0 & 1 \times 0.4 & 0 \\ 0 & 0.3 \times 0.3 & 0.7 \times 0.3 \end{bmatrix}$$

$$P(x, y) = 0 \quad \begin{array}{c} \\ x_1 \\ x_2 \\ x_3 \end{array} \begin{array}{c} \begin{array}{ccc} y_1 & y_2 & y_3 \end{array} \\ \begin{bmatrix} 0.24 & 0.06 & 0 \\ 0 & 0.4 & 0 \\ 0 & 0.09 & 0.21 \end{bmatrix} \end{array}$$

(As a check, it can be seen that the sum of all the entires in P (x, y) is 1)

The probabilities P (y_1), P (y_2) and P (y_3) can be obtained by adding the column of P (x, y) giving

$$P(y_1) = 0.24$$

$$P(y_2) = 0.06 + 0.4 + 0.09 = 0.55$$

$$P(y_3) = 0.21$$

The conditional probability matrix $P(x/y)$ can be obtained by dividing the columns of P (x, y) by P (y_1), $P(y_2)$ and P (y_3) respectively, giving

$$P(x/y) = \begin{array}{c} \\ x_1 \\ x_2 \\ x_3 \end{array} \begin{array}{c} \begin{array}{ccc} y_1 & y_2 & y_3 \end{array} \\ \begin{bmatrix} 1 & 0.109 & 0 \\ 0 & 0.727 & 0 \\ 0 & 0.164 & 1 \end{bmatrix} \end{array}$$

(Check : The sum of all the columns of $P(x/y)$ is 1)

The entropies can now be calculated as follows

$$H(x) = -\sum_{j=1}^{3} P(x_j) \log P(x_j) = -(0.3 \log 0.3 + 0.4 \log 0.4 + 0.3 \log 0.3)$$

$$= 1.571 \text{ bits/message}$$

$$H(x) = -\sum_{j=1}^{3} P(y_k) \log P(y_k)$$

$$= -[0.24 \log 0.24 + 0.55 \log 0.55 + 0.21 \log 0.21) = 1.44 \text{ bits/message}$$

$$H(x, y) = -\sum_{j=1}^{3}\sum_{k=1}^{3} P(x_j, y_k) \log (x_j, y_k)$$

$$= -[0.24 \log 0.24 + 0.06 \log 0.06 + 0.4 \log 0.4 + 0.09 \log 0.09 + 0.21 \log 0.21]$$

$$= 2.053 \text{ bits/message}$$

$$H(x/y) = -\sum_{j=1}^{3}\sum_{k=1}^{3} P(x_j, y_k) \log P(x_j/y_k)$$

$$= -[0.24 \log 1 + 0.06 \log 0.109 + 0.4 \log 0.727 + 0.09 \log 0.164 + 0.21 \log 1]$$

$$= 0.610 \text{ bit/message}$$

$$H(x, y) = -\sum_{j=1}^{3}\sum_{k=1}^{3} P(x_j, y_k) \log P(y_k, x_j)$$

$$= -[0.24 \log 0.8 + 0.06 \log 0.2 + 0.4 \log 1 + 0.09 \log 0.03 + 0.21 \log 0.7]$$

$$= 0.7799 \text{ bit/message}$$

REVIEW QUESTIONS

1. What do you understand by information ? What are the various units of information ?
2. What do you understand by average information ?
3. Write short notes on :

   (*a*) Joint Entropy . (*b*) Conditional Entropy.
4. What do you understand by channel capacity ?
5. Explain :

   Kraft inequality and Prefix coding.
6. Explain Shannon-Hartely law for capacity of continuous channel.
7. (a) State and prove Shannon-Hartley theorem. How it can give trade off between bandwidth and channel capacity ? What is the maximum value of channel capacity ?

   (*b*) Describe source coding. How Shannon-fano coding can give maximum efficiency.
8. (*a*) What is Entropy, prove that—

   $$H(X, Y) = H(X/Y) + H(Y) = H(X/Y) + H(Y)$$

   (*b*) A Gaussian channel has 1MHz bandwidth. Calculate the channel capacity if the signal power to noise spectrum density ratio $\left(\frac{S}{\eta}\right)$ is 105 Hz. Also find the maximum information rate.
9. A DMS X has five symbol x_1, x_2, x_3, x_4 and x_5 with $p(x_1) = 0.4$, $p(x_2) = 0.19$, $p(x_3) = 0.16$, $p(x_4) = 0.15$ and $p(x_5) = 0.1$.

   (*a*) Construct a Shannon-fano code for X and calculate the efficiency of the code.

   (*b*) Repeat for the Huffman code and compare the result.
10. (*a*) Explain the term Entropy as average uncertainty.

    (*b*) For a source generating M number of messages, Entropy H is limited as :

    $$0 \le H \le \log_2 M.$$
11. Explain Shanon theorem and Shano limit.
12. (*a*) Define Shannon's Entropy taking the suitable properties as axioms, derive the Mathematical form of Shannon's Entropy.

(*b*) Let H (p_1, p_2 ... p_n) be measure of uncertainty defined on a completer probability distribution

$$P = \{p_1, p_2, ..., p_n\} \text{ then prove}$$

$$H(p_1, p_2, ... p_n) < \log M$$

with equality if and only if all $p_i = 1/M$.

13. (*a*) Define conditional and marginal entropies for the random variates X and Y. Also define mutual information I (X,Y) and prove,

$$I(X,Y) = H(X) + H(Y) - H(X,Y)$$

$$= H(Y) - H(y/x).$$

14. (*a*) What is Entropy of a zero-memory source? Show that the value of Entropy is maximum and equal to log n if and only if all the source symbols are equiprobable.

(*b*) An analog signal is bandlimited to BHz, sampled at the nyquist rate and are quantized into 4 levels. The quantization levels Q_1, Q_2, Q_3 and Q_4 are assumed independent and occur with probabilites $p_1 = p_4 = \frac{1}{8}$ and $p_2 = p_3 = \frac{3}{8}$. Find the information rate of the source.

15. Define differential Entropy. Write expression for the same.

16. Write short notes :

(*i*) BW-SNR Tradeoff in the light of Shannon's theorem.

17. (*a*) Explain source coding with its importance.

(*b*) Explain the term mutual information.

18. For a continuous random variable X, given a mean square value, the maximum entropy is obtained when the distribution of X is gaussian. Find the maximum value of Entropy.

19. A voice-grade channel of the telephone network has a bandwidth of 3.4 kHz.

(*a*) Calculate the information capacity of the telephone channel for a signal-to-noise ratio of 30 dB.

(*b*) Calculate the minimum signal-to-noise ratio required to support information transmission through the telephone channel at the rate of 9,6000 b/s.

20. Alphanumeric data are entered into a computer from a remote terminal through a voice-grade telephone channel. The channel has a bandwidth of 3.4 kHz and output signal-to-noise ratio of 20dB. The terminal has a total of 128 symbols. Assume that the symbols are equiprobable and the successive transmissions are statistically independent.

(*a*) Calculate the information capacity of the channel.

(*b*) Calculate the maximum symbol rate for which error-free transmission over the channel is possible.

21. What you understand by 'Bandwidth compression'? Discuss how it can be achieved and at what cost and also its implications using Shannon-Hartley theorem.

22. A TV picture consists of 525 lines of picture information and each line consists of 525 picture elements and each element has 256 brightness levels. How much information is contained in each picture? If the transmission rate is 30 pictures per sec., what is the bit rate r_b per sec.?

[**Ans:** 2.2 Megabits, 66Mbs.]

23. (*a*) Plot channel capacity C versus B, with S/η = constant, for the gaussian channel.

(*b*) If the channel bandwidth B = 5 kHz and a message is being transmitted with R = 10^6 bits per sec., find S/η for $R \leq C$.

24. An analog signal has a 4-kHz bandwidth. The signal is sampled at 3 times the Nyquist rate and quantized using 256 quantization levels. The S_i/η ratio at the receiver is $S_i/\eta = 1$ MHz.

(*a*) Calculate the time t between samples.

(*b*) The quantized and sampled signal is encoded into a binary PCM waveform. Calculate the number of bits N and find P_e.

(*c*) The quantized and sampled signal is encoded into 1 of 256 orthogonal signals. Find P_e.

9

ERROR CORRECTING CODES

9.1 ERROR CONTROL CODING

In chapter 6, we described Modulation Techniques for transmitting digital information over noisy channels. Basically Modulation is a signal processing operation that is used to provide for the efficient transmission of digital information over the channel.

Now in this chapter, we study another important signal processing operation, namely channel coding, which is used to provide for reliable transmission of digital information over the channel.

Coding is a techniques that adds some redundant bits in real data so that two basic objective are performed at the receiver side; **Error Detection** and **Error Correction**. These codes are called **Error Controlling Codes**.

9.1.1 Type of Error Controlling Codes

There are basically two types of error controlling codes

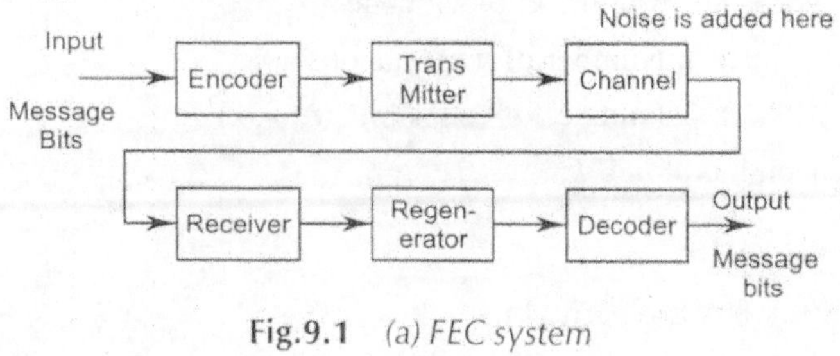

Fig.9.1 *(a) FEC system*

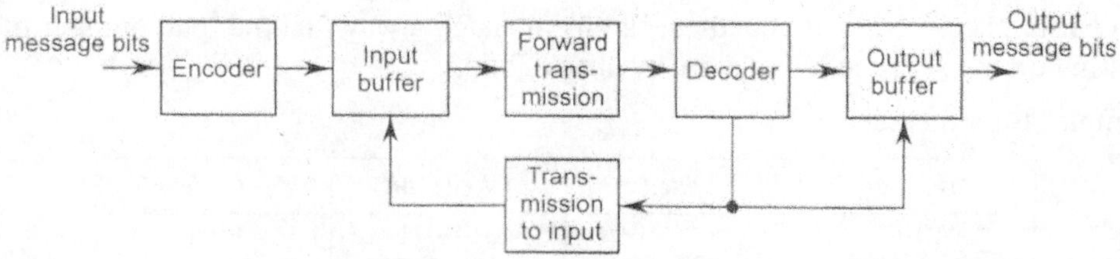

Fig.9.1 *(b) ARQ system*

9.1.1.1 *Forward Error Correction Codes (FEC)*

In the FEC codes basic conept of error detection and correction is performed at receiver side. FEC codes are basically *block codes* and *convolution codes*.

In Block codes ; we a code (n, k) is generated. Here k data bits are accepted by the encoder and it will add $(n - k)$ reduant bit in this data and that will produce a code of n bits simply $n > k$. For example (7, 4) block codes ; the data bit are 4 ; now 7 – 4 = 3 bits are added to the data bit by encoder. Thus totally,

4 bits of data + 3 (bits reduant bits)

= 7 bit code is generated :

These 3 bit added will help us to detect and correct the error when this code is received at receiver side and will help us to produce a correct data.

In convolutional codes, we have a memory space. Now this time coder will accept the data bits as a continuous sequence and thereby will produce a code of continuous sequence.

9.1.1.2 *Automatic Error Request Codes (AER)*

In automatic request codes, the receiver will decode the code ; and if it found some error ; it will ask to transmitter to retransmit the data (coded form) again.

In general case, we use block codes mostly because if we have a single transmitter and a number of receiver, then if any single receiver found some error (due to its channel between transmitter and this receiver) then it will be impractical that due to a single receiver among a number of receiver will say to transmitter to Retransmit the message (data) again while other receivers are working satisfactory.

Now we will study the FEC codes in detail.

9.2 BLOCK CODE (ARITHMATIC CODES/GROUP CODES)

In the block codes, each block of k message bits in encoded into a block of n bits (n > K). The check bits are derived from the message bits and are added to them. The n-bit block of a channel encoder output is called a code word. So for a (n, k) block code

$$k = \text{Number of message bits}$$

$$n = \text{Number of codes bits } (n > k)$$

Then added bits as check bits = $n - k$.

9.2.1 Parity Check Bits

The simplest form of check bits are parity bits.

If added check bits are such a way that the total number of 1's in the code word is even then it is called even parity check code ; if added check bits are such a way that the total number of 1's in the code words in odd then it is called odd parity check code.

For example, for a message

Message	*Added*	*Check bit*	*Code*
010101	1	0101011	Even
010101	0	0101010	Odd

9.2.2 Linear and Systematic Code

A block code is called systematic code. If the message bits in unaltered form, are transmitted along with code

If we have message bits = 1101

Add check bits = 101

Then 1101 : 101 'or'

101 : 1101 is called the systematic block code;

Now if check bits $(n - k)$ of the code word are generated from k message bits according to some predetermined rule :

$$C_{k+1} = P_{11}\, d_1 \oplus P_{21}\, d_2 \oplus \ldots \oplus P_{k1}\, d_k$$
$$C_{k+2} = P_{12}\, d_1 \oplus P_{22}\, d_2 \oplus \ldots \oplus P_{k2}\, d_k$$
$$\ldots$$
$$C_n = P_{1(n-k)}\, d_1 \oplus P_{2(n-k)}\, d_2 \oplus \ldots \oplus P_{k(n-k)}\, d_k \qquad (9.1)$$

where co-efficient P_{ij} in equation are 0's and is so that values of k's are 0's and 1's.

Now if we choose the co-efficients P_{ij} in such a way that all equations represented. Here are linearly independent (means to say that no equation in the set of complete equations can be represented by a linear combination of the remaining equations) then it is called *linear block code*.

Mathematically it can be shown as that A block code is linear if

b_1 and b_2 are in a code c and

$b_1 \oplus b_2$ is also in the code c, then it is called linear block code ;

For example,

$$C = \{[0\ 0\ 0]\ [1\ 1\ 0]\ [1\ 0\ 1]\ [0\ 1\ 1]\}$$

is a linear code because

$$110 \oplus 101 = 011 \text{ etc.}$$

9.2.3 Mathematical Structure of Linear Systematic Block Codes

Now to accelerate the speed of operation ; a linear systematic block code (n, k), can be represented in the form of matrix notation. So again write the equations in the matrix form.

Message (data) can be represented by a row vector

$$D = [d_1, d_2, d_3, \ldots, d_k] \text{ total } k \text{ bits}$$
$$C = [c_1, c_2, c_3, \ldots, c_n] \text{ total } n \text{ bits}$$

Now again,

$$C = [d_1, d_2, d_3, \ldots, d_k, c_3 \ldots c_{n-k}]$$

$$[c_1, c_2, c_3, \ldots, c_n] = [d_1, d_2, \ldots, d_n] \begin{bmatrix} 1 & 0 & \ldots & 0 \\ 0 & 1 & \ldots & 0 \\ \vdots & \vdots & \ldots & \vdots \\ 0 & 0 & \ldots & 1 \end{bmatrix} \begin{bmatrix} P_{11} & P_{12} & \ldots & P_{1n-k} \\ P_{21} & P_{22} & \ldots & P_{2n-k} \\ \vdots & \vdots & \ldots & \vdots \\ P_{k1} & P_{k2} & \ldots & P_{kn-k} \end{bmatrix} \qquad (9.2)$$

9.2.4 Generator Matrix

Representing equation in the form

$$C = dG \tag{9.3}$$

then G is called generator matrix

$$G = \begin{bmatrix} 1 & 0 & \dots & 0 & \vdots & P_{11} & P_{12} & \dots & P_{1n-k} \\ 0 & 1 & \dots & 0 & \vdots & P_{21} & P_{22} & \dots & P_{2n-k} \\ \vdots & \vdots & \dots & \vdots & & \vdots & \vdots & \dots & \vdots \\ 0 & 0 & \dots & 1 & \vdots & P_{k1} & P_{k2} & \dots & P_{kn-k} \end{bmatrix} \tag{9.4}$$

$$= [I_k \vdots P]_{k \times n} \tag{9.5}$$

where I_k is identity matrix of order $k \times k$. P is the co-efficient matrix of order $k\ (n - k)$

$$P = \begin{bmatrix} P_{11} & P_{12} & \dots & P_{1n-k} \\ P_{21} & P_{22} & \dots & P_{2n-k} \\ \vdots & \vdots & \dots & \vdots \\ P_{k1} & P_{k2} & \dots & P_{kn-k} \end{bmatrix} \tag{9.6}$$

Thus for a given generator matrix ; a data word can be used to generate equivalent code.

9.2.5 Parity Check Matrix

There is another way to represent the relation between parity check co-efficien• and that is denoted by H.

$$H = [P^T : I_{n-k}]_{(n-k) \times n} \tag{9.7}$$

where P^T = Transpose of P that can be obtained by equation — interchanging row by columb

$$P^T = \begin{bmatrix} P_{11} & P_{21} & \dots & P_{k1} \\ P_{12} & P_{22} & \dots & P_{k2} \\ \vdots & \vdots & \dots & \vdots \\ P_{1n-k} & P_{2n-k} & \dots & P_{kn-k} \end{bmatrix} \tag{9.8}$$

of order $(n - k) \times (n - k)$.

Now identity matrix is of order $k \times (n - k)$.

9.2.6 Parity Check Equation

Let us see that performing the following matrix multiplication :

$$HG^T = [P^T : I] \begin{bmatrix} I \\ P^T \end{bmatrix} \tag{9.9}$$

$$= P^T \oplus P^T$$

$$\boxed{HG^T = 0} \tag{9.10}$$

Now take the equation

$$C = dG$$

Again multiply this equation by H^T

$$CH^T = dG\,H^T$$

But $GH^T = 0$

$$\boxed{CH^T = 0} \quad (9.11)$$

This is called the parity check matrix. This is very important quality that will help us during decoding of the code at receiver side.

9.3 DECODING OF LINEAR BLOCK CODE USING SYNDROME MATRIX

At the transmitter side ; with use of a generator matrix ; we generate a code c. Now at the receiver side we will have a received code r ; simply received code 'r' will contain transmitted code ; and a error due to channel that is e.

So, $$r = c \oplus e \quad (9.12)$$

Note that this addition is also *module-2 addition.*

Here e = error vector : If error is zero then,

$$r = c \quad (9.13)$$

If any co-efficient of error vector is one ; that means at the place of that one an error has occurred.

Say code transmitted was $c = \overset{C_6}{1}\,1\,0\,1\,1\,0\,\overset{C_0}{1}$

and error generated will be definitely in the order of $(1 \times n)$ that

$$e = 0\,1\,0\,0\,0\,0\,0$$

Then received data will be

$$r = 1\,1\,0\;\;1\,1\,0\,1 \oplus 0\;\;1\,0\,0\,0\,0\,0 = c \oplus e$$

$$= 100110\,1$$

So we see that at C_5 an error has been occurred ; so

$$e_i = \begin{cases} 1 & \text{If at } i^{\text{th}} \text{ location any error has been occurred} \\ 0 & \text{Otherwise} \end{cases}$$

Now decoding starts with the computation of $1 \times (n - k)$ vector called **syndrome matrix.**

It is given by : $$\boxed{S = rH^T} \quad (9.14)$$

Now we will study some essential properties of syndrome matrix.

Property 1 : *"The syndrome depends upon only the error pattern and not on the transmitted code word"* or

$$S = rH^T$$

But $$r = c \oplus e$$

$$S = (c \oplus e)\, H^T$$

$$S = C\, H^T \oplus e\, H^T \tag{9.15}$$

But $\boxed{CH^T = 0}$ called Parity Check Matrix

$$\boxed{S = eH^T} \tag{9.16}$$

This for our code system, we know the value of syndrome and have information of parity check matrix, then we can easily find the error pattern occurred at the path.

Property 2 : *All the error pattern that differ at most by a code word have the same syndrome.*

For k-message bits we have 2^k distinct code vectors denoted by c_i ; $i = 0, 1, \ldots, 2^{k-1}$, correspondingly, for any error pattern e, we define the 2^k distinct vector e_i as

$$e_i = e + c_i \tag{9.17}$$

The set of vector $\{e_i, i = 0, 1, \ldots, 2^{k-1}\}$

So defined is called *a coset of the code*. In other word, we may write the following conclusion :

(*a*) Coset has extactly 2^k elements that differ at most by code vector.

(*b*) Thus an (n, k) linear block code has 2^{n-k} possible cosets.

$$e_i = e + c_i$$

Now multiply both sides by H^T

$$e_i\, H^T = (e + c_i)\, H^T$$

$$= eH^T + c_i\, H^T$$

But $\qquad e_i\, H^T = 0$

$$\boxed{e_i\, H^T = eH^T} \tag{9.18}$$

So we can say that any coset e_i corresponding to common error pattern has a unique and same syndrome.

Property 3 : *The syndrome S is the sum of those columns of the matrix H corresponding to the error location.*

To prove this property, let the parity check matrix H be expanded in terms of its column as follows :-

$$H = [h_1, h_2, h_3, \ldots, h_n]$$

The substituting equation in equation

$$S = eH^T$$

$$= [e_1, e_2, \ldots, e_n] \begin{bmatrix} h_1^T \\ h_2^T \\ \vdots \\ h_n^T \end{bmatrix} \tag{9.19}$$

$$\boxed{S = \sum_{i=1}^{n} e_i\, h_i^T} \tag{9.20}$$

where h_i^T is the ith row of matrix H^T and e_i is the ith element of the error pattern.

Property 4 : *If we have n-bit code word, there are* $\binom{n}{i}$ *multiple error patterns, where i is the number of error locations in the error pattern.*

Where $$\binom{n}{i} = \frac{n!}{n-i!.i!} \quad (9.21)$$

Let us consider (7, 4) codes then if we find the possible error pattern of single bit error. Then,

$$n = 7$$
$$i = 1$$

So, $$\binom{7}{1} = \frac{7!}{7-1!.1!} = \frac{7!}{6!.1!} = 7$$

So we have totally of error pattern having single bit error that are :

1 0 0 0 0 0 0
0 1 0 0 0 0 0
0 0 1 0 0 0 0
0 0 0 1 0 0 0
0 0 0 0 1 0 0
0 0 0 0 0 1 0
0 0 0 0 0 0 1 Total 7 error patterns generated.

This property also shows that if we are able to find the syndrome for every possible error pattern, then we can detect all the errors in the code received.

So if any value of (n and k) that satisfies the Hamming bound with equality sign

$$2^{n-k} \geq \sum_{i=0}^{t} \binom{n}{i} \quad (9.22)$$

Then this code is called perfect code.

$$2^{n-k} = \sum_{i=0}^{t} \binom{n}{i} \text{ for perfect code.} \quad (9.23)$$

9.3.1 Practical Method to Decode the Message Using Syndroms Matrix

(*a*) **Formation of Cosets :** Standard array for (n, k) linear code can be formed as

(*i*) The 2^k code words are placed in a row with all zero code words C_o as the leftmost one.

(*ii*) An error pattern e is picked up and placed under C_o.

(*iii*) Now form a second row by adding e to each of the remaining code words in the first row: (It should be noted that every e should not be appeared before and has minimum Hamming weight).

(*iv*) Step '2' is repeated until all the possible error patterns have been account for.

(*v*) The addition of each column has a unique value.

Each row in standard array is called a coset. The left most element of a row is called coset leader of the coset has a unique syndrome as discussed in property 2.

For example, we have drawn here standard array for (0, 3) codes shown in example.

Now to decode a code received we follow the following example.

(*a*) Suppose we have received code

$$r = 101001.$$

(*b*) Place this received code in the standard array as shown in example. Now we can see that for this received code error pattern for single bit error

$$= 000100.$$

(*c*) The top element of this coloumb shows the transmitted code that is

$$\boxed{C = 101101}$$

Thus we have decoded data.

Example 1. *Consider the (6, 3) linear code generated by the following matrix :*

$$G = \begin{bmatrix} 1 & 0 & 0 & 0 & 1 & 1 \\ 0 & 1 & 0 & 1 & 0 & 1 \\ 0 & 0 & 1 & 1 & 1 & 0 \end{bmatrix}$$

We then have the parity check matrix

$$H = \begin{bmatrix} 0 & 1 & 1 & 1 & 0 & 0 \\ 1 & 0 & 1 & 0 & 1 & 0 \\ 1 & 1 & 0 & 0 & 0 & 1 \end{bmatrix}$$

Find the Standard array for the (6, 3) linear code.

Solution. Standard array for the (6, 3) linear code :

Coset leader							
000000	100011	010101	001110	110110	101101	011011	111000
100000	000011	110101	101110	010110	001101	111011	011000
010000	110011	000101	011110	100110	111101	011011	101000
001000	101011	011101	000110	111110	100101	010011	110000
000100	100111	010001	001010	110010	101001	011111	111100
000010	100001	010111	001100	110100	101111	011001	111010
000001	100010	010100	001111	110111	101100	011010	111001
100100	000111	110001	101010	010010	001001	111111	011100
110000	010011	100101	111110	000110	011101	101011	001000

But this type of decoding method has one disadvantage ; that the memory requirement is very much high to store the standard array. For example here for (6, 3) codes we have totally $2^6 = 64$ elements in standard array. If we use codes with block length $n = 7, 9, 12, \ldots$ Then total number of elements in standard array will be high.

Now so will find most easy way to find the transmitted code.

(*i*) For a given parity check matrix, H and for a particular error pattern find the syndromes by using property '3' of syndrom.

Now we have all possible error vectors and corresponding syndromes.

(*ii*) Now for a received code '*r*', calculate syndrom by formula

$$S = rH^T$$

(*iii*) Now see the step '1' ; now for this syndrom find the error vector.

(*iv*) Now we have '*e*' and '*r*', then find code by relation

$$C = r \oplus e$$

This process will require smaller space to mostly preferred.

9.4 HAMMING CODE

Hamming codes are a special case of linear block codes ; having these parameters

$$\begin{aligned} &\text{Block length} && n = 2^m - 1 \\ &\text{Number of message bits} && k = 2^m - m - 1 \\ &\text{Number of parity bits} && n - k = m. \end{aligned} \tag{9.24}$$

Where $m \geq 3$, these are called **Hamming codes**. Suppose that (7, 4) code is a Hamming code. Remaining process is same as discussed in the case of linear block code.

We will keep some points in the mind about Hamming codes :

(*a*) Hamming codes have the property that the minimum distance $d_{min} = 3$ (independent to the value assigned *m*.

(*b*) An important property of hamming block codes is that it follows the equation of perfect codes.

$$2^{n-k} = \sum_{i=0}^{t} \binom{n}{i}$$

For $t = 1$ and $n = 7$, $k = 4$; then

$$2^{7-4} = \sum_{i=0}^{1} \binom{7}{4}$$

Solve with

$$2^{n-k} = \sum_{i=0}^{1} \binom{n}{i} = \frac{n!}{n - i_1!.i_1!} + \frac{n!}{n - i_2!.i_2!}$$

where $i_1 = 0, i_2 = 1$

So

$$2^{7-4} = \frac{7!}{7-0!.0!} + \frac{7!}{7-1!.1!}$$

But $0! = 1$

$$= 1 + \frac{7!}{6!.1!} = 1 + \frac{7.6!}{6!} = 1 + 7$$

$$\boxed{2^3 = 8.}$$

R.H.S. = L.H.S.

So called perfect code for a single bit error. Again it is an block code so the decoding easily can be done by syndrom decoding method.

9.4.1 Encoder of Hamming Code

It is very easy to design a encoder for Hamming code ; again it should be clear that Hamming codes are a special class of linear block code ; they have identical representation of matrix form as linear block code ; but they are perfect linear block code for a single bit error.

Basic principle of designing a encoder for a systematic linear block code is :

(*a*) It should generate the data bits as it is because every systematic code has data (message) bits in unaltered form.

(*b*) Now it should generate check bits by the addition of different message bits because a check bit is given by

$$C_{k+1} = P_{11}\, d_1 \oplus P_{21}\, d_2 \oplus \ldots \oplus P_{k,1}\, d_k.$$

Here don't confuge by $p_{11}, p_{21}, \ldots$ because they are co-efficient '0' or 1,.

For example, if all co-efficients = 1, then

$$c_{k+1} = d_1 \oplus d_2 \oplus \ldots \oplus d_k.$$

Thus by adding these message bits we can generate the check bits.

Now place these check bit along with data (message) bit and thus a systematic linear block code (can also be Hamming code) is generated.

Now by example, given below, we will study the designing of encoder.

So encoder can be designed as

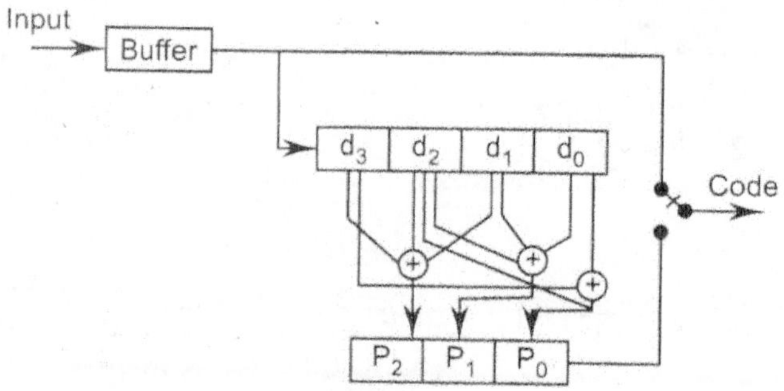

Fig. 9.2

where
$$P_1 = d_3 \oplus d_2 \oplus d_1$$
$$P_2 = d_2 \oplus d_1 \oplus d_0$$
$$P_3 = d_3 \oplus d_2 \oplus d_0$$

called Parity Bits.

Example 2. *For the generator matrix*

$$G = \begin{bmatrix} 1 & 0 & 0 & 0 & : & 1 & 0 & 1 \\ 0 & 1 & 0 & 0 & : & 1 & 1 & 1 \\ 0 & 0 & 1 & 0 & : & 1 & 1 & 0 \\ 0 & 0 & 0 & 1 & : & 0 & 1 & 1 \end{bmatrix}$$

design the encoder for (7, 4) code for the data vector

$$= [1\ 1\ 1\ 1]$$

Solution. We have generator matrix :

$$G = \begin{bmatrix} 1 & 0 & 0 & 0 & : & 1 & 0 & 1 \\ 0 & 1 & 0 & 0 & : & 1 & 1 & 1 \\ 0 & 0 & 1 & 0 & : & 1 & 1 & 0 \\ 0 & 0 & 0 & 1 & : & 0 & 1 & 1 \end{bmatrix}$$

Now $\quad d = \begin{matrix} d_1 & d_2 & d_3 & d_4 \\ [\,1 & 0 & 1 & 0\,] \end{matrix}$

Now, $\quad c = dG = \begin{matrix} d_3 & d_2 & d_1 & d_0 \\ [\,1 & 1 & 1 & 1\,] \end{matrix} \begin{bmatrix} 1 & 0 & 0 & 0 & : & 1 & 0 & 1 \\ 0 & 1 & 0 & 0 & : & 1 & 1 & 1 \\ 0 & 0 & 1 & 0 & : & 1 & 1 & 0 \\ 0 & 0 & 0 & 1 & : & 0 & 1 & 1 \end{bmatrix}$

$$= \begin{bmatrix} d_3 & d_2 & d_1 & d_0 & d_3 \oplus d_2 \oplus d_1 & d_2 \oplus d_1 \oplus d_0 & d_3 \quad d_2 \quad d_0 \\ 1 & 0 & 1 & 0 & 1 \oplus 1 \oplus 1 & 1 \oplus 1 \oplus 1 & 1 \oplus 1 \oplus 1 \end{bmatrix}$$

$= [1\,0\,1\,0\,1\,1\,1]$

So encoder can be designed as

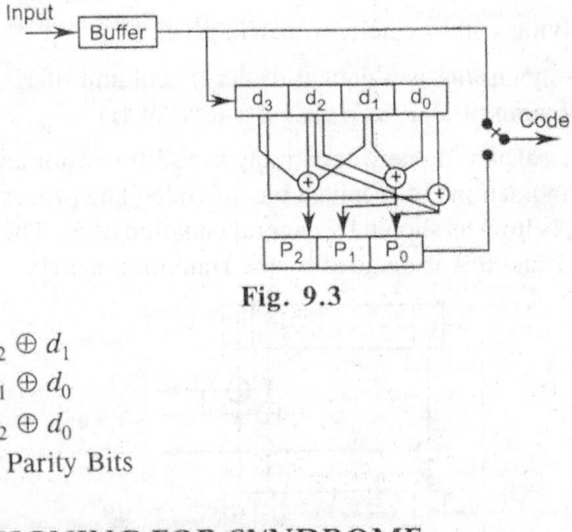

Fig. 9.3

where $\quad P_1 = d_3 \oplus d_2 \oplus d_1$

$P_2 = d_2 \oplus d_1 \oplus d_0$

$P_3 = d_3 \oplus d_2 \oplus d_0$

Called Parity Bits

9.5 DECODER DESIGNING FOR SYNDROME

We know that $\quad S = rH^T \quad$ But $\quad H^T = \begin{bmatrix} P \\ I \end{bmatrix}$

So, $$S = r \begin{bmatrix} P \\ I \end{bmatrix} \tag{9.25}$$

Now take the general expression for

$S = [S_1, \ldots, S_k]$ (Syndrome)

$r = [r_1, r_2, \ldots, r_n]$ (Received code)

and $$P = \begin{bmatrix} P_{11} & P_{12} & \ldots & P_{1n-k} \\ P_{21} & P_{22} & \ldots & P_{2n-k} \\ \vdots & \vdots & \ldots & \vdots \\ P_{k1} & P_{k2} & \ldots & P_{kn-k} \end{bmatrix}$$

Now,
$$[S_1, \ldots S_k] = [r_1, r_2, \ldots, r_n] \begin{bmatrix} P_{11} & P_{12} & \ldots & P_{1n-k} \\ P_{21} & P_{22} & \ldots & P_{2n-k} \\ \vdots & \vdots & \ldots & \vdots \\ P_{k1} & P_{k2} & \ldots & P_{kn-k} \\ 1 & 0 & \ldots & 0 \\ 0 & 1 & \ldots & 0 \\ 1 & \vdots & \ldots & 0 \\ 0 & 0 & \ldots & 1 \end{bmatrix} \tag{9.26}$$

So
$$S_1 = P_{11}\, r_1 \oplus P_{21}\, r_2 \oplus \ldots \oplus P_{k1}\, r_k \oplus r_{k+1} \oplus 0 \oplus 0 \oplus 0 \ldots$$
that represent the first bit of the syndrome matrix.

'or' for a general 'j' bit of syndrome
$$\boxed{S_j = r_1\, P_{1j} \oplus r_2\, P_{2j} \oplus \ldots \oplus r_k\, P_{kj} \oplus r_{k+j}} \tag{9.27}$$
we can verify it by substituting $j = 2$ mean
$$S_2 = r_1\, P_{12} + r_2\, P_{22} \oplus \ldots \oplus r_k\, P_{k2} \oplus r_{k+2}$$
that can be verified by solving with the help of matrix given :

If we see that 'j' bit of syndrome is identical to the 'j' column of H^T : for example, first bit of syndrome S is the first coloumn of H^T, 'or' (the First Row of H)

So to design a syndrome calculating network simply use shift resistor and generator $[S_1, S_2, S_3, \ldots, S_k]$ by adding different combinations of received bits of code (The process will be more clear by the numerical example solved below) as shown by general equation of S_j. The values of co-efficients P_{1j}, P_{2j}, ..., P_{kj} is each '0' or 1 and this is decided by the Hamming matrix.

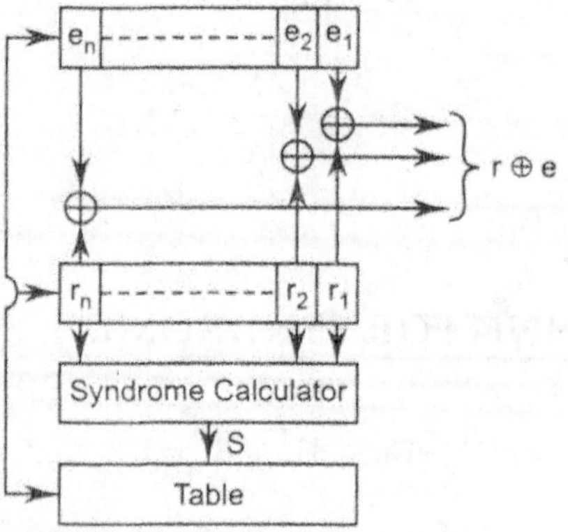

Fig. 9.4 *Decoding of hamming code using syndrome calculator*

9.5.1 Decoding Circuit (Using Syndrome Calculator)

Now the circuit diagram has been drawn here to decode a received word and find the final result that is the transmitted code.

The operation of circuit is very easy.

(*i*) First of all we receive a word r.

(*ii*) With the help of 'r', syndrome calculator circuit (explained in previous topic), generate a syndrome.

(*iii*) Now by this generated syndrome, we see the error pattern.

This is simply a software that is used to search a file, folder etc., in computer. Say we have syndrome 111 then by table we get the error pattern 0100000 searched by the software.

(*iv*) Now this error pattern is loaded in the error loading resistor.

(*v*) Now simply perform adding (Module - 2) between received word resistor and this error loading resistor ; that generate transmitted code

as $$c = r \oplus e.$$

9.6 CYCLIC CODES

Cyclic codes is also a special case of linear block code. They are mostly used code in the field of communication, because they can be easily encoded and very efficiently decoded. They have a well defined mathematical structure so easily processed by a designer.

A binary code is said to be cyclic code if that exhibits two fundamental properties:

1. *Linearity Property* : The sum of two codes words is also a code word.
2. *Cyclic Property* : Any cyclic shift of a code word is also a code word.

By Property 1, it can be easily said that it is a linear block code : By property second for a n bit cyclic code; If first code is

$$(x_0, x_1, x_2, \ldots, x_{n-1}) \text{ then by cyclic shifting}$$
$$(x_{n-1}, x_0, x_1, x_2, \ldots, x_{n-2}) \text{ and}$$
$$(x_{n-2}, x_{n-1}, x_0, \ldots, x_{n-3})$$

are all codes.

This formulation of the cyclic property suggests that we may treat the elements of a code word of length n as the co-efficients of a polynomial of degree $(n - 1)$. So code word polynomial can be represented as

$$x\,(D) = x_0 + x_1\,D + \ldots + x_{n-1}\,D^{n-1} \qquad (9.28)$$

where D is a arbitrary real variable.

Note that for binary codes, the co-efficients are 1s 'or' 0s.

Here we can say that every multiplication of D represent one cyclic shift to the right direction. For example,

$$x\,(D) = 010101, \text{ then its polynomial}$$

D^5	D^4	D^3	D^2	D^1	D^0
0	1	0	1	0	1

$$x\,(D) = D^4 + D^2 + 1$$

Now multiply x (D), by D

$$x_2\,(D) = D\,x(D) = D\,(D^4 + D^2 + 1)$$
$$= D^5 + D^3 + D$$

again decode the code

x_2 (D) =

D^5	D^4	D^3	D^2	D^1	D^0
1	0	1	0	1	0

Thus x (D) = 010101 is shift by one

x_2 (D) = 101010 (is the shifted version of x (D) by one bit in right hand direction).

9.6.1 Mathematical Analysis

(*i*) Again the general expression for cyclic code

$$x(D) = x_0 + x_1(D) + \ldots + x_{n-1} D^{n-1} \tag{9.29}$$
$$= x_{n-1} D^{n-1} + x_{n-2} D^{n-2} + \ldots + x_1 D + x_0$$

Now take a shifted version of x (D) ; recall that this new code will also be a Cyclic code.

$$D\,x(D) = x_{n-1} D^{n} + x_{n-2} D^{n-1} + \ldots + x_1 D^2 + x_0 D \tag{9.30}$$

But according to definition of Cyclic code : The order of polynomial of any cyclic code of n bits should be $n - 1$. That means (x_{n-1}) D^n is not a part of code because it is the order of n so

$$D\,x(D) = x_{n-1} \cdot D^{n} - x_{n-1} + \underline{x_{n-1} + x_{n-2} D^{n-1} + \ldots + x_1 D^2 + x_0 D} \tag{9.31}$$

Underlined equation shows the cyclic code containing all terms $(x_0, x_1, \ldots, x_{n-1})$.

$$D\,x(D) = x_{n-1} D^n - x_{n-1} + \text{Cyclic code}$$
$$\text{Cyclic code} = Dx(D) - x_{n-1} D^n - x_{n-1}$$

$$\boxed{\text{Cyclic code} = Dx(D) + x_{n-1}(D^n + 1)} \tag{9.32}$$

(Due to Module-2 operation)

(*ii*) But cyclic code polynomial in terms of message polynomial and generator polynomial is given by :

$$C(D) = d(D)\,.\,G(D), \text{ so}$$
$$d(D)\,G(D) = Dx(D) + x_{n-1}(D^n + 1) \tag{9.33}$$

It means that G (D) will always have a factor of $(D^n + 1)$.

There may be a number of factors of $(D^n + 1)$ but, we will choose only the factor that is the order of $(n - k)$ so.

Order of g (D) = $(n - k)$ = number of check bits to find a polynomial of g (D) solve $(D^n + 1)$ and collect the polynomial of order of $(n - k)$ from the different factor of $(D^n + 1)$.

For example, for a (7, 4) cyclic code we have $D^n + 1 = D^7 + 1$

Now, find the factor $D^7 + 1$

$$D^7 + 1 = (D + 1)(D^3 + D^2 + 1)(D^3 + D + 1)$$

No the two factors are of order '3' so we may have two generator polynomials

$$\boxed{\begin{aligned} g_1(D) &= D^3 + D^2 + 1 \\ g_2(D) &= D^3 + D + 1 \end{aligned}} \tag{9.34}$$

(*iii*) The relation discussed

$$GH^T = 0$$

also shows that if g (D) is the one of factors of $(1 + D^n)$, h (D) is will also be the factor of $(1 + D^n)$. But now h (D) will be the degree of k.

For example, discussed in point (*ii*) for (7, 4) code.

$$D^7 + 1 = (D + 1)(D^3 + D^2 + 1)(D^3 + D + 1).$$

Then one possible parity check polynomial

$$h(d) = (D + 1)(D^3 + D^2 + 1) = D^4 + D^3 + D^3 + D^2 + 1 = D^4 + D^2 + 1$$

that has the degree = k = 4.

So

Order of $h(d) = k$ = number of message bit ; to find a polynomial $g(D)$, solve $D^n + 1$, and collect the polynomials of order of k from the factor of $D^n + 1$.

(*iv*) The generator polynomial of an (n, k) cyclic code is unique, and it is also a code word of degree $(n - k)$.

It is unique because any code (n, k) arrays have only one polynomial of degree $(n - k)$.

(*v*) Any polynomial of order $(n + 1)$ 'or' less in code word if and only if it is multiple of generator polynomial. In other words.

In (n, k) codes
(*i*) One polynomial of degree $(n - k)$ is called generator polynomial.
(*ii*) Remaining polynomial of degree $(n - 1)$ 'or' less are code words if they are multiple of $g(D)$.

9.6.2 Systematic Cyclic Code

In a systematic cyclic code, message bit must be included in un-altered form.

In mathematical form, they are represented by

$$\boxed{C(D) = \underbrace{D^{n-k} \cdot d(D)}_{\text{I}} + \underbrace{\text{Rem}\left[\frac{D^{n-k} \cdot d(D)}{g(D)}\right]}_{\text{II}}} \tag{9.35}$$

Here, $d(D)$ = Data polynomial.

$$= m_0 + m_1 D + \ldots + m_{k-1} + D^{k-1}$$

$$\boxed{x^{n-k} d(D) = m_0 D^{n-k} + m_1 D^{n-k+1} + \ldots + m_{k-1} D^{n-1}} \tag{9.36}$$

(*i*) So first term of equation denotes the message polynomial of maximum degree that this $(n - 1)$ as discussed above (point V).

(*ii*) Now by dividing $\dfrac{D^{n-k} \cdot d(D)}{g(D)}$ we find the polynomial of x^{n-k} degree that is generator polynomial.

(*iii*) So by term II,

Remainder of $\dfrac{D^{n-k} \cdot d(D)}{g(D)}$, we have found the polynomial that don't have data polynomial and generator polynomial. Now this Remainder will have only terms that are neither the generator nor the multiple of generator (that are called code). So this remaining part of code is called parity polynomial.

So parity check polynomial

$$P(D) = \frac{D^{n-k} . d(D)}{g(D)} \tag{9.37}$$

9.6.3 Encoder Designing for the Generation of Systematic Cyclic Code :

For a systematic cyclic code generation, are need

(*i*) First k bits of message it self that can be easily transmitted by switching the input switch to the message bit position.

(*ii*) Now to generate the check bit, we have to perform the divisions find the remainder by equation

$$\text{Rem}\left[\frac{x^{n-k} . d(D)}{g(D)}\right]$$

and this can be easily performed by using shift resister encoder ; the encoding of check bit starts when the feedback switch is closed and position of input switch is at message bit position. Start with taking the contained of each resistor initially 000 ... Now k bits of message bits are shifted in to the shift resister.

After the k shift ; the content of shift resistor shows the $(n - k)$ check bit. so,

(*a*) At a time we shift input message bit at transmitter line and generate the check bit.

(*b*) After k shifts the all message bits are transmitted on the transmitting line and at the same time. Now the contained of shift resister are check bit.

(*c*) Now move input switch at check bit position and feedback switch is opened. So now check bit are transferred on the transmitting line.

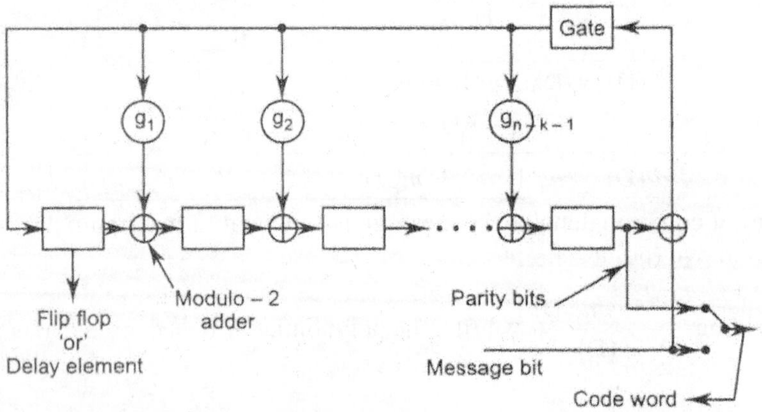

Fig. 9.5 *Encoder for an (n, K) cyclic code*

Example 3. *The generator matrix for a (6, 3) block code is shown below. Obtain all the code vectors of this code vector.*

$$G = \begin{bmatrix} 1 & 0 & 0 & : & 0 & 1 & 1 \\ 0 & 1 & 0 & : & 1 & 0 & 1 \\ 0 & 0 & 1 & : & 1 & 1 & 0 \end{bmatrix}$$

Solution. Here we have

Total bits in codes = 6

Total bits in data = 3

Hence we have totally 2^3 data vectors.

S. No.	*Message/Data*		
1.	0	0	0
2.	0	0	1
3.	0	1	0
4.	0	1	1
5.	1	0	0
6.	1	0	1
7.	1	1	0
8.	1	1	1

So we can get codes by the formula C = dG

For [0, 0, 0] ;

$$C = [0 \quad 0 \quad 0]\begin{bmatrix} 1 & 0 & 0 & : & 0 & 1 & 1 \\ 0 & 1 & 0 & : & 1 & 0 & 1 \\ 0 & 0 & 1 & : & 1 & 1 & 0 \end{bmatrix}$$

$$= [0 \quad 0 \quad 0 \quad 0 \quad 0 \quad 0]$$

For $d = [0 \quad 0 \quad 1]$;

$$C = [0 \quad 0 \quad 1]\begin{bmatrix} 1 & 0 & 0 & : & 0 & 1 & 1 \\ 0 & 1 & 0 & : & 1 & 0 & 1 \\ 0 & 0 & 1 & : & 1 & 1 & 0 \end{bmatrix}$$

$$= [0 \quad 0 \quad 1 \quad 1 \quad 1 \quad 0]$$

For $d = [0 \quad 1 \quad 0]$;

$$C = [0 \quad 1 \quad 0]\begin{bmatrix} 1 & 0 & 0 & : & 0 & 1 & 1 \\ 0 & 1 & 0 & : & 1 & 0 & 1 \\ 0 & 0 & 1 & : & 1 & 1 & 0 \end{bmatrix}$$

$$= [0 \quad 1 \quad 0 \quad 1 \quad 0 \quad 1]$$

For $d = [0 \quad 1 \quad 1]$;

$$C = [0 \quad 1 \quad 1]\begin{bmatrix} 1 & 0 & 0 & : & 0 & 1 & 1 \\ 0 & 1 & 0 & : & 1 & 0 & 1 \\ 0 & 0 & 1 & : & 1 & 1 & 0 \end{bmatrix}$$

$$= [0 \oplus 0, 1 \oplus 0, 0 \oplus 1, 1 \oplus 1, 0 \oplus 1, 1 \oplus 0]$$

$$= [0, 1, 1, 0, 1, 1]$$

And thus we can get all the code words. These table shows all the possible code words for data, we should see that last $n - k = 6 - 3 = 3$ bits of each code word shows the parity bits.

S. No.	Message			Codeword					
	m_1	m_2	m_3	C_1	C_2	C_3	C_4	C_5	C_6
1	0	0	0	0	0	0	0	0	0
2	0	0	1	0	0	1	1	1	0
3	0	1	0	0	1	0	1	0	1
4	0	1	1	0	1	1	0	1	1
5	1	0	0	1	0	0	0	1	1
6	1	0	1	1	0	1	1	0	1
7	1	1	0	1	1	0	1	1	0
8	1	1	1	1	1	1	0	0	0

Parity Bits

Example 4. *For (7, 4) Hamming codes, parity matrix is given by :*

$$H = \begin{bmatrix} 0 & 1 & 1 & 1 & : & 1 & 0 & 0 \\ 1 & 1 & 0 & 1 & : & 0 & 1 & 0 \\ 1 & 0 & 1 & 1 & : & 0 & 0 & 1 \end{bmatrix}$$

Calculate codes for

$$d = [1\ 0\ 1\ 0],\ [0\ 1\ 1\ 0],\ [1\ 1\ 1\ 0]$$

Solution. Given that :

$$\mathrm{H} = \begin{bmatrix} 0 & 1 & 1 & 1 & : & 1 & 0 & 0 \\ 1 & 1 & 0 & 1 & : & 0 & 1 & 0 \\ 1 & 0 & 1 & 1 & : & 0 & 0 & 1 \end{bmatrix}$$

But $\mathrm{H} = [\mathrm{P}^T : \mathrm{I}]$

Thus we can say

$$\mathrm{P}^T = \begin{bmatrix} 0 & 1 & 1 & 1 \\ 1 & 1 & 0 & 1 \\ 1 & 0 & 1 & 1 \end{bmatrix}$$

$$\mathrm{P} = \begin{bmatrix} 0 & 1 & 1 \\ 1 & 1 & 0 \\ 1 & 0 & 1 \\ 1 & 1 & 1 \end{bmatrix}$$

Thus now we can have generator matrix

$$\mathrm{G} = [\mathrm{I} : \mathrm{P}]$$

$$= \begin{bmatrix} 1 & 0 & 0 & 0 & : & 0 & 1 & 1 \\ 0 & 1 & 0 & 0 & : & 1 & 1 & 0 \\ 0 & 0 & 1 & 0 & : & 1 & 0 & 1 \\ 0 & 0 & 0 & 1 & : & 1 & 1 & 1 \end{bmatrix}$$

Thus for $d = [1\ 0\ 1\ 0]$,

$$C = dG = [1\ 0\ 1\ 0]\begin{bmatrix} 1 & 0 & 0 & 0 & : & 0 & 1 & 1 \\ 0 & 1 & 0 & 0 & : & 1 & 1 & 0 \\ 0 & 0 & 1 & 0 & : & 1 & 0 & 1 \\ 0 & 0 & 0 & 1 & : & 1 & 1 & 1 \end{bmatrix}$$

$$= [1 \oplus 0, 0 \oplus 0, 0 \oplus 1, 0 \oplus 0, 0 \oplus 1, 1 \oplus 0, 1 \oplus 1]$$

$$= [1\ 0\ 1\ 0\ 1\ 1\ 0]$$

for $d = [0\ 1\ 1\ 0]$

$$C = [0\ 1\ 1\ 0]\begin{bmatrix} 1 & 0 & 0 & 0 & : & 0 & 1 & 1 \\ 0 & 1 & 0 & 0 & : & 1 & 1 & 0 \\ 0 & 0 & 1 & 0 & : & 1 & 0 & 1 \\ 0 & 0 & 0 & 1 & : & 1 & 1 & 1 \end{bmatrix}$$

$$= [0 \oplus 0, 1 \oplus 0, 0 \oplus 1, 0 \oplus 0, 1 \oplus 1, 1 \oplus 0, 0 \oplus 1]$$

$$= [0\ 1\ 1\ 0\ 0\ 1\ 1]$$

$$C = [1\ 1\ 1\ 0]\begin{bmatrix} 1 & 0 & 0 & 0 & : & 0 & 1 & 1 \\ 0 & 1 & 0 & 0 & : & 1 & 1 & 0 \\ 0 & 0 & 1 & 0 & : & 1 & 0 & 1 \\ 0 & 0 & 0 & 1 & : & 1 & 1 & 1 \end{bmatrix}$$

again for $d = [1\ 1\ 1\ 0]$

$$= [1 \oplus 0 \oplus 0, 0 \oplus 1 \oplus 0, 0 \oplus 0 \oplus 1, 0 \oplus 0 \oplus 0, 0 \oplus 1 \oplus 1, 1 \oplus 1 \oplus 0, 1 \oplus 0 \oplus 1]$$

$$= [1\ 1\ 1\ 0\ 0\ 0\ 0]$$

Example 5. *For a (7, 4) Hamming code, generator matrix is given by :*

$$G = \begin{bmatrix} 1 & 0 & 0 & 0 & : & 1 & 1 & 0 \\ 0 & 1 & 0 & 0 & : & 0 & 1 & 1 \\ 0 & 0 & 1 & 0 & : & 1 & 1 & 1 \\ 0 & 0 & 0 & 1 & : & 1 & 0 & 1 \end{bmatrix}$$

We have received code [0 1 1 0 1 1 0] with a single bit error. Calculate the correct transmitted code word.

Solution. Here we have :

$$G = \begin{bmatrix} 1 & 0 & 0 & 0 & : & 1 & 1 & 0 \\ 0 & 1 & 0 & 0 & : & 0 & 1 & 1 \\ 0 & 0 & 1 & 0 & : & 1 & 1 & 1 \\ 0 & 0 & 0 & 1 & : & 1 & 0 & 1 \end{bmatrix}$$

But $G = [I : P]$

So,
$$P = \begin{bmatrix} 1 & 1 & 0 \\ 0 & 1 & 1 \\ 1 & 1 & 1 \\ 1 & 0 & 1 \end{bmatrix}$$

So we can calculate

$$P^T = \begin{bmatrix} 1 & 0 & 1 & 1 \\ 1 & 1 & 1 & 0 \\ 0 & 1 & 1 & 1 \end{bmatrix}$$

Now we may write parity matrix

$$H = [P^T : I]$$

$$= \begin{bmatrix} 1 & 0 & 1 & 1 & : & 1 & 0 & 0 \\ 1 & 1 & 1 & 0 & : & 0 & 1 & 0 \\ 0 & 1 & 1 & 1 & : & 0 & 0 & 1 \end{bmatrix}$$

Now calculate sydnrome for single bit error patterns given below :

$$\begin{bmatrix} 0 & 0 & 0 & 0 & 0 & 0 & 0 \\ 0 & 0 & 0 & 0 & 0 & 0 & 1 \\ 0 & 0 & 0 & 0 & 0 & 1 & 0 \\ 0 & 0 & 0 & 0 & 1 & 0 & 0 \\ 0 & 0 & 0 & 1 & 0 & 0 & 0 \\ 0 & 0 & 1 & 0 & 0 & 0 & 0 \\ 0 & 1 & 0 & 0 & 0 & 0 & 0 \\ 1 & 0 & 0 & 0 & 0 & 0 & 0 \end{bmatrix}$$

By the formula,

$$S = eH^T$$

So for example,

$$\text{for } e = 0\,0\,0\,0\,0\,0\,1$$

$$H^T = \begin{bmatrix} 1 & 1 & 0 \\ 0 & 1 & 1 \\ 1 & 1 & 1 \\ 1 & 0 & 1 \\ 1 & 0 & 0 \\ 0 & 1 & 0 \\ 0 & 0 & 1 \end{bmatrix}$$

$$S = eH^T$$

$$= [0\ 0\ 1]$$ (By using property 3 of syndrome]

Then, finally we have,

Syndrome	*Error pattern*
0 0 0	0 0 0 0 0 0 0
0 0 1	0 0 0 0 0 0 1
0 1 0	0 0 0 0 0 1 0
1 0 0	0 0 0 0 1 0 0
1 0 1	0 0 0 1 0 0 0
1 1 1	0 0 1 0 0 0 0
0 1 1	0 1 0 0 0 0 0
1 1 0	1 0 0 0 0 0 0

Now we have,

$$r = [0\ 1\ 1\ 0\ 1\ 1\ 0]$$

But by formula,

$$S = rH^T = [0\ 1\ 1\ 0\ 1\ 1\ 0] \begin{bmatrix} 1 & 1 & 0 \\ 0 & 1 & 1 \\ 1 & 1 & 1 \\ 1 & 0 & 1 \\ 1 & 0 & 0 \\ 0 & 1 & 0 \\ 0 & 0 & 1 \end{bmatrix}$$

$$[0 \oplus 1 \oplus 1 \oplus 0,\ 1 \oplus 1 \oplus 0 \oplus 1,\ 1 \oplus 1 \oplus 0 \oplus 0]$$

$$S = [0,\ 1\ 0]$$

By the table for

$$S = [0\ 1\ 0]$$

error, $e. = [0\ 0\ 0\ 0\ 0\ 1\ 0]$

So we may say that transmitted code

$$C = r \oplus e$$

$$= 0\ 1\ 1\ 0\ 1\ 1\ 0 \oplus 0\ 0\ 0\ 0\ 0\ 1\ 0$$

$$C = 0\ 1\ 1\ 0\ 1\ 0\ 0$$

Example 6. *For a (7, 4) cyclic code ; calculate cyclic code for message polynomial* $(x^3 + x + 1)$.

Solution. We may start with

$$g\ (D) = (D^7 + 1) \text{ or}$$

$$g\ (x) = (x^7 + 1)$$

[as g (D) is a $(n - k)$ order polynomial solved $(D^n + 1)$]

$$g(x) = (1 + x)(1 + x^2 + x^3)(1 + x + x^3)$$

The generator polynomial of $(n - k)$ power

$$g(x) = (x^3 + x^2 + 1) \text{ 'or'}$$

$$g(x) = (x^3 + x + 1)$$

Now taking any one of these two ; we may choose

$$g(x) = (x^3 + x^2 + 1)$$

So cyclic code,

$$C(x) = g(x) \,.\, d(x)$$

$$= (x^3 + x^2 + 1)(x^3 + x + 1)$$

$$= x^6 + x^5 + x^3 + x^4 + x^3 + x + x^3 + x^2 + 1$$

(additions are module 2 addition)

$$= x^6 + x^5 + (x^3 + x^3 + x^3) + x^4 + x^2 + x + 1$$

$$= x^6 + x^5 + x^3 + x^4 + x^2 + x + 1 \qquad [\because\ x^3 \oplus x^3 = 0]$$

$$C = \begin{pmatrix} x^6 & x^5 & x^4 & x^3 & x^2 & x & x^0 \\ 1 & 1 & 1 & 1 & 1 & 1 & 1 \end{pmatrix}$$

Thus, finally we have cyclic code,

$$= [1\ 1\ 1\ 1\ 1\ 1\ 1]$$

Example 7. *Consider a (7, 4) Hamming code generated by*

$$g(D) = 1 + D^2 + D^3$$

Construct the code words for data polynomial.

$$d(D) = D^3 + 1$$

Solution. Given that

$$g(D) = 1 + D^2 + D^3$$

If we have to find code words we have to find generator matrix first.

So proceed with

$$D^0\, g(D) = D^3 + D^2 + 1$$

$$D\, g(D) = D^4 + D^3 + D$$

$$D^2\, g(D) = D^5 + D^4 + D^2$$

$$D^3\, g(D) = D^6 + D^5 + D^3$$

Thus, we may write

$$G = \begin{bmatrix} D^6 & D^5 & D^4 & D^3 & D^2 & D^1 & D^0 \\ 0 & 0 & 0 & 1 & 1 & 0 & 1 \\ 0 & 0 & 1 & 1 & 0 & 1 & 0 \\ 0 & 1 & 1 & 0 & 1 & 0 & 0 \\ 1 & 1 & 0 & 1 & 0 & 0 & 0 \end{bmatrix}$$

If we proceed with this form of generator matrix, we will not get code in the form of (m, P) called systematic form. So convert it by simply matrix operations in the form $G = [I : P]$ that will be

$$G = \begin{bmatrix} 1 & 0 & 0 & 0 & : & 1 & 0 & 1 \\ 0 & 1 & 0 & 0 & : & 1 & 1 & 1 \\ 0 & 0 & 1 & 0 & : & 1 & 1 & 0 \\ 0 & 0 & 0 & 1 & : & 1 & 1 & 1 \end{bmatrix}$$

Now we have,

$$d = (D^3 + 1)$$

In the form of matrix

$$d = \begin{matrix} d^3 & d^2 & d^1 & d^0 \\ [1 & 0 & 0 & 1] \end{matrix}$$

So, $C = d \,.\, G$

$$= [1 \;\; 0 \;\; 0 \;\; 1] \begin{bmatrix} 1 & 0 & 0 & 0 & : & 1 & 0 & 1 \\ 0 & 1 & 0 & 0 & : & 1 & 1 & 1 \\ 0 & 0 & 1 & 0 & : & 1 & 1 & 0 \\ 0 & 0 & 0 & 1 & : & 1 & 1 & 1 \end{bmatrix}$$

$$= [1 \oplus 0, 0 \oplus 0, 0 \oplus 0, 0 \oplus 1, 1 \oplus 1, 0 \oplus 1, 1 \oplus 1]$$

$$C = [1\,0\,0\,1\,0\,1\,0]$$

Code polynomial can be $\boxed{C = [d^6 + d^3 + d]}$

Example 8. *For a (7, 4) cyclic code ; calculate systematic cyclic code for*

$$d(D) = D^3 + 1$$

Solution. Now again for (7, 4) cyclic code $g(D) = D^7 + 1$

$$= (1 + D)(1 + D^2 + D^3)(1 + D + D^3)$$

Again choosing,

$$g(D) = 1 + D + D^3$$

Now for systematic cyclic code

$$C(d) = \text{Rem}\,\frac{D^{n-k} \,.\, d(D)}{g(D)} + D^{n-k} \,.\, d(D)$$

So start with $= \text{Rem of } g(D) \overline{\big)\, D^{n-k} \,.\, d(D)}$

$$= \text{Rem of } g(D) \overline{\big)\, D^3 (D^3 + 1)}$$

$$= \text{Rem of } g(D) \overline{\big)\, D^6 + D^3}$$

$$\begin{array}{r} D^3 + D \\ 1 + D + D^3 \overline{\big)\, D^6 + D^3} \\ D^6 + D^3 + D^4 \\ D^4 \\ \hline D^4 + D^2 + D \\ \hline \text{Rem} \rightarrow D^2 + D \end{array}$$

So, $$C(D) = D^{7-4}(D^3 + 1) + (D^2 + D) = D^3(D^3 + 1) + (D^2 + D)$$

$$\boxed{C(D) = D^6 + D^3 + D^2 + D}$$

Here we can draw the encoder.

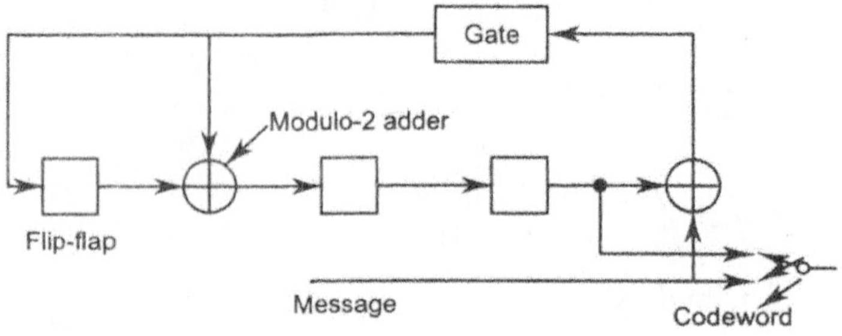

Fig. 9.6

Example 9. *Consider a (7, 4) block code generated by*

$$G = \begin{bmatrix} 1 & 0 & 0 & 0 & \vdots & 1 & 1 & 1 \\ 0 & 1 & 0 & 0 & \vdots & 1 & 1 & 0 \\ 0 & 0 & 1 & 0 & \vdots & 1 & 0 & 1 \\ 0 & 0 & 0 & 1 & \vdots & 0 & 1 & 1 \end{bmatrix}$$

Find out the error vector ?

Solution. Step 1: $$H = \left[P^T : I_{n-k}\right]_{(n-k)\times n}$$

$$H = \underbrace{\begin{bmatrix} 1 & 1 & 1 & 0 \\ 1 & 1 & 0 & 1 \\ 1 & 0 & 1 & 1 \end{bmatrix}}_{P^T}\underbrace{\begin{matrix} 1 & 0 & 0 \\ 0 & 1 & 0 \\ 0 & 0 & 1 \end{matrix}}_{I_3}$$

Step 2: We have $k = 4$

$n = 7$

$\therefore$ $2^k = 2^4$ codewords for 2^4 messages (0 0 0 0) … (1 1 1 1).

Step 3: Choose a specific value of d from the 16 combinations for example 1011.

$\therefore$ $C = dG$

$$dG = [1011]\begin{bmatrix} 1 & 0 & 0 & 0 & 1 & 1 & 1 \\ 0 & 1 & 0 & 0 & 1 & 1 & 0 \\ 0 & 0 & 1 & 0 & 1 & 0 & 1 \\ 0 & 0 & 0 & 1 & 0 & 1 & 1 \end{bmatrix} = [1\ 0\ 1\ 1\ 0\ 0\ 1]$$

Step 4: Calculate syndrome

$$S = CH^T$$

$$S = CH^T = (1011001)\begin{bmatrix} 1 & 1 & 1 & 0 & \vdots & 1 & 0 & 0 \\ 1 & 1 & 0 & 1 & \vdots & 0 & 1 & 0 \\ 1 & 0 & 1 & 1 & \vdots & 0 & 0 & 1 \end{bmatrix}^T$$

$$= (1011001)\begin{bmatrix} 1 & 1 & 1 \\ 1 & 1 & 0 \\ 1 & 0 & 1 \\ 0 & 1 & 1 \\ 1 & 0 & 0 \\ 0 & 1 & 1 \\ 0 & 0 & 1 \end{bmatrix} = [0\,0\,0]$$

Syndrome is zero, it means it is a valid code word.

Step 5: If R = 1 0 0 1 0 0 1 is given, find

$S = RH^T$

$$S = (1001001)\begin{bmatrix} 1 & 1 & 1 \\ 1 & 1 & 0 \\ 1 & 0 & 1 \\ 0 & 1 & 1 \\ 1 & 0 & 0 \\ 0 & 1 & 0 \\ 0 & 0 & 1 \end{bmatrix} = [1\,0\,1]$$

S = 101

Step 6: Compare value of S by H^T. Now 101 is equal to third row of H^T. It means third bit from the left is in error. Make it 1 if 0 and 0 if it is 1.

So the transmitted word is

C = 1 0 1 1 0 0 1

Error vector E = R – C

E = 0 0 1 0 0 0 0

Example 10. *A parity check code has the parity check matrix*

$$H = \left[\begin{array}{ccc|ccc} 1 & 0 & 1 & 1 & 0 & 0 \\ 1 & 1 & 0 & 0 & 1 & 0 \\ 0 & 1 & 1 & 0 & 0 & 1 \\ \hline & P^T & & & I_{n-k} & \end{array}\right]$$

(i) Determine the generator matrix G.

(ii) Find the code word that beings 1 0 1………

(iii) Suppose that the recieved word is 1 1 0 1 1 0

Decode this received word

Solution. (*i*) Since H is a 6 × 3 matrix $n = 6$ and $k = 3$

$$P = \begin{bmatrix} 1 & 1 & 0 \\ 0 & 1 & 1 \\ 1 & 0 & 1 \end{bmatrix}$$

Then the generator matrix G is

$$G = [I_k \vdots P] = \begin{bmatrix} C_1 & C_2 & C_3 & \vdots & C_4 & C_5 & C_6 & \\ 1 & 0 & 0 & \vdots & 1 & 1 & 0 & R_1 \\ 0 & 1 & 0 & \vdots & 0 & 1 & 1 & R_2 \\ 0 & 0 & 1 & \vdots & 1 & 0 & 1 & R_3 \end{bmatrix}$$

(ii)

$$C = dG = [101] = \begin{bmatrix} C_1 & C_2 & C_3 & \vdots & C_4 & C_5 & C_6 & \\ 1 & 0 & 0 & \vdots & 1 & 1 & 0 & R_1 \\ 0 & 1 & 0 & \vdots & 0 & 1 & 1 & R_2 \\ 0 & 0 & 1 & \vdots & 1 & 0 & 1 & R_3 \end{bmatrix}$$

$$= [1\ 0\ 1\ 0\ 1\ 1]$$

(iii)

$$R = [1\ 1\ 0\ 1\ 1\ 0]$$

$$S = RH^T$$

$$= [110110] \begin{bmatrix} C_1 & C_2 & C_3 & \\ 1 & 1 & 0 & R_1 \\ 0 & 1 & 1 & R_2 \\ 1 & 0 & 1 & R_3 \\ 1 & 0 & 0 & R_4 \\ 0 & 1 & 0 & R_5 \\ 0 & 0 & 1 & R_6 \end{bmatrix} = [011]$$

since S is equal to the second row of H^T, an error is at the second bit, the correct code word is 1 0 0 1 1 0, and the data bits are 1 0 0.

1 0 0   1 0 1 0 1 1
1 0 1   1 1 0 1 0 1

Example 11. *Consider (7, 4) linear code whose generator matrix is*

$$G = \begin{bmatrix} 1 & 0 & 0 & 0 & \vdots & 1 & 0 & 1 \\ 0 & 1 & 0 & 0 & \vdots & 1 & 1 & 1 \\ 0 & 0 & 1 & 0 & \vdots & 1 & 1 & 0 \\ 0 & 0 & 0 & 1 & \vdots & 0 & 1 & 1 \end{bmatrix}$$

(i) Find all the code vectors of this code.
(ii) Find the parity check matrix for this code.
(iii) Find the minimum weight of this code.
(iv) Prove equation $CH^T = 0$ [Raj. Univ. 2007]

Solution. Given (7, 4) linear code

$$G = \begin{bmatrix} 1 & 0 & 0 & 0 & \vdots & 1 & 0 & 1 \\ 0 & 1 & 0 & 0 & \vdots & 1 & 1 & 1 \\ 0 & 0 & 1 & 0 & \vdots & 1 & 1 & 0 \\ 0 & 0 & 0 & 1 & \vdots & 0 & 1 & 1 \end{bmatrix}$$

(*i*) Here k = 4   n = 7

K denotes no. of bits in a message

There are 16 possible combinations of 4 - bits ranging from 0000 - 1111.

for each combination, the codeword is formed by

$$C = dG$$

$$C = [1111]\left[\begin{array}{cccc|ccc} 1 & 0 & 0 & 0 & 1 & 0 & 1 \\ 0 & 1 & 0 & 0 & 1 & 1 & 1 \\ 0 & 0 & 1 & 0 & 1 & 1 & 0 \\ 0 & 0 & 0 & 1 & 0 & 1 & 1 \end{array}\right]$$

$$\therefore \quad C = 1\,1\,1\,1\,1\,1\,1$$

So on for other combinations codewords can be formed

(*ii*) Parity check matrix

$$H = [\,P^T/I_{n-k}]$$

$$H = \left[\begin{array}{cccc|ccc} 1 & 1 & 1 & 0 & 1 & 0 & 0 \\ 0 & 1 & 1 & 1 & 0 & 1 & 0 \\ 1 & 1 & 0 & 1 & 0 & 0 & 1 \end{array}\right]$$

(*iii*) Minimum weight

First construct all codewords by

$$C = dG$$

Weight (No. of nonzero)

0000000	0
0001011	3
0010110	3
0011101	4
0100111	4
0101100	3
0110001	3
0111010	4
1000101	3
1001110	4
1010011	4
1011000	3
1100010	3
1101001	4
1110100	4
1111111	7

Minimum weight = 3

(*iv*) Equation $CH^T = 0$

Choose any $C = 0\,0\,1\,0\,1\,1\,0$

$$H^T = \begin{bmatrix} 1 & 0 & 1 \\ 1 & 1 & 1 \\ 1 & 1 & 0 \\ 0 & 1 & 1 \\ 1 & 0 & 0 \\ 0 & 1 & 0 \\ 0 & 0 & 1 \end{bmatrix}$$

Now $$CH^T \Rightarrow [0\,0\,1\,0\,1\,1\,0] \Rightarrow [0010110]\begin{bmatrix} 1 & 0 & 1 \\ 1 & 1 & 1 \\ 1 & 1 & 0 \\ 0 & 1 & 1 \\ 1 & 0 & 0 \\ 0 & 1 & 0 \\ 0 & 0 & 1 \end{bmatrix}$$

$$\Rightarrow [0\ 0\ 0]$$ **Hence Proved.**

Example 12. *The parity check bits of a (8, 4) block code are generated by*

$$C_5 = d_1 + d_2 + d_4$$
$$C_6 = d_1 + d_2 + d_3$$
$$C_7 = d_1 + d_3 + d_4$$
$$C_8 = d_1 + d_3 + d_4$$

where d_1, d_2, d_3 and d_4 are the message digits.

(i) Find the generator matrix and the parity check matrix for this code.
(ii) Find the minimum weight of this code.
(iii) Find the error-detecting capabilities of this code.
(iv) Show through an example that this code can detect three errors/codeword. [Raj. Univ. 2003]

Solution. Given
$$C_5 = d_1 + d_2 + d_4$$
$$C_6 = d_1 + d_2 + d_3$$
$$C_7 = d_1 + d_3 + d_4$$
$$C_8 = d_1 + d_3 + d_4$$

Here
$$n = 8$$
$$k = 4$$

Comparing the given equation by the standard equations as described in theory

$$P_{11} = 1 \quad P_{21} = 1 \quad P_{31} = 0 \quad P_{41} = 1$$
$$P_{12} = 1 \quad P_{22} = 1 \quad P_{32} = 1 \quad P_{42} = 0$$
$$P_{13} = 1 \quad P_{23} = 0 \quad P_{33} = 1 \quad P_{43} = 1$$
$$P_{14} = 0 \quad P_{24} = 1 \quad P_{34} = 1 \quad P_{44} = 1$$

(*i*) Now generator matrix

$$G = [I_K/P]$$

Where
$$P = \begin{bmatrix} P_{11} & P_{12} & \cdots\cdots & P_{1n-k} \\ P_{21} & P_{22} & \cdots\cdots & P_{2n-k} \\ \vdots & \vdots & \vdots & \vdots \\ P_{k1} & P_{k2} & \cdots\cdots & P_{k,n-k} \end{bmatrix}$$

$$P = \begin{bmatrix} 1 & 1 & 1 & 0 \\ 1 & 1 & 0 & 1 \\ 0 & 1 & 1 & 1 \\ 1 & 0 & 1 & 1 \end{bmatrix}$$

Hence Proved.

Generator matrix
$$G = \left[\begin{array}{cccc|cccc} 1 & 0 & 0 & 0 & 1 & 1 & 1 & 0 \\ 0 & 1 & 0 & 0 & 1 & 1 & 0 & 1 \\ 0 & 0 & 1 & 0 & 0 & 1 & 1 & 1 \\ 0 & 0 & 0 & 1 & 1 & 0 & 1 & 1 \end{array}\right]$$

Parity check matrix $H = [P^T/I_{n-k}]$

$$H = \left[\begin{array}{cccc|cccc} 1 & 1 & 0 & 1 & 1 & 0 & 0 & 0 \\ 1 & 1 & 1 & 0 & 0 & 1 & 0 & 0 \\ 1 & 0 & 1 & 1 & 0 & 0 & 1 & 0 \\ 0 & 1 & 1 & 1 & 0 & 0 & 0 & 1 \end{array}\right]$$

(*ii*) To find out the minimum weight, first codewords are constructed weight.

0	0	0	0	0	0	0	0	0
0	0	0	1	1	0	1	1	4
0	0	1	0	0	1	1	1	4
0	0	1	1	1	1	0	0	4
0	1	0	0	1	1	0	1	4
0	1	0	1	0	1	1	0	4
0	1	1	0	1	0	1	0	4
0	1	1	1	0	0	1	1	4
1	0	0	0	1	1	1	0	4
1	0	0	1	0	1	0	1	4
1	0	1	0	1	0	0	1	4
1	0	1	1	0	0	1	0	4
1	1	0	0	0	0	1	1	4
1	1	0	1	1	0	0	0	4
1	1	1	0	0	1	0	0	4
1	1	1	1	1	1	1	1	8

∴ Minimum weight = 4

(*iii*) Error detection capability

$$= d_{\min} - 1$$
$$= 4 - 1 = 3$$

(*iv*) Take any two adjacent codewords, their distance = 4

∴ can detect 3 errors/codeword

Example 13. *The generator matrix for a (7, 4) linear code is*

$$G = \begin{bmatrix} 1 & 0 & 0 & 0 & \vdots & 1 & 0 & 1 \\ 0 & 1 & 0 & 0 & \vdots & 1 & 1 & 1 \\ 0 & 0 & 1 & 0 & \vdots & 1 & 1 & 0 \\ 0 & 0 & 0 & 1 & \vdots & 0 & 1 & 1 \end{bmatrix}$$

find (i) All the code vector of this code

(ii) The parity check matrix for this code.

(iii) The minimum weight of this code.

Solution. (*i*) Since $k = 4$, there are $M = 2^4 = 16$ messages, and C = MG. Thus code vectors are tabulated as below:

Message				*Code vectors C*							*Weight*
0	0	0	0	0	0	0	0	0	0	0	0
0	0	0	1	0	0	0	1	0	1	1	3
0	0	1	0	0	0	1	0	1	1	0	3
0	0	1	1	0	0	1	1	1	0	1	4
0	1	0	0	0	1	0	0	1	1	1	4
0	1	0	1	0	1	0	1	1	0	0	3
0	1	1	0	0	1	1	0	0	0	1	3
–	–	–	–	–	–	–	–	–	–	–	
1	1	1	1	1	1	1	1	1	1	1	–

The above code vector has been obtained using the relation.

$$C = MG = [0\,1\,0\,1]\begin{bmatrix} 1 & 0 & 0 & 0 & : & 1 & 0 & 1 \\ 0 & 1 & 0 & 1 & : & 1 & 1 & 1 \\ 0 & 0 & 1 & 0 & : & 1 & 1 & 0 \\ 0 & 0 & 0 & 1 & : & 0 & 1 & 1 \end{bmatrix} = [0\,1\,0\,1\,1\,0\,0]$$

This is one vector the other row vectors have been obtained using similar way.

(*ii*) Parity check matrix H for this code is using $G = [I_4/P]$.

$$H = \begin{bmatrix} 1 & 1 & 1 & 0 & : & 1 & 0 & 0 \\ 0 & 1 & 1 & 1 & : & 0 & 1 & 0 \\ 1 & 1 & 0 & 1 & : & 0 & 0 & 1 \\ & & & & : & & & \\ & & P^T & & : & & I_3 & \end{bmatrix}$$

(*iii*) Associated with each (n, k) block code is a parity check matrix H. We find the minimum weight = 3.

Example 14. *For a (6, 3) systematic linear block code the three - parity check bits C_4, C_5 and C_6 are formed from the following equations :*

$$C_4 = d_1 \oplus d_3$$
$$C_5 = d_1 \oplus d_2 \oplus d_3$$
$$C_6 = d_1 \oplus d_2$$

(i) Write down the generator matrix G.

(ii) Construct all possible code words.

(iii) Suppose that the received word is 010111. Decode this received word by finding the location of the error and the transmitted data bits.

Solution. (*i*) From the given equation we obtain

$$p^T = \begin{bmatrix} d_1 & d_2 & d_3 \\ 1 & 0 & 1 \\ d_1 & d_2 & d_3 \\ 1 & 1 & 1 \\ d_1 & d_2 & d_3 \\ 1 & 1 & 0 \end{bmatrix}$$

$$G = [I_3 \mid P] = \left[\begin{array}{ccc|ccc} 1 & 0 & 0 & 1 & 1 & 1 \\ 0 & 1 & 0 & 0 & 1 & 1 \\ 0 & 0 & 1 & 1 & 1 & 0 \end{array}\right]$$

(*ii*) Since K = 3, we have $2^3 = 8$ data words, thus, of d = [101], then using, we obtain

$$C = dG = [101]\left[\begin{array}{ccc|ccc} 1 & 0 & 0 & 1 & 1 & 1 \\ 0 & 1 & 0 & 0 & 1 & 1 \\ 0 & 0 & 1 & 1 & 1 & 0 \end{array}\right]$$

$$= [1\,0\,1\,0\,0\,1]$$

in a similar manner, the other code words can be constructed. They are listed in table

d	*c*	*d*	*c*
000	000000	100	100111
001	000110	101	101001
010	010011	110	110100
011	011101	111	111010

(*iii*)

$$H^T = \begin{bmatrix} P \\ I_3 \end{bmatrix} = \begin{bmatrix} 1 & 1 & 1 \\ 0 & 1 & 1 \\ 1 & 1 & 0 \\ 1 & 0 & 0 \\ 0 & 1 & 0 \\ 0 & 0 & 1 \end{bmatrix}$$

Now, $r = [0\ 1\ 0\ 1\ 1\ 1]$

The syndrome S of r is

$$S = rH^T = [010111] = \begin{bmatrix} 1 & 1 & 1 \\ 0 & 1 & 1 \\ 1 & 1 & 0 \\ 1 & 0 & 0 \\ 0 & 1 & 0 \\ 0 & 0 & 1 \end{bmatrix} = [100]$$

Since S is equal to the fourth row of H^T, an error is at the fourth bit, the correct code word is 0 1 0 0 1 1, and the data bits are 0 1 0.....

Example 15. *The Generator polynomial of a (7, 4) cyclic code is* $g(X) = 1 + X + X^3$. *Find codewords for message vector 1011 in the following ways :*

(*i*) *By forming the code polynomial using*

$$V(X) = g(X)\ m(X) \qquad \text{where } m(X) = \text{message polynomial}$$

(*ii*) *By using systematic encoding method.*

Solution. (*i*) Message vector

$$m = (m_0, m_1, m_2, m_3) = 1011$$

so message polynomial

$$m(X) = 1 + X^2 + X^3$$

Now code polynomial

$$\begin{aligned} V(X) &= m(X)\,g(X) \\ &= (1 + X^2 + X^3)(1 + X + X^3) \\ &= 1 + X + X^3 + X^2 + X^3 + X^5 + X^3 + X^4 + X^6 \\ &= 1 + X + X^2 + X^4 + X^5 + X^6 \qquad [\text{using } X \oplus X = 0] \end{aligned}$$

and $V = (1\ 1\ 1\ 0\ 1\ 1\ 1\ 0)$

(*ii*) In systematic form

$$m(X) = 1 + X^2 + X^3,\ n = 7,\ k = 4,\ n - k = 3$$

$$X^{n-k}\,m(X) = X^3\,(1 + X^2 + X^3) = X^3 + X^5 + X^6$$

Dividing $X^{n-k}\,m(X)$ by $g(X)$ using polynomial division, we can write

$$\frac{(X^3 + X^5 + X^6)}{(1 + X + X^3)} = \underbrace{(1 + X + X^2 + X^3)}_{\substack{\text{Quotient} \\ q(X)}} + \underbrace{(1)}_{\substack{\text{Remainder} \\ r(X)}}$$

so message polynomial

$$\begin{aligned} V(X) &= r(X) + X^3 m(X) \\ &= 1 + X^3 + X^5 + X^6 \end{aligned}$$

and code $V = \underbrace{100}_{\text{parity}}\ \underbrace{1011}_{\text{message bits}}$

Example 16. *A cyclic (7, 3) code has the generator polynomial*

$$g(X) = X^4 + X^2 + X + 1 \qquad \text{...}(i)$$

Find the generator matrix for this code. Further find the minimum distance for this code.

Solution. The generator matrix for the given (7,3) cyclic code with the generator polynomial $g(X) = X^4 + X^2 + X + 1$, is given by

$$G = \begin{bmatrix} x^{k-1}\,g(x) \\ x^{k-1}\,g(x) \\ \vdots \\ g(x) \end{bmatrix} = \begin{bmatrix} x^{3-1}\,g(x) \\ x^{3-2}\,g(x) \\ x^{3-3}\,g(x) \end{bmatrix} = \begin{bmatrix} x^2\,g(x) \\ x\,g(x) \\ g(x) \end{bmatrix}$$

$$\Rightarrow \qquad G = \begin{bmatrix} (x^6 + x^4 + x^3 + x^2) \\ x^5 + x^3 + x^2 + x \\ x^4 + x^2 + x + 1 \end{bmatrix}$$

$$\Rightarrow \qquad G = \begin{bmatrix} 1011100 \\ 0101110 \\ 0010111 \end{bmatrix}$$

The above matrix is not in its systematic form. The systematic form of this matrix can be obtained by performing elementary row transformations as follows :

$$R_1 \rightarrow R_1 + R_3$$

$$\Rightarrow \qquad G = \begin{bmatrix} 1001011 \\ 0101110 \\ 0010111 \end{bmatrix}$$

The above matrix can be written as

$$G = [I_k \mid P] = [I_3 \mid P]$$

where
$$I_3 = \begin{bmatrix} 100 \\ 010 \\ 001 \end{bmatrix} \quad \text{and} \quad P = \begin{bmatrix} 1011 \\ 1110 \\ 0111 \end{bmatrix}$$

Equation (*i*) gives the systematic form of the generator matrix for the given (7,3) cyclic code. Since $k = 3$, the data words are (000), (100), (010), (001), (110), (011), (101), (111).

The corresponding codewords are obtained by using the relation $C = dG$ as follows

$$m = (000) \Rightarrow C = (000)\begin{bmatrix} 1001011 \\ 0101110 \\ 0010111 \end{bmatrix} = (0000000)$$

$$m = (100) \Rightarrow C = (100)\begin{bmatrix} 1001011 \\ 0101110 \\ 0010111 \end{bmatrix} = (1001011)$$

$$m = (010) \Rightarrow C = (010)\begin{bmatrix}1001011\\0101110\\0010111\end{bmatrix} = (0101110)$$

$$m = (001) \Rightarrow C = (001)\begin{bmatrix}1001011\\0101110\\0010111\end{bmatrix} = (0010111)$$

$$m = (110) \Rightarrow C = (110)\begin{bmatrix}1001011\\0101110\\0010111\end{bmatrix} = (1100101)$$

$$m = (011) \Rightarrow C = (011)\begin{bmatrix}1001011\\0101110\\0010111\end{bmatrix} = (0111001)$$

$$m = (101) \Rightarrow C = (101)\begin{bmatrix}1001011\\0101110\\0010111\end{bmatrix} = (1110010)$$

$$m = (111) \Rightarrow C = (111)\begin{bmatrix}1001011\\0101110\\0010111\end{bmatrix} = (1110010)$$

Since the minimum distance between any two codewords is 4,therefore the minimum distance of this code is 4.

Example 17. *Construct a systematic (7,4) cyclic code using the generator polynomial*

$$g(x) = x^3 + x^2 + 1.$$

Solution. We have to construct a cyclic code in which the no of data bits are 4 and no of check bits are 3, so that the no of code bits are 7 *i.e.*

$$n = 7 \text{ and } k = 4$$

The generator polynomial is $g(x) = x^3 + x^2 + 1$. The total no of data words possible are $2^4 = 16$. We consider the following data word

$$m = [1010]$$

The corresponding data polynomial is

$$m(x) = x^3 + 0 \cdot x^2 + x + 0 \cdot (1) = x^3 + x$$

Further $\quad x^{n-k}\, m(x) = x^{7-4}\,(x^3 + x)$

$$= x^3\,(x^3 + x) \qquad [\because\ x^3 \cdot x^3 = x^3]$$

$$= x^3 + x^4$$

Hence

$$
\begin{array}{r|l|l}
x^3+x^2+1 & x^6+x^4 & x^3+x^2+1 \leftarrow q(x) \\
 & x^6+x^5+x^3 & \\ \hline
 & \quad x^5+x^4+x^3 & \\
 & \quad x^5+x^4+x^2 & \\ \hline
 & \qquad x^3+x^2 & \\
 & \qquad x^3+x^2+1 & \\ \hline
 & \qquad\qquad 1\ r(x) &
\end{array}
$$

Therefore the code polynomial corresponding to the data polynomial $m(x) = x^3 + x$ is given by

$$V(x) = x^{n-k}\, m(x) + r(x)$$

where $r(x)$ is the remainder obtained on dividing $x^{n-k}\, m(X)$ by the generator polynomial $g(x)$.

$$\therefore \quad V(x) = x^3(x^3 + x) + 1 = x^6 + x^4 + 1$$

The code polynomial can also be obtained by using

$$V(x) = g(x)\, q(x)$$

Where $g(x)$ is the generator polynomial and $q(x)$ is the quotient obtained on dividing $x^{n-k}\, m(X)$ by $g(x)$

$$\therefore \quad V(X) = (x^3 + x^2 + 1)(x^3 + x^2 + 1)$$

$$
\begin{aligned}
\Rightarrow \quad V(X) &= x^3(x^3 + x^2 + 1) + x^2(x^3 + x^2 + 1) + 1(x^3 + x^2 + 1) \\
&= x^6 + x^5 + x^3 + x^5 + x^4 + x^2 + x^3 + x^2 + 1 \\
&= x^6 + (x^5 + x^5) + x^4 + (x^3 + x^3) + (x^2 + x^2) + 1 \\
&= x^6 + x^4 + 1
\end{aligned}
$$

Now the code vector corresponding to the code polynomial

$$V(X) = x^6 + x^4 + 1 = x^6 + 0x^5 + x^4 + 0x^3 + 0x^2 + 0x + 1 \text{ is } (1010001).$$

We can construct the entire code in this manner. As an alternative we can use the generator matrix G for constructing the entire code, the generator matrix G for a (n, k) cyclic code is given by

$$
G = \begin{bmatrix} x^{k-1}\, g(x) \\ x^{k-2}\, g(x) \\ \vdots \\ g(x) \end{bmatrix} = \begin{bmatrix} g_0 g_1 \cdots\cdots g_{n-k}\ 0\,0 \cdots\cdots\cdots 0 \\ 0 g_0 g_1 \cdots\cdots g_{n-k}\ 0 \cdots\cdots\cdots 0 \\ \cdots\cdots\cdots\cdots\cdots\cdots\cdots\cdots \\ 000 \cdots\cdots\cdots g_0 g_1 \cdots\cdots\cdots g_{n-k} \end{bmatrix}
$$

Here $g(x) = x^3 + x^2 + 1$ and $k = 4$, therefore

$$
\Rightarrow \quad G = \begin{bmatrix} x^3 g(x) \\ x^2 g(x) \\ x g(x) \\ g(x) \end{bmatrix} = \begin{bmatrix} x^3(x^3 + x^2 + 1) \\ x^2(x^3 + x^2 + 1) \\ x(x^3 + x^2 + 1) \\ x^3 + x^2 + 1 \end{bmatrix} = \begin{bmatrix} x^6 + x^5 + x^3 \\ x^5 + x^4 + x^2 \\ x^4 + x^3 + x \\ x^3 + x^2 + 1 \end{bmatrix}
$$

$$\Rightarrow \qquad G = \begin{bmatrix} x^6 + x^5 + 0x^4 + x^3 + 0x^2 + 0x + 0 \\ 0x^6 + x^5 + x^4 + 0x^3 + x^2 + 0x + 0 \\ 0x^6 + 0x^5 + x^4 + x^3 + 0x^2 + x + 0 \\ 0x^6 + 0x^5 + 0x^4 + x^3 + x^2 + 0x + 1 \end{bmatrix}$$

$$\Rightarrow \qquad G = \begin{bmatrix} 1\,1\,0\,1\,0\,0\,0 \\ 0\,1\,1\,0\,1\,0\,0 \\ 0\,0\,1\,1\,0\,1\,0 \\ 0\,0\,0\,1\,1\,0\,1 \end{bmatrix}$$

The above matrix is not in its systematic form. To convert it into the standard form. We perform the following operations.

$$R_1 \to R_1 + R_2 + R_3 \qquad R_2 \to R_2 + R_3 \qquad R_3 \to R_3 + R_4$$

These manipulations result in the desired generator matrix.

$$\Rightarrow \qquad G = \begin{bmatrix} 1000 \vdots 110 \\ 0100 \vdots 011 \\ 0010 \vdots 111 \\ 0001 \vdots 101 \end{bmatrix}$$

Now all the code vectors can be obtained using the relation C = dG as follows :

$$m = (0000) \Rightarrow C = (0000) \begin{bmatrix} 1000110 \\ 0100011 \\ 0010111 \\ 0001101 \end{bmatrix} = (0000000)$$

$$m = (0001) \Rightarrow C = (0001) \begin{bmatrix} 1000110 \\ 0100011 \\ 0010111 \\ 0001101 \end{bmatrix} = (0001101)$$

$$m = (0010) \Rightarrow C = (0010) \begin{bmatrix} 1000110 \\ 0100011 \\ 0010111 \\ 0001101 \end{bmatrix} = (0010111)$$

$$m = (0100) \Rightarrow C = (0100) \begin{bmatrix} 1000110 \\ 0100011 \\ 0010111 \\ 0001101 \end{bmatrix} = (0100011)$$

$$m = (1000) \Rightarrow C = (1000)\begin{bmatrix}1000110\\0100011\\0010111\\0001101\end{bmatrix} = (1000110)$$

$$m = (0011) \Rightarrow C = (0011)\begin{bmatrix}1000110\\0100011\\0010111\\0001101\end{bmatrix} = (0011010)$$

$$m = (0101) \Rightarrow C = (0101)\begin{bmatrix}1000110\\0100011\\0010111\\0001101\end{bmatrix} = (0101110)$$

$$m = (0110) \Rightarrow C = (0110)\begin{bmatrix}1000110\\0100011\\0010111\\0001101\end{bmatrix} = (0110100)$$

$$m = (1001) \Rightarrow C = (1001)\begin{bmatrix}1000110\\0100011\\0010111\\0001101\end{bmatrix} = (1001011)$$

$$m = (1010) \Rightarrow C = (1010)\begin{bmatrix}1000110\\0100011\\0010111\\0001101\end{bmatrix} = (1010001)$$

$$m = (0111) \Rightarrow C = (0111)\begin{bmatrix}1000110\\0100011\\0010111\\0001101\end{bmatrix} = (0111001)$$

$$m = (1011) \Rightarrow C = (1011)\begin{bmatrix}1000110\\0100011\\0010111\\0001101\end{bmatrix} = (1011100)$$

$$m = (1100) \Rightarrow C = (1100)\begin{bmatrix}1000110\\0100011\\0010111\\0001101\end{bmatrix} = (1100101)$$

$$m = (1101) \Rightarrow C = (1101)\begin{bmatrix}1000110\\0100011\\0010111\\0001101\end{bmatrix} = (1101000)$$

$$m = (1110) \Rightarrow C = (1110)\begin{bmatrix}1000110\\0100011\\0010111\\0001101\end{bmatrix} = (1110010)$$

$$m = (1111) \Rightarrow C = (1111)\begin{bmatrix}1000110\\0100011\\0010111\\0001101\end{bmatrix} = (1111111)$$

The Hamming minimum distance between any two codewords is 3. Hence this is a single error correcting code and 14 of these codewords can be obtained by successive cyclic shifts of two codewords 1110010 and 1101000. The remaining two codewords 1111111 and 0000000 remain unchanged under cyclic shift.

Example 18. *The generator polynomial for a (15, 7) cyclic code is*

$$g(x) = 1 + x^4 + x^6 + x^7 + x^8$$

(i) Find the code vector (in systematic form) for the message polynomial $D(x) = x^2 + x^3 + x^4$.

(ii) Assume that the first and last bits of the code vector $V(x)$ *for* $D(x) = x^2 + x^3 + x^4$ *is in error.*

Solution. Given (15, 7) cyclic code

$$g(x) = 1 + x^4 + x^6 + x^7 + x^8$$

(*i*) V = ?

$$\frac{x^{n-k}\,D(x)}{g(x)} \Rightarrow \frac{x^8\,(x^2 + x^3 + x^4)}{1 + x^4 + x^6 + x^7 + x^8}$$

$$\Rightarrow \frac{x^{10} + x^{11} + x^{12}}{1 + x^4 + x^6 + x^7 + x^8}$$

$$
\begin{array}{r}
x^4 \quad +1 \\
1 + x^4 + x^6 + x^7 + x^8 \overline{\big) x^{10} + x^{11} + x^{12}} \\
x^4 + x^8 + x^{10} + x^{11} + x^{12} \quad \left[\because \quad x^{10} \oplus x^{10} = 0\right] \\
\hline
x^4 + x^8 \quad \left[\because \quad x^{11} \oplus x^{11} = 0\right] \\
1 + x^4 + x^6 + x^7 + x^8 \quad \left[\because \quad x^{12} \oplus x^{12} = 0\right] \\
\hline
1 + x^6 + x^7 \\
\hline
r
\end{array}
$$

Now $\quad V = \dfrac{(1000\,0011\,0011100)}{4 \qquad D}$

(*ii*) Assume that last and first bits suffer from transmission errors.

By changing last and first bit

$$V = 0\,0\,0\,0\,0\,0\,1\,1\,0\,0\,1\,1\,1\,0\,1$$

is $\rightarrow$

$$\therefore \quad V(x) = x^6 + x^7 + x^{10} + x^{11} + x^{12} + x^{14}$$

$$
\begin{array}{r}
x^{10} + x^9 + x^8 + x^7 + x^6 + x^4 \\
x^2 + x^3 + x^4 \overline{\big) x^6 + x^7 + x^{10} + x^{11} + x^{12} + x^{14}} \\
x^{12} + x^{13} + x^{14} \\
\hline
x^6 + x^7 + x^{10} + x^{13} + x^{11} \\
x^{11} + x^{12} + x^{13} \\
\hline
x^6 + x^7 + x^{10} + x^{12} \\
x^{10} + x^{11} + x^{12} \\
\hline
x^6 + x^7 + x^{11} \\
x^9 + x^{10} + x^{11} \\
\hline
x^6 + x^7 + x^9 + x^{10} \\
x^8 + x^9 + x^{10} \\
\hline
x^6 + x^7 + x^8 \\
x^6 + x^7 + x^8 \\
\hline
\end{array}
$$

Remainder = 0

9.7 CONVOLUTIONAL CODING

Now we will discuss one more class of forward error request codes. In the case of block codes ; we take input message bits as a block ; for example (7, 4) encoder take (4) bits of message at a time so we use buffer to take them in the encoder. In the case of convolutional coding ; the message bits are taken serially.

This type of coding is very much important and preferred for space and satellite communication.

9.7.1 Convolutional Encoding at Code Rate = 1

Here we have drawn a fundamental unit for convolutional encoding using a tapped shift resistor.

If we see the diagram, the shown present state is of L bits then the length of shift resistor must be (L + 1). Each tap gain g is a binary digit representing a short circuit connection 'or' an open circuit. The message bits in the resister are combined by module 2 addition to for a encoded coded output.

$$x_j = m_{j-L}\, g_L \oplus \ldots \oplus m_j - 1\, g_1 \oplus m_j\, g_0 \tag{9.38}$$

$$x_j = \sum_{i=0}^{L} m_{j-i}\, g_i \tag{9.39}$$

The name convolutional encoding comes from the fact that equation has a form of digital (discrete) convolutional. Now it should be cleared that for a incoming message bit ; this bit will influence the encoded output for (L + 1) shifts totally for a single output bit at a time).

For example, consider the case given below.

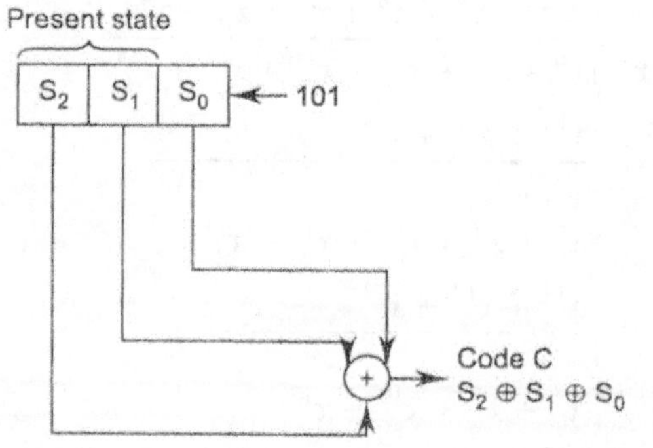

Fig. 9.7 *(a) Convolutional encoding at code rate = 1*

at rate = 1

take initially $S_0, S_1, S_2 = 0\ 0\ 0$;

Cycle / Shifts	S_2	S_1	S_0	Code
0	0	0	0	0
1	0	0	1	1
2	0	1	0	1
3	1	0	1	0
4	0	1	0	1
5	1	0	0	1

So, we can say that if we have totaly present state = L.

(*a*) Any message bit will influence output code for (L + 1) cycle.

In our example for L = 2

Every message bit influence the output encoded code for = L + 1

= 2 + 1

= 3 states as shown in figure.

(*b*) Point (*a*) discussed also shows that we should encode data for complete output untill the LSB bit of data (message) has arrived at the MSB bit of shift resister.

(After that we will arrive again at 0 shift state).

9.7.2 Convolution Encoding for Code Rates = $\frac{1}{n}$

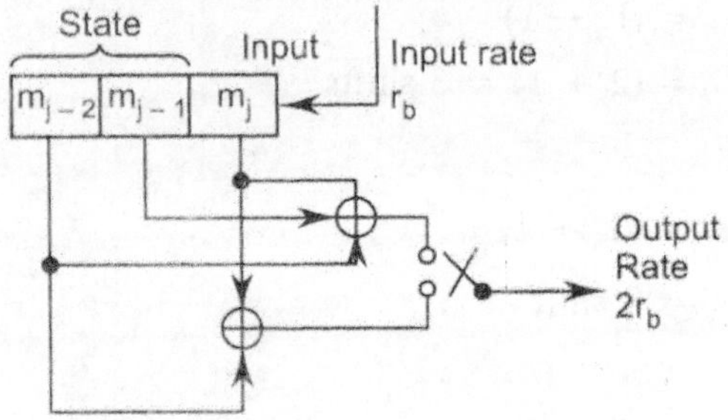

Fig. 9.8 *Convolutional encoder for code rate* = $\frac{1}{n}$

In our example,

$$n = \text{Output code bit} = 1 \text{ at a shift}$$

$$k = \text{Input code bit} = 1 \text{ at a shift}$$

$$\text{So code rate} = \frac{k}{n} = 1.$$

Now to provide extra bits needed for error control ; a complete convolutional encoder must generate output bits at a higher rate than the message rate. suppose that at output in a shift (one second) we can get 2 bits, and at input in a shift (one second), we can get only 1 bit, then we have flexibility to add 1 bit for error control along with 1 bit of message so we can transfer data at 2 bits/second speed without any problem.

To achieve higher output bit rate two 'or' more Module-2 summers to the register. Now we will get more than one output encoded bits as shown in the following figure so they are interleaved by using a commutator switch.

For example, for the encoder shown in the following figure; generate (n = 2) encoded output bits.

$$\left.\begin{aligned} x_j &= m_{j-2} \oplus m_{j-1} \oplus m_j \\ x'_j &= m_{j-2} \oplus m_j \end{aligned}\right\} \qquad (9.40)$$

Now, the interleaved output stream produces code

$$X = x_1\ x'_1\ x_2\ x'_2\ x_3\ x'_3 \ldots$$

at one cycle output $X = x_1\ x'_1$

So output code rate $= 2r_b$ for input data rate r_b.

So we have code rate $= \dfrac{k}{n} = \dfrac{r_b}{2r_b} = \dfrac{1}{2} = \dfrac{1}{n}$.

In this case each input bit will effect the each output coded bit for (L + 1) shifts as discussed above ; but now the output does remain one.

So x_j will be effect by $= (L + 1)$

$= (2 + 1) = 3$ shifts

due to input data bits.

and x'_j will also be effected by $= (L + 1)$

$= 3$ shift

due to input data bit.

Thus, now a single bit of input data / message, will effect the output encoded code for n (L + 1) codes shifts. In n (L + 1) shifts we will get n (L + 1) output codes.

So for a shift resister having total present state = L.

The total shifts influenced by a single bit of message = Encoded output codes influenced by a single bit of message = n (L + 1)

Now we have to perform the coding for n (L + 1) shifts. Here n (L + 1) is called constraint length

Constraint length = n (L + 1)

Now for single output $n = 1$

Influence of input messasge bit at output code $= n$ (L + 1)

$= (L + 1)$

as discussed in the point (a). So point (a) is a special case of this general formulation, that is for a single bit output at a time.

9.7.3 Encoding for Code Rate Higher than $\left(\frac{1}{n}\right)$

Consider the case when code rate was required $= \left(\dfrac{1}{n}\right)$

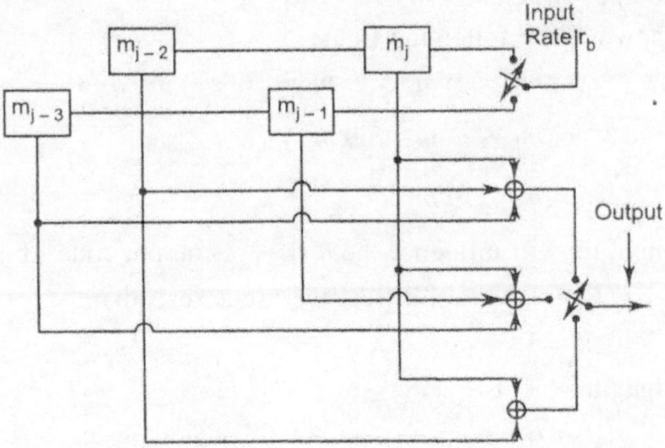

Fig. 9.9 *Code rate > 1/n*

Then, input bit rate $k = r_b$

and output bit rate $n = 2r_b$, output bit rate was achieved by taking two output interleaved by commutator switch at output. So,

$$\text{Input data} = r_b.$$

$$\text{Number of Input} = 1 \qquad \text{Input rate} = 1 \,.\, r_b = r_b$$

$$\text{Number of Output} = n \qquad \text{Output rate} = n \,.\, r_b$$

$$\text{Then, code rate} = \frac{r_b}{nr_b} = \frac{1}{n}$$

Again for

$$\text{Number of input} \geq 2 \qquad \text{Input rate} \geq 2 \,.\, r_b$$

$$\text{Number of output} = n \qquad \text{Output rate} = n \,.\, r_b$$

$$\text{Then code rate} \geq \frac{2r_b}{nr_b}$$

or $$\text{code rate} \geq \frac{2}{n}$$

So now code rate is greater than $\frac{1}{n}$.

So we can say that to get code rate higher than $\frac{1}{n}$; we have to apply a commulator switch at input also so that input is distributed in more than one path. Now follow the following procedure for $\frac{2}{3}$ code rate.

(*a*) Apply a commutator switch at input that distribute input between 2 input resistor.

(*b*) Now along with each input resister, arrange the shift resistor of length L + 1. (One shift resister + one input resister).

(*c*) Take 3 output by output switch.

In our example, shown in the following figure.

$$x_j = m_{j-3} \oplus m_{j-2} \oplus m_j$$

$$x'_j = m_{j-3} \oplus m_{j-1} \oplus m_j \qquad (9.41)$$

$$x''_j = m_{j-2} \oplus m_j.$$

In this case each input bit will influence the n (L + 1) output code ; but now

L = 1 (One shift resistor / state resistor)

$n = 3.$

So constraint length = 3 (1 + 1)

= 6.

9.8 CODE TREE

Three different but related graphical representations has been divised for the study of convolutional encoding. These three graphical representations are named as code tree, code trellis and state diagrams.

Now we will study first graphical representation named by code tree. For study of code tree of our example, we will follow the following standard method :

(*a*) Each branch of code tree represents the input symbol.

(*b*) As in binary system, input symbol can be either '1' or'0'. So the convolutional code tree distinguish the binary symbols '0' and '1' as follow —

"A input '0' is specifies the upper branch of a bifurcation, while input 1 specifies the lower branch".

(*c*) Code tree propagate from left to right as the input sequence is serially applied.

(*d*) At the each branch, we have to indicate the output binary symbol.

Now keeping all these points in our mind we will proceed for our example :—

Let at the initial all the bits of convolutional encoder are reset. Now as indicated in the following figure ; we have total 2 bits that are represented by the state of code tree ;

So, we have states

0 0, 0 1, 1 0, 1 1 now for simplicity, we take a abbreviation of these 4 codes (states)

0 0	*a*
0 1	*b*
1 0	*c*
1 1	*d*

Now we have to take each state and discuss about the output of encoder along with the next state of the encoder.

Supposing at starting all the register value of encoder is 0 0 0 ; it should be noted that location of S_2 S_1 shows the value of present state ; now if we want to find the next state we have to shift resister contained in left hand direction by one so we can say that now

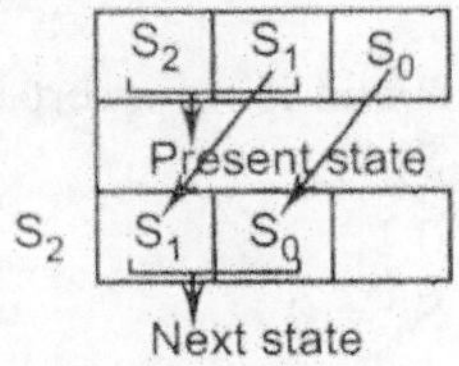

For present state $S_2\ S_1$. The next state will be $S_1\ S_0$;

Now we will discuss the all possible présent state ; for these present state will calculate output for binary inputs '0' and '1' and for here binary input the next state.

(*a*) When the present state = 0 0 (*a*)

(*i*) The present state ($S_2\ S_1$) is 0 0,.now if we insert the input bit (S_0) '0'. Then the content of shift resister is total

$$\text{Present state } S_2\ S_1 = 0\ 0\ (a)$$

$$\text{Input bit } S_0 = 0$$

$$\text{So content of shift resister} = S_2\ S_1\ S_0$$

$$= 0\ 0\ 0$$

$$\text{So output bit } C_1 = S_2 \oplus S_1 \oplus S_0 = 0$$

$$C_2 = S_2 \oplus S_0$$

$$= 0.$$

$$\text{So output encoded code} = C_1\ C_2$$

$$= 0\ 0$$

$$\text{and next state} = S_1\ S_0 = 0\ 0\ (a).$$

So, if present state is 0 0 and if we insert input bit '0', the output of encoder is 0 0 and next state itself 0 0.

(*ii*) The present state is 0 0, now if we insert the input bit '1', then the

$$\text{Present state } S_2\ S_1 = 0\ 0$$

$$\text{Input bit } S_0 = 1$$

$$\text{So the content of shift resister} = S_2\ S_1\ S_0$$

$$= 0\ 0\ 1$$

$$\text{So output bit } C_1 = S_2 \oplus S_1 \oplus S_0$$

$$= 1$$

$$C_2 = S_2 \oplus S_0$$

$$= .1$$

$$\text{So output encoded} = C_1\ C_2$$

$$= 1\ 1$$

$$\text{and next state} = S_1\ S_0$$

$$= 0\ 1\ (b)$$

So, if present state is '0 0' and if we insert input bit '1', the output of encoder is '1 1' and next state 0 1.

(*b*) When the present state is 0 1 (*b*) :

(*i*) If the present state (S_2 S_1) is '0 1' and if we insert input bit '0', then the

Present state = S_2 S_1
= 0 1
Input bit = S_0 = 0
Content of shift resister = S_2 S_1 S_0
= 0 1 0
So the output bit C_1 = 1
C_2 = 0
So output encoded code = C_1 C_2
= 10
and next state of shift resistor = S_1 S_2
= 10 (*c*)

> So, if the present state is 0 1 and if we insert input bit '0', the output of encoder is '1 0' and next state is 1 0.

(*ii*) If for present state S_2 S_1 = 0 1 ; we insert input bit '1', then the

Present state = S_2 S_1
= 0 1
Input bit = S_0 = 0 1
Content of shift resistor = S_2 S_1 S_0
= 0 1 1
So the output bit C_1 = 0
C_2 = 1
So encoded output = C_1 C_2
= 0 1
and next state of shift resistor = S_1 S_0
= 1 1 (*d*)

> So, if the present state is 0 1 and if we insert input bit '1', the output of encoder is '0 1' and next state is 1 1.

(*c*) When the present state = 10 (*c*) :

(*i*) If the present state is S_2 S_1 = 10 and if we insert input bit S_0 = 0, then the

Present state = S_2 S_1
= 10
Input bit = S_0 = 0
Total content of shift resister = S_2 S_1 S_0
= 1 0 0
So the output bit C_1 = 1
C_2 = 1
So output encoded = C_1 C_2 = 1 1
and next state = S_1 S_0
= 0 0 (*a*)

So, if the value of present state is '10', if we insert input bit '0', the output of encoder is 1 1 and next state is 0 0.

(*ii*) If the present state is $S_2\ S_1 = 10$ and if we insert input bit $S_0 = 1$; then the

$$\text{Present state} = S_2\ S_1 = 10$$

$$\text{Input bit} = S_0 = 0\ 1$$

$$\text{Total content of shift resister} = S_2\ S_1\ S_0 = 1\ 0\ 1$$

$$\text{So the output bit } C_1 = 0$$

$$C_2 = 0$$

$$\text{So output encoded} = C_1\ C_2 = 0\ 0$$

$$\text{and next state} = S_1\ S_0 = 0\ 1\ (b)$$

So, if the value of present state is '10' and if we insert input bit '1' the output of encoder is '0 0' and next state is '0 1'.

(*d*) When the present state 1 1 (*d*) :

(*i*) If the present state is $S_2\ S_1 = 1\ 1$ and if we insert input bit $S_0 = 0$, then the

$$\text{Present state} = S_2\ S_1 = 1\ 1$$

$$\text{Input bit} = S_0 = 0$$

$$\text{Total content of shift resister} = S_2\ S_1\ S_0 = 1\ 1\ 0$$

$$\text{So the output bit } C_1 = 0$$

$$C_2 = 1$$

$$\text{So output encoded} = C_1\ C_2 = 0\ 1$$

$$\text{and next state} = S_1\ S_0 = 1\ 0\ (c).$$

So, if the present state is 11 and if we insert input bit '0', then the output of encoder is '01' and next state is '1 0'.

(*ii*) If the present state is $S_2\ S_1 = 11$ and if we insert the input bit $S_0 = 0$, then

$$\text{Present state} = S_2\ S_1 = 1\ 1$$

$$\text{Input bit} = S_0 = 1$$

$$\text{Total content of shift resister} = S_2\ S_1\ S_0 = 1\ 1\ 1$$

$$\text{So output bit } C_1 = 1$$

$$C_2 = 0$$

$$\text{So output encoded} = C_1\ C_2 = 10$$

$$\text{Next state} = S_1\ S_0 = 11\ (d).$$

If the present state is '11' and if we insert input bit '1', then the output of encoder is '10', and the next state is '11'.

The all process has been shown in tabular form. Now combined all the parts and make a code tree as shown in Fig. 9.10.

Table 9.1 Convolution coding

	S_2 S_1 S_0 0 0 0 Present state	$C_1 = S_2 \oplus S_1 \oplus S_0$ $C_2 = S_2 \oplus S_0$ Output = C_1 C_{12}	S_2 S_1 S_0 [ ] [ ] [ ] Next state = S_1 S_0	
a = 0 0	0 ← 0 ← 0 ← 0 S_2 S_1 S_0 0 0 0	$C_1 = 0 \oplus 0 \oplus 0$ $= 0$ $C_2 = 0 \oplus 0$ $= 0$ Output = C_1 C_2 = 0 0	Next state = S_1 S_0 = 0 0 0 0 0 (= a)	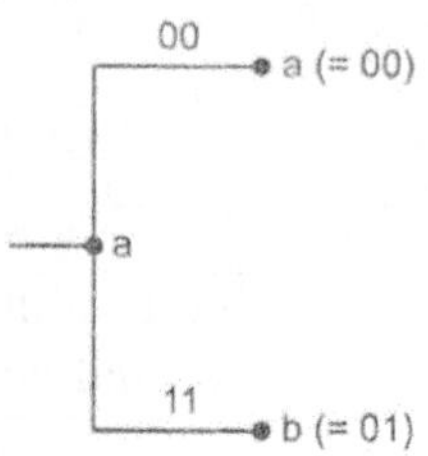
	0 ← 0 ← 0 ← 1 S_2 S_1 S_0 0 0 1	$C_1 = 0 \oplus 0 \oplus 1$ $= 1$ $C_2 = 0 \oplus 1$ $= 1$ Output = 1 1	Next state = S_1 S_0 = 01(= b) 0 0 1	
b = 0 0	0 ← 0 ← 1 ← 0 0 1 0	$C_1 = 1$ $C_2 = 0$ Output = 1 0	Next state = (10 = c) 0 1 0	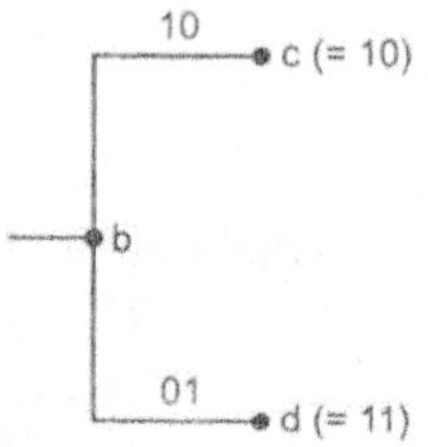
	0 ← 0 ← 1 ← 1 0 1 1	$C_1 = 0$ $C_2 = 1$ Output = 0 1	Next state = 11 (= d) 0 1 1	
c = 0 0	0 ← 1 ← 0 ← 0 1 0 0	$C_1 = 1$ $C_2 = 1$ Output = 1 1	Next state = 00 (= a) 1 0 0	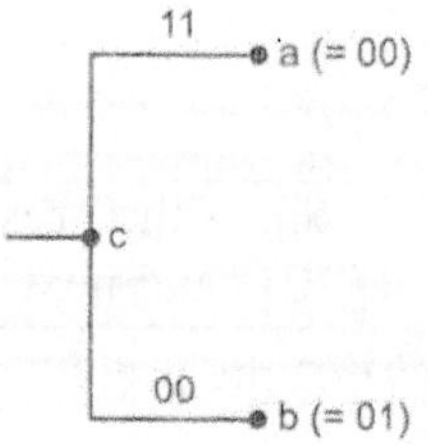
	0 ← 1 ← 0 ← 1 1 0 1	$C_1 = 0$ $C_2 = 0$ Output = 0 0	Next state = 01 (= b) 1 0 1	
d = 1 1	0 ← 1 ← 1 ← 0 1 1 0	$C_1 = 0$ $C_2 = 1$ Output = 0 1	Next state 10 (= c) 0 1 0	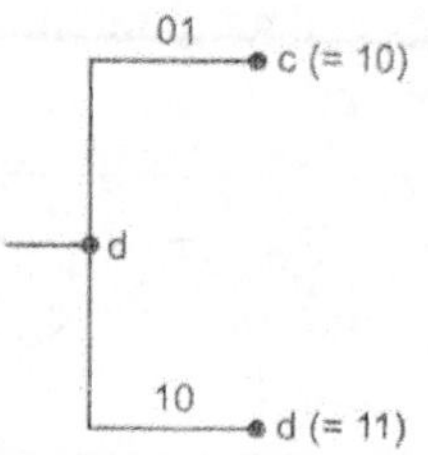
	0 ← 1 ← 1 ← 1 1 1 1	$C_1 = 1$ $C_2 = 0$ Output = 1 0	Next state 11 (= d) 0 1 1	

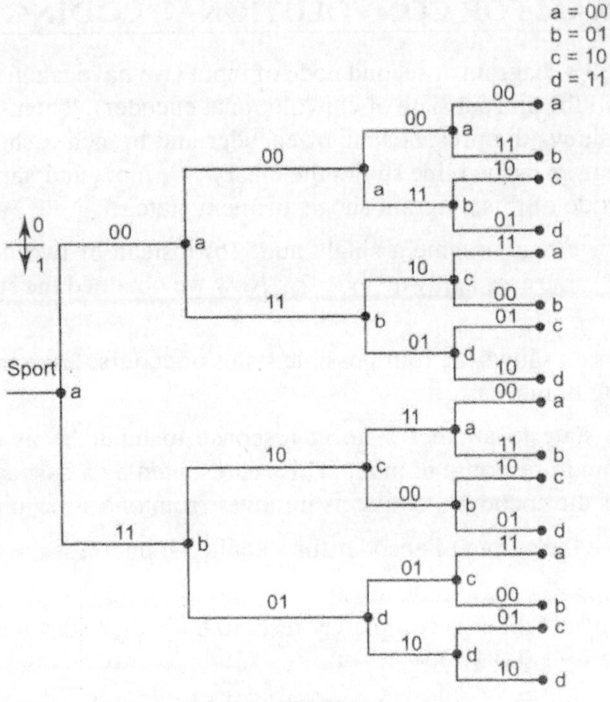

Fig. 9.10 *Code tree*

9.9 CODE TRELLIS FOR CONVOLUTIONAL ENCODER

Code trellis is nothing but more compact representation of code tree. In the code tree, for differentiate the input bit '0' and '1', the output due to inserting input bit '0', we have drawn upper branch at the point of bifercation. If we applied input bit '1', we have drawn lower branch at the point of bifercation as discussed in the case of code tree. But now, as the complete diagram is place on a planer surface, so we have differentiate the input (0) and input (1) as :

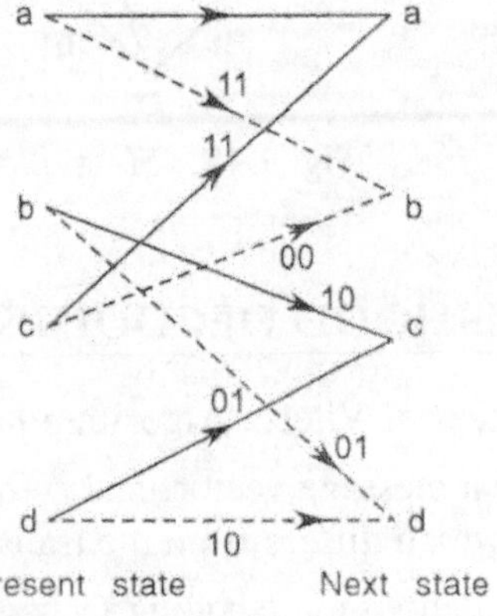

Fig. 9.11 *Code trellis*

(*i*) Input bit '0' by drawing a solid line.

(*ii*) Input bit '1' by drawing a dashed line.

9.10 STATE DIAGRAM FOR CONVOLUTIONAL CODING

Now consider a part of trellis diagram at second node of input (we have taken second node 'or' upper node so that it can cover all the present state of convolutional encoder). Then see the figure, it is clear that left side of this past shows the present state of encoder and branches shows the binary input as discussed in the trellis diagram dashed line shows the binary '1', input and hard line shows the binary '0' input. Now the right side of this diagram shows the next state.

It should be clear that we may assume a single node (*b*) instead of two nodes '*b*' (because both nodes are identical). Now same case apply at node '*c*'. Now we obtained the state diagram of encoder shown in Fig. 9.12.

The nodes of state diagram shows the four possible states of coders. Each node have two incoming branches and two outgoing branches.

A transition from one state to another state in response to input '0' is represented by a solid branches, while a transition in response to input '1' is represented by a dashed line. The binary label on each branch represents the encoder's output as it moves from one state to another state.

We can easily determine the output of encoder for a applied input message sequence by using state diagram.

If we assume that initially encoder is completely reset (0 0 0), then start with the node (*a*). Now if the applied input message is (1 0 0 1 1).

Now from state '*a*' we will go to state (*b*) through dashed line due to applied input '1'.

Again go from state '*b*' to state '(*c*)' due to applied signal '0'. Thus for message sequence (10011) followed path is '*abcabd*'. That causes output sequence (11, 10, 11, 11, 01) which agree results that encoded easier using code tree 'or' trellis diagram. Thus, input output relation of convolutional encoder is completely described by state diagram.

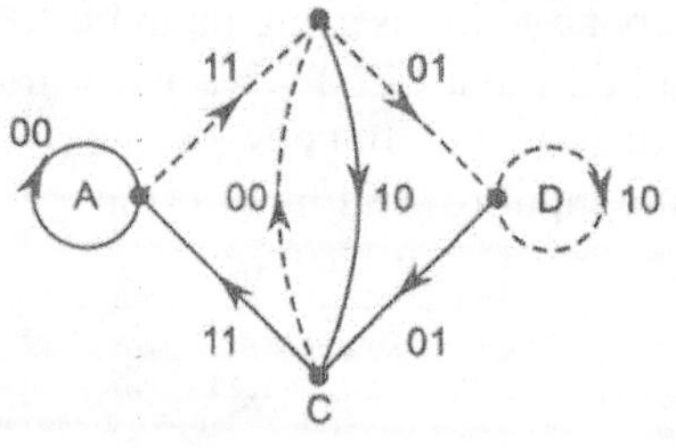

Fig. 9.12 *State diagram*

9.11 MAXIMUM-LIKELIHOOD DECODING OF CONVOLUTIONAL CODE

Here we will develop the concept of Viterbi Algorithm for the decoding of convolutional code.

Now start with let m denote a message vector, and r - denote the corresponding code vectors. Let y denote the received vector, which differs from the transmitted code vector due to channel noise.

Now our main purpose is to choose a decoding technique such that decoded y is very closely to transmittetd x.

Thus the probability of getting y, when x has been transmitted.

$$= P(y/x) \tag{9.42}$$

If we assume that all the equiprobable messages, then the minimum decoding error will occur, if we choose maximum likehood function that is given by :

$$= \log_e P(y/x) \tag{9.43}$$

If we have totally message sequence of N length ; let x_i and y_i denote the ith elements of x and y respectively then we have :

$$P(y/x) = \prod_{i=1}^{N} P(y_i/x_i) \tag{9.44}$$

Correspondingly, the log likelihood function is given by

$$\log_e P(y/x) = \prod_{i=1}^{N} \log_e (y_i/x_i) \tag{9.45}$$

Now suppose that 'y' differs from 'x' (due to error) in exactly d positions, where d is the hamming distance between vector 'x' and 'y'. Then likelihood function becomes,

$$\boxed{\log_e P(y/x) = d \log_e P + (N - d) \log_e (1 - P)} \tag{9.46}$$

$$= d \log_e \left(\frac{P}{1-P}\right) + N \log_e (1 - P) \tag{9.47}$$

So we can say that N $\log_e$ (1 – P) is a constant term, if we want to decrease the error and increase the maximum likelihood probability, choose 'y' for a particular 'x' in such a way that decoding of 'y', provide you the code that has minimum hamming distance with x.

So we can say that maximum likelihood decoder is working as minimum distance decoder.

This is the basic concept of Viterbi Algorithm, that now for a particular received code y, compare 'y' with each possible transmitted code 'x' and choose 'closest' one (so that hamming distance is minimum).

So now take code tree 'or' trellis diagram of a particular code pattern ; and perform the following step.

(*a*) At each node of code tree / trellis diagram, you will get more than one path.

(*b*) So compute the matrix for each possible path.

(*c*) The matrix for a particular path is defined as the hamming distance between the coded sequence represented by the path and the received sequence.

(*d*) Now discard the path of maximum hamming distance and choose only the minimum hamming distance path.

Only problem may arise when the paths entering a states are compared and they have identical matrix ; then it's up to you to choose one and discard other.

9.12 SEQUENTIAL DECODING

The greatest advantage of Viterbi Algorithm is that for a constraint length k, code rate $\frac{k}{n}$ convolutional code, the number of operation performed in decoding L bits is L $2^{n\,(k-1)}$.

So it is linear with length L. But on the other hand disadvantage is that it is expotenitally depends upon constraint length k ; so it cannot be good for ($k \geq$ for 1 1).

To overcome this disadvantage, (basically to reduce computation complexicity) we use sequential decoding technique. Here we will see that sequential decoding does not depend upon the constraint length k so can be used for large k.

Sequential decoding is an trial and error technique for searching out the correct path in a code tree. During the course of this search, the decoder moves forward 'or' backward in code tree, one node at a time.

Fano Algorithm : Several algorithms have been developed for decoding using sequential decoding. But fano algorithm is important one and also requires less memory space.

To perform this algorithm, first of all take a node called (origin node) at the code tree. From this node, the decoder moves forward in the tree decoder has maintained a thresold (that is integer multiple of thresold spacing Δ). If at that node, the value of path matrix is greater than this thresold value then decoder except that node. Now if thresold value is less than path matrix, than it moves backward.

Now just take a example.

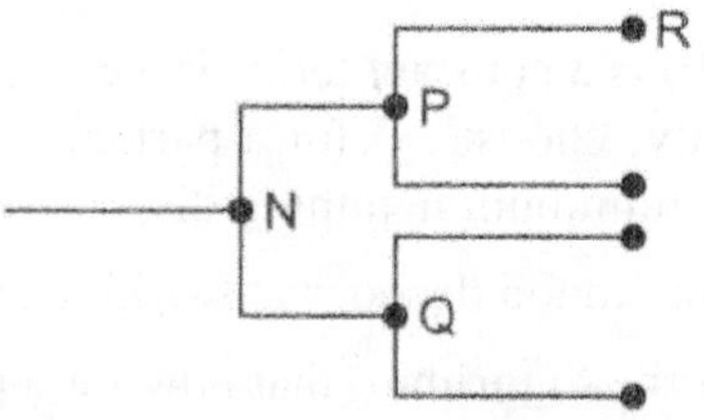

Fig. 9.13 *Sequential coding*

Suppose that the decoder has traced a path up to Node P and T_e denote the current running thresold. Firstly, decoder tries to make a forward move. In particular, it loops forward to the 'best' of the two succeeding nodes. Let R be the best node, viewed from node P. If the path matrix of node R is greater than the thresold value T_e, the decoder moves forward to node R. If the node 'R' is not satisfying this condition, then decoder loops backward to the proceeding node N. Again at node 'N' it will check the path matrix of node 'N' is greater than thresold or not, if yes then OK ; otherwise if will move side node Q.

At a special case if at node 'R' and 'N' both have path matrix below then thresold, then it is said that thresold is loosened ; and now decoder reduces its thresold by reducing the thresold value.

REVIEW QUESTIONS

1. What are the various concepts of error controlling codes?
2. What do you understand by block codes ? Differeniate between linear Block codes & Hamming codes.
3. What is the syndrome calculator ? Define & explain the technique of decoding of Block codes using syndrome calculator.
4. What is systematic cyclic code ?
5. Take your own example and draw the code tree for that.
6. What do you understand by
   (*a*) Code tree
   (*b*) State diagram for decoding of convolutional code.
   (*c*) Code trellis for decoding of convolutional code.
7. Write the sequential decoding for convolutional code.
8. How does the convolutional code differs from generally used other codes?
9. The polynomial $1 + X^7$ has $1 + X + X^3$ and $1 + X^2 + X^3$ as primitive factors. We used $1 + X + X^3$ as the generator polynomial for a (7, 4) Hamming polynomial. This should lead to a (7,4) Hamming code that is different from the code analyzed. Develop the encoder and syndrome calculator for the generator polynomial:
$$g(X) = 1 + X^2 + X^3$$
Compare your results with those in Example ?
10. Consider the (7, 4) Hamming code defined by the generator polynomial
$$g(X) = 1 + X + X^3$$
The codeword 0111001 is sent over a noisy channel, producing the received word 0101001 that has a single error. Determine the syndrome polynomial $s(X)$ for this received word, and show that it is identical to the error polynomial $e(X)$.
11. For a(n, k) linear block code, how would you distinguish between Hamming distance and parity check code?
12. Suppose we know H-matrix, can we find the Code-matrix? If yes, which relationship should we use?
13. Consider a (7, 4) block code, the generator matrix is,
$$G = \begin{bmatrix} 1000 & 111 \\ 0100 & 110 \\ 0010 & 101 \\ 0001 & 011 \end{bmatrix}$$
Find: (*a*) all the code vectors of this code,
   (*b*) the parity check matrix for this code,
   (*c*) the minimum weight of this code,
   (*d*) verify whether a codeword C is generated by G and find the Syndrome S in the process.
14. Design a rate ½ Convolutional encoder with a constraint length $v = 4$ and $d = 6$.
   (*i*) Construct the State Diagram for this encoder.
   (*ii*) Construct the Trellis Diagram for this encoder.
   (*iii*) What is the d_{free} for this code?
   (*iv*) Give the Generator Matrix, G.
   (*v*) Is this code Non-Catastrophic? Why?
15. Design a (12, 3) systematic convolutional encoder with a constraint length $v = 3$ and $d \geq 8$.

III B.E. (VI Semester) (Main/Back) EXAMINATION, 2007

(New four year Semester Scheme)
DIGITAL COMMUNICATION
[Branch: Electronics and Comm. Engineering]

Time: Three Hours **Max. Marks:** 80

Attempt any five questions. All questions carry equal marks.

Q1. (*a*) What is meant by noise figure and how can it be calculated for an amplifier or a receiver?

(*b*) Explain the concept of effective noise temperature of a network. An antenna is connected to a receiver having a noise figure of 3dB. The available power gain of the receiver is 80 dB and the noise-equivalent-bandwidth is 10 MHz. If the available output noise power is 10u.w, determine the equivalent antenna (noise) temperature.

(*c*) Determine the equivalent noise temperature of the following low noise receiving system. Assume room temperature of the following low noise receiving system. Assume room temperature to be 290°K, low noise amplifier gain = 25 dBs.

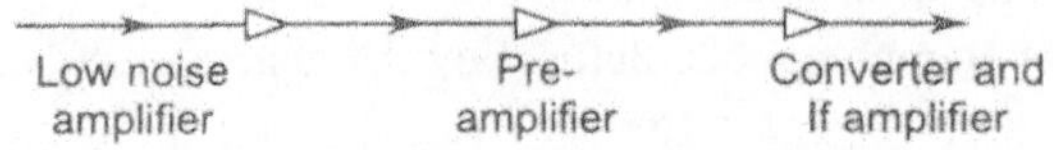

Fig.1 *Low noise amplifier noise temp. = 4° K, preamplifier gain = 17 dBs. preamp noise figure = 6 dBs. converter and if amplifier noise figure = 12dBs.*

Ans. (*a*) **Noise Figure :** The calculation of the equivalent noise resistance of an amplifier, receiver or temperature may have either of two purposes. The first purpose is comparison of two kinds of evaluating their performance. The second purpose is comparison of two kinds of evaluating their performance. The second is comparison of noise and signal at the same point to ensure that the noise is not excessive, in the second instance, and also when equivalent noise resistance

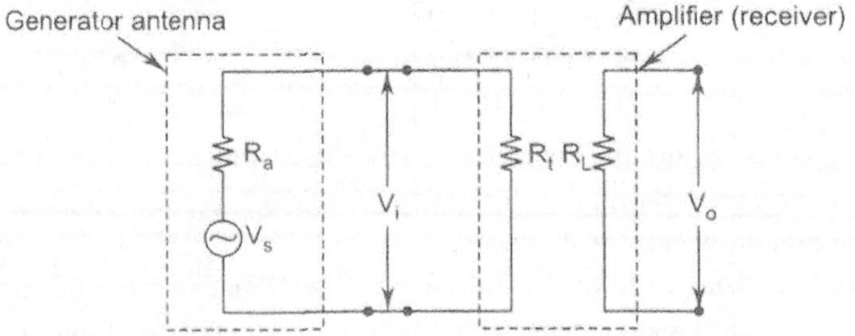

Fig. 2 *Block diagram for noise figure calculation*

is difficult to obtain the signal to noise ratio $\left(\frac{S}{N}\right)$ is very often used

$$\frac{S}{N} = \frac{X_s}{X_n} = \frac{V_s^2/R}{V_s^2/R} = \left(\frac{V_s}{V_n}\right)^2 \text{ or } \left[F = \frac{(S/N)_{in}}{(S/N)_{o/p}}\right]$$

Calculation of Noise Figure

The calculation procedure may be broken down into a number of general steps.

- Determine the signal input power P_{si}
- Determine the noise input power P_{ni}
- Calculate the input signal to noise ratio $\frac{S}{N_i}$ from the ratio of P_{si} and P_{ni}
- Determine the signal output power P_{si}
- Write P_{no} for the noise output power to be determined later.
- Calculate the output signal to noise ratio (S/N_o) from the ratio of P_{so} and P_{no}.
- Calculate the generalized from of Noise figure from step (*iii*) and (*vi*).
- Calculate P_{no} from R_{eq} of possible and substitute into the general will be

$$V_{si} = \frac{V_s R_t}{R_a + R_t}$$

$$R_{si} = \frac{V_{si}^2}{R_f} = \left(\frac{V_s R_f}{R_a + R_t}\right)^2 \times \frac{1}{R_t} = \frac{V_s^2 R_t}{(R_a + R_t)^2}$$

similarly the noise input voltage and power will be $V_{ni}^2 = 4kT\,\Delta f\,\frac{R_a R_t}{R_a + R_t}$

$$P_{ni} = \frac{V_{ni}^2}{R_t} = 4kT\,\Delta f\,\frac{R_a R_t}{R_a + R_t} \times \frac{1}{R_t} = \frac{4kT\,\Delta f\,R_a}{R_a + R_t}$$

The input signal to noise ratio will be

$$\frac{S}{N_i} = \frac{P_{si}}{P_{ni}} = \frac{V_s^2 R_t}{(R_a + R_t)^2} \div \frac{4kT\,\Delta f\,R_a}{R_a + R_t} = \frac{V_s^2 R_t}{4kT\,\Delta f\,(R_a + R_f)}$$

The output signal power will be

$$P_{so} = \frac{V_{so}^2}{R_L} = \frac{(AV_{si})^2}{R_L} = \left(\frac{AV_s R_t}{R_a + R_t}\right)^2 \times \frac{1}{R_L} = \frac{A^2 V_s^2 R_t^2}{(R_a + R_t)^2 R_L}$$

P_{no} = noise output power.

Thus, the output signal-to-noise ratio will be

$$\frac{S}{N_o} = \frac{P_{so}}{P_{no}} = \frac{A^2 V_2^2 R_t^2}{(R_a + R_t)^2 R_L P_{no}}$$

The general expression for the noise figure is

$$F = \frac{S/N_i}{S/N_o} = \frac{V_s^2}{4kT\,\Delta f\,(R_a + R_t)} \div \frac{A^2 V_s^2 R_t^2}{(R_a + R_t)^2 R_L \times P_{no}}$$

$$= \frac{R_L P_{no}(R_a + R_t)}{4kT\,\Delta f\,A^2 R_a R_t}$$

Noise Figure from Equivalent Noise Resistance :

$$R'_{eq} = R_{eq} - R_t$$

The total equivalent noise resistance for this receiver will now be

$$R = R'_{eq} + \frac{R_a R_t}{R_a + R_t}$$

Thus the equivalent noise voltage generated at the input of the receiver will be

$$V_{ni} = \sqrt{4kT\,\Delta f R}$$

Since the amplifier has an overall voltage gain A and may now be treated as through it were noiseless. The noise output will be

$$P_{no} = \frac{V_{no}^2}{R_L} = \frac{(AV_{ni})^2}{R_L} = \frac{A^2 \times 4kT\,\Delta f R}{R_L}$$

The noise figure in terms of the equivalent noise resistance,

$$F = \frac{R_L(R_a + R_f)}{4kT\,\Delta f\,A^2 R_a R_t} P_{no} = \frac{R_L(R_a + R_t)}{4kT\,\Delta f\,A^2 R_a R_t} \times \frac{A^2\,4kT\,\Delta f R}{R_L}$$

$$= R\frac{R_a + R_t}{R_a R_t} = \left[R'_{eq} + \frac{R_a R_t}{R_a + R_t}\right] \times \frac{R_a + R_t}{R_a R_t} = 1 + \frac{R'_{eq}(R_a + R_t)}{R_a R_t}$$

$$F = 1 + \frac{R_{eq}'}{R_a}$$

$$\textbf{Noise Figure } F = \frac{\text{S/N at the input}}{\text{S/N at the output}}$$

$$F = 1 + R_{eq}'\frac{R_a + R_t}{R_a R_t} = 1 + \frac{R_{eq}}{R_a} \text{ if } R_t >> R_a$$

$$F = 1 + \frac{R_{eq}'}{R_a} = 1 + \frac{T_{eq}}{R_a}$$

Noise Temperature $T_{eq} = T_0 (F - 1)$

T_{eq} = equivalent noise temperature of the receiver or amplifier under consideration.

Noise Figure of Cascaded Amplifier

$$F = F_1 + \frac{F_2 - 1}{G_{a_1}} + \frac{F_3 - 1}{G_{a_1} G_{a_2}} + \frac{F_4 - 1}{G_{a_1} G_{a_2} G_{a_3}} + \ldots + \frac{F_n}{G_{a_1} G_{a_2} \ldots\ldots G_a(n-1)}$$

Equivalent Noise Temperature of Cascaded Stages

$$F = 1 + \frac{T_{eq}}{T_0}$$

$$F = F_1 + \frac{F_2 - 1}{G_{a_1}} + \frac{F_3 - 1}{G_{a_1} G_{a_2}} + \frac{F_4 - 1}{G_{a_1} G_{a_2} G_{a_3}} + \ldots + \frac{F_n}{G_{a_1} G_{a_2} \ldots\ldots G_a(n-1)}$$

$$1+\frac{T_{eq}}{T} = 1+\frac{T_{eq_1}}{T_0}+\frac{1+\frac{T_{eq_2}}{T_0}-1}{G_{a_1}}+\frac{1+\frac{T_{eq_3}}{T_0}-1}{G_{a_1}G_{a_{12}}}+\ldots$$

$$T_{eq} = T_{eq_1}+\frac{T_{eq_2}}{G_{a_1}}+\frac{T_{eq13}}{G_{a_2}}+\frac{T_{eq_4}}{G_{a_1}G_{a_2}G_{a_3}}+\ldots$$

Noise in Frequency Mixers: Except at microwave frequency, mixers are more noisy than amplifiers rising identical devices because (*i*) $g_0 < g_m$, (*ii*) with inadequate image rejection. The noise associated with the image frequency also gets accepted.

Ans. (*b*) **Noise Temperature:** Noise temperature is a measure of noise in the signal. It is advantageous to use noise temperature for measurement of noise in case of low frequency wave signals and also when the noise is low because the effective variation in case of noise temperature is better.

It is given by

$$T_{eq} = T_0\,(F-1) \text{ where F : Noise Figure}$$

T_o : Absolute temperature = 290°K.

Numerical : NF = 3dB. = $10^{0.3}$

$$\text{Power gain} = 10\log_{10}\left(\frac{P_o}{P_i}\right) = 80 \text{ dB or} \Rightarrow \frac{P_o}{P_i} = 10^8$$

Noise equation B.W. = 10 MHz

$$P_0 = 10\times10^{-6}\, w$$

Determine T_{eq}. by $T_{eq} = T_0\,(F-1)$

$$= 290\,(10^{0.3}-1)\ °K.$$

Ans. (*c*) Low noise amplifier gain G_{a1} = 25dB = 316

pre – amplifier gain G_{a2} = 17dB = 50

pre – amplifier noise Figure F_2 = 6dB = 4

Converter and IF noise figure F_3 = 12dB = 16

Equivalent Noise Temperature

$$T_{eq} = T_{eq_1}+\frac{T_{eq_2}}{G_{a_1}}+\frac{T_{eq_3}}{G_{a_1}G_{a_2}} = T_{eq_1}+\frac{(F_2-1)\,290}{G_{a_1}}+\frac{(F_3-1)\,290}{G_{a_1}G_{a_2}}$$

$$= 4+\frac{(4-1)\times290}{316}+\frac{(16-1)\times290}{316\times50} = 7.025\text{ K}.$$

Q2. (*a*) State and prove Sampling theorem. Explain natural and top sampling.

(*b*) Consider a sine wave of frequency f_m and amplitude A_m, which is applied to a delta modulation of step size Δ, show that slop overload distortion will occur If: $A_m > \Delta/2\pi f_m T_s$ where T_s is the sampling period. What is the maximum power that may be transmitted without slope over load distortion?

(*c*) Five signal channels are sampled and time multiplexed. The multiplexed signal is then passed through a low pass filter. Three of the channels handle signals covering the frequency range 300 to 3300 Hz. Other two carry signals covering the range 50 Hz. to

10 KHz include 0 is a synchronizing signal: (*i*) What is the minimum sampling ratio? (*ii*) For this sampling rate what is the minimum bandwidth of the low pass filter? (*iii*) What would be the maximum bandwidth if these channels were frequency multiplexed, using normal AM techniques.

Ans. (*a*) Refer to Chapter No. - 2,

Ans. (*b*) Refer to Chapter No. - 3, Example - 24

Ans. (*c*) (*i*) Smapling rate must be $\geq 2f_m = 2 \times 10$ KHz = 20 KHz

$$T_s = \frac{1}{f_s} = \frac{1}{2\text{KHz}} = 50 \ \mu \text{ sec in this time 5 channel are multiplexed. Thus } \tau = \frac{50}{5} = 10 \ \mu \text{ sec}$$

(*ii*) Bandwidth $BW = \frac{1}{\tau} = \frac{10^6}{10} = 100\text{KHz}$

(*iii*) DSB amplitude modulation require

3 channel requires 3×3.3 KHz = 10 KHz

2 channel require 2×10 KHz = 20 KHz

Minimum BW requires therefore = 2×20 KHz = 40 KHz

Thus example illustrate the digital techniques require much larger bandwidth per channel making TDM unavailable.

Q3. (*a*) What are the different type of errors in Delta modulator ? How can these be removed?

(*b*) Write short note on Matched filter.

(*c*) A computer puts out binary data at the rate of 56 kb/ sec. The computer output is transmitted using a base band binary PAM system that is designed to have a raised cosine spectrum. Determine the transmission bandwidth required for each of the following roll off factors: (*i*) $\alpha = 0.25$; (*ii*) $\alpha = 0.5$; (*iii*) $\alpha = 0.75$; (*iv*) $\alpha = 1.0$.

Ans. (*a*) There are two types of quantization error.

(*a*) **Slope overload distortion** (*b*) **Granular noise.**

(*a*) **Slope overload distrotion**, if the step size Δ is too small for the stair case approximation to follow the input waveform $m(t)$. Then stair case approximation fall behind $m(t)$ and this condition is called **slope overload** and resulting **quantization** error is **called slope overlaod distortion**.

(*b*) **Granular noise :** This occurs when the step size is two relative to the local slope characteristic of the input waveform $x(t)$. Thereby causing the stair case approximation to hunt around a relatively flat segment of the input waveform.

Hence it is clear that a large step size required to accomodate wide **dynamic range**. Whereas a small step size is required for the acurate representation of relatively low level signals. Therefore choice of optimum step size minimize the mean square value of quantization error in a **linear delta modualtion** and this will be the result of the compromise between slope **overload distortion** and **granular noise**.

Ans. (*b*) Matched Filter.

(*i*) The signal pass through a filter/intergrator so that the sampling time, the signal voltage must be emphasised, correlated in comparison with noise voltage. The integration is optimum filter for the purpose of minimizing the probability error P_e.

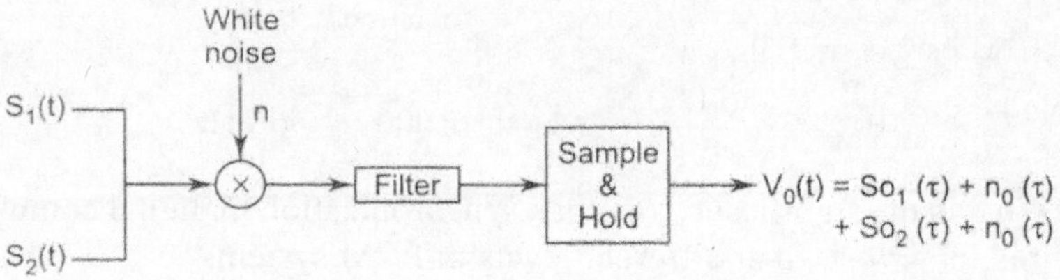

Fig. 3

(*ii*) The received signal is binary wavefoṫms.

(*iii*) One binary digit is represented by $S_1(t)$ for time interval T and the other bits $S_2(t)$.

$$S_1(t) = A\cos\omega_c t;\ S_2(t) = -A\cos\omega_c t$$

(*iv*) The *i*/P which is $S_1(t)$ or $S_2(t)$ includes noise $n_O(t)$.

(*v*) The noise is guassian and has a spectrum intensity of nf^2.

(*vi*) The signal and noise are filtered and when sampled at a end of each bit interval

$$V_O(T) = S_{O1}(T) + n_O(T)$$

$$V'_O(T) = S_{O2}(T) + n_O(T)$$

when noise is omit this '*a*' absent $V_O(T) = S_{O1}(T)$ and $V'_O(T) = S_{O2}(T)$

$$\text{Avg} = \frac{S_{O1}(T) + S_{O2}(T)}{2};\ n_O(T) \geq \frac{S_{O1}(T) - S_{O2}(T)}{2}$$

$$P_e = \int_{\infty}^{\infty} \frac{e^{\frac{-n_0^2(T)}{2\sigma_0^2}}}{\sqrt{2\pi}\,\sigma_0}\, d\eta_0(T) = \frac{S_{O1}(T) - S'_{O2}(T)}{2}$$

$$P_e = \frac{1}{2}\,\text{erfc}\left|\frac{S_{O1}(T) - S_{O2}(T)}{2\sqrt{2\sigma_0}}\right|$$

Ans. (*c*) Given data is r_b = 56 kb/sec = 56,000 b/sec

Now for filter with raised cosine spectrum we have the BW;

$$\text{BW} = \text{Nyquist BW} \times (1 + \alpha)$$

Where α is the roll off factor

Now the Bandwidth B will be : $B = \frac{r_b}{2}(1 + \alpha)$

(*i*) **For α = 0.25**

$$B_{0.25} = \frac{56,000}{2}(1 + 0.25) = 35,000 = 35 \text{ kHz}$$

(ii) **For α = 0.5**

$$B_{0.5} = \frac{56,000}{2}(1 + 0.5) = 42,000 = 42 \text{ kHz}$$

(*iii*) **For α = 0.75**

$$B_{0.75} = \frac{56,000}{2}(1 + 0.75) = 49,000 = 49 \text{ kHz}$$

(*iv*) **For $\alpha = 1.0$**

$$B_{1.0} = \frac{56,000}{2}(1+1) = 56,000 = 56 \text{ kHz}.$$

Q4. (*a*) What is the significance of clock synchronization in digital communication systems? How it is incorporated at different levels in PCM systems?

(*b*) Consider an audio signal with spectral components limited to the frequency band of 300 to 3300 Hz. A PCM signal is generated with a sampling rate of 8000 sample/sec. The required output signal-to-quantizing noise ratio is 30 dB.

(*i*) Calculate the minimum number of uniform quantizing level needed, and what is the minimum number of bits per sample needed.

(*ii*) Calculate the minimum system bandwidth required.

(*iii*) Repeat parts: (*i*) and (*ii*) when a μ law compander is used with $\mu = 255$.

Ans. (*a*) Refer to Chapter No. - 3.

Ans. (*b*) (*i*) Using $\left(\frac{S}{N_q}\right)_{0dB} = 10\log\left(\frac{S}{N_q}\right)_0 = 1.76 + 20\log L$, we have $\left(\frac{S}{N_q}\right)_{0\,dB_0} = 1.76 +$

$20\log L \geq 30$

$$\log L \geq \frac{1}{20}(30 - 1.76) = 1.412 \rightarrow L \geq 25.82$$

Thus the minimum number of uniform quantizing levels needs is 26.

$$n = \lceil \log_2 L \rceil = \lceil \log_2 26 \rceil = \lceil 4.7 \rceil = 5\text{b per sample}$$

The minimum number of bits per sample is 5.

(*ii*) From equation $f_{PCM} = \frac{n}{2} f_s \geq n f_m$ Hz, the minimum required system bandwidth is

$$f_{PCM} = \frac{n}{2} f_s = \frac{5}{2}(8000) = 20000\text{Hz} = 20\text{kHz}$$

(*iii*) **Using equation**

$$\left(\frac{S}{N_q}\right)_{0dB} = 20\log L - 10.1 \geq 30$$

$$\log L \geq \frac{1}{20}(30 + 10.1) = 2.005 \rightarrow L \geq 101.2$$

Thus the minimum number of quantizing levels needed is 102.

$$n = \lceil \log_2 L \rceil = \lceil 6.67 \rceil = 7$$

The minimum number of bits per sample is 7. The minimum bandwidth required for this case is $f_{PCM} = \frac{n}{2} f_s = \frac{7}{2}(8000) = 28000\text{Hz} = 28\text{kHz}$.

Q5. (*a*) With the help of waveform and mathematical expressions, describe the ASK, FSK, Q- PSK, 64 -QPSK, with respect to: (*i*) Spectrum of modulated signal (*ii*) The detection using coherent and non-coherent receiver.

(*b*) What is QPSK? How is QPSK modified for use with satellite? What are the advantages of QPSK?

(*c*) Determine the probability error of the PSK, ASK and FSK system and the required B.W. of the each system.

Ans. (*a*) Refer to Chapter No - 6.

Ans. (*b*) Refer to Chapter No - 6.

Ans. (*c*) Refer to Chapter No - 6.

Q6. (*a*) An analog signal band limited to 10 KHz quantized is 8 levels of PCM system with probability of 1/4, 1/5, 1/5, 1/10, 1/20, 1/20 and 1/20 respectively. Find the entropy and rate of information.

(*b*) Show that:

(*i*) $H(x, y) = H(x) + \left(\frac{y}{x}\right) = H(y) + \left(\frac{x}{y}\right)$

(*ii*) $H(x, y) = H(x) + H\left(\frac{y}{x}\right)$

Ans. (*a*) As per sampling theorem, the signal should be sampled at a frequency given by

$$f_s = 10 \times 2 \text{ KHz} = 20 \text{ KHz}$$

Considering each of the eight quantized level as a message, the entropy of the source can written as

$$H = \frac{1}{4}\log_2 4 + \frac{1}{5}\log_2 5 + \frac{1}{10}\log_2 10 + \frac{1}{10}\log_2 10 + \frac{1}{20}\log_2 20 + \frac{1}{20}\log_2 20 + \frac{1}{20}\log_2 20$$

$$= \frac{1}{4}\log_2 4 + \frac{1}{5}\log_2 5 + \frac{2}{20}\log_2 10 + \frac{3}{20}\log_2 20$$

$$= 2.84 \text{ bits/messages}$$

As the sampling frequency is 20 KHz. then the rate at which message is produced is

$$= 2000 \text{ message/sec}$$

Therefore the information rate is

$$R = rH = 20000 \times 2.84 = 56800 \text{ bits/sec.}$$

Ans. (*b*) (*i*) $H(X, Y) = -\sum_{i=1}^{m}\sum_{j=1}^{m} P(X=i, Y=j)\log_2 P(X=i, Y=j)$

$$= -\sum_{i=1}^{m}\sum_{j=1}^{m} P(X=i, Y=j)\left[\log_2 P(X=i)\,P(Y=j/X=i)\right]$$

$$\left[\therefore P(X=i, Y=j) = P(X=i)\,P(Y=j/X=i)\right]$$

$$= -\sum_{i=1}^{m}\sum_{j=1}^{m} P(X=i, Y=j)\left[\log_2 P(X=i) + \log_2 P(Y=j/X=i)\right]$$

$$\left[\because \log XY = \log X + \log Y\right]$$

$$= -\sum_{i=1}^{m}\sum_{j=1}^{m} P(X=i, Y=j)\log_2 P(X=i)$$

$$-\sum_{i=1}^{m}\sum_{j=1}^{m} P(X=i, Y=j)\log_2 P(Y=j/X=i)$$

$$= -\sum_{i=1}^{m}\log_2 \frac{(X=i)}{(X=j)}\sum_{j=1}^{m} P(Y=j/X=i)$$

$$-\sum_{i=1}^{m}\sum_{j=1}^{m} P(X=i, Y=j)\log_2 P(Y=j/X=i)\quad\left[\because \sum_{j=1}^{m} P(Y=i/X=i)=1\right]$$

$$= \left[-\sum_{i=1}^{m} P(X=i)\log_2 P(X=i)\right] + \left[-\sum_{i=1}^{m}\sum_{j=1}^{m} P(X=i, Y=j)\log_2 P(Y=j/X=i)\right]$$

$$= H(X) + H(Y/X)$$

(ii) $H(X, Y) = H(Y) + H(X/Y)$

$$= -\sum_{i=1}^{m}\sum_{j=1}^{m} P(X=i, Y=j)\log_2 P(X=i, Y=j)$$

$$= -\sum_{i=1}^{m}\sum_{j=1}^{m} P(X=i, Y=j)\left[\log_2 P(Y=j)\,P(X=i/Y=j)\right]$$

$$\left[\because P(X=i, Y=j) = P(Y=i)\,P(X=i/Y=j)\right]$$

$$= -\sum_{i=1}^{m}\sum_{j=1}^{m} P(X=i, Y=j)\log_2 P(Y=j)$$

$$-\sum_{i=1}^{m}\sum_{j=1}^{m} P(X=i, Y=j)\log_2 P(X=i/Y=j)$$

$$= -\sum_{i=1}^{m}\sum_{j=1}^{m} P(Y=j)\,P(X=i/Y=j) + H(X/Y)\log_2 P(Y=j)$$

$$= \sum_{j=1}^{m} P(Y=j)\log_2 P(Y=j)\sum_{i=1}^{m} P(X=i/Y=j) + H(X/Y)$$

$$\left[\because \sum_{i=1}^{m} P(X=i/Y=j)=1\right]$$

$$= \sum_{j=1}^{m} P(Y=j)\log_2 P(Y=j) + H(X/Y)$$

$$H(X, Y) = H(Y) + H(X/Y).$$

Q7. (*a*) Find the capacity of the discrete channel shown in Fig. 4.

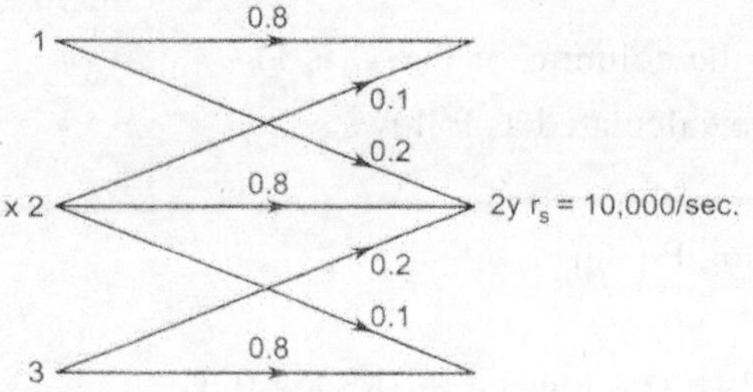

Fig. 4

Ans. Figure gives the conditional probability matrix P(*y*/*x*) as

$$p(y/x) = \begin{array}{c} \\ x_1 \\ x_2 \\ x_3 \end{array}\begin{array}{c} \begin{array}{ccc} y_1 & y_2 & y_3 \end{array} \\ \begin{bmatrix} 0.8 & 0.2 & 0 \\ 0 & 1.0 & 0 \\ 0 & 0.3 & 0.7 \end{bmatrix}\end{array}$$

$$P[x] = [0.3, 0.4, 0.3]$$

The joint probability matrix P (*x*, *y*) can be obtained by multiplying the rows of P(*y*/*x*) by P (x_1), P (x_2) and P (x_3) *i.e.*, by 0.3, 0.4 and 0.3 respectively, giving

$$P(x, y) = \begin{bmatrix} 0.8\times 0.3 & 0.2\times 0.3 & 0 \\ 0 & 1\times 0.4 & 0 \\ 0 & 0.3\times 0.3 & 0.7\times 0.3 \end{bmatrix}$$

$$P(x, y) = 0 \begin{array}{c} \\ x_1 \\ x_2 \\ x_3 \end{array}\begin{array}{c} \begin{array}{ccc} y_1 & y_2 & y_3 \end{array} \\ \begin{bmatrix} 0.24 & 0.06 & 0 \\ 0 & 0.4 & 0 \\ 0 & 0.09 & 0.21 \end{bmatrix}\end{array}$$

(As a check, it can be seen that the sum of all the entries in P (*x*, *y*) is 1)

The probabilities P (y_1), P (y_2) and P (y_3) can be obtained by adding the column of P (*x*, *y*) giving

$$P(y_1) = 0.24$$

$$P(y_2) = 0.06 + 0.4 + 0.09 = 0.55$$

$$P(y_3) = 0.21$$

The conditional probability matrix P(*x*/*y*) can be obtained by dividing the columns of P (*x*, *y*) by P (y_1), P(y_2) and P (y_3) respectively, giving

$$P(x/y) = \begin{array}{c} \\ x_1 \\ x_2 \\ x_3 \end{array} \begin{array}{c} \begin{array}{ccc} y_1 & y_2 & y_3 \end{array} \\ \begin{bmatrix} 1 & 0.109 & 0 \\ 0 & 0.727 & 6 \\ 0 & 0.164 & 1 \end{bmatrix} \end{array}$$

(Check : The sum of all the columns of P(x/y) is 1)

The entropies can now be calculated as follows

$$H(x) = -\sum_{j=1}^{3} P(x_j) \log P(x_j)$$

$$= -(0.3 \log 0.3 + 0.4 \log 0.4 + 0.3 \log 0.3)$$

$$= 1.571 \text{ bits/message}$$

$$H(x) = -\sum_{j=1}^{3} P(y_k) \log P(y_k)$$

$$= -[0.24 \log 0.24 + 0.55 \log 0.55 + 0.21 \log 0.21)$$

$$= 1.44 \text{ bits/message}$$

$$H(x, y) = -\sum_{j=1}^{3}\sum_{k=1}^{3} P(x_j, y_k) \log (x_j, y_k)$$

$$= -[0.24 \log 0.24 + 0.06 \log 0.06 + 0.4 \log 0.4 + 0.09 \log 0.09 + 0.21 \log 0.21]$$

$$= 2.053 \text{ bits/message}$$

$$H(x, y) = -\sum_{j=1}^{3}\sum_{k=1}^{3} P(x_j, y_k) \log P(x_j/y_k)$$

$$= -[0.24 \log 1 + 0.06 \log 0.109 + 0.4 \log 0.727 + 0.09 \log 0.164 + 0.21 \log 1]$$

$$= 0.612 \text{ bit/message}$$

$$H(x, y) = -\sum_{j=1}^{3}\sum_{k=1}^{3} P(x_j, y_k) \log P(y_k, x_j)$$

$$= -[0.24 \log 0.5 + 0.06 \log 0.2 + 0.4 \log 1 + 0.09 \log 0.03 + 0.21 \log 0.7]$$

$$= 0.482 \text{ bit/message.}$$

Q7. (*b*) An Information source produces 8 different symbols with probabilities 1/2, 1/4, 1/8, 1/16, 1/32, 1/64, 1/128, 1/256 respectively. These symbols are encoded as 000,001, 010,011, 100, 110, and 111 respectively:

(*i*) What is the amount of information symbol?

(*ii*) What are the probabilities of occurring for P (0) and $P_1(1)$.

(*iii*) Which is the efficiency of the code so obtained?

(*iv*) Give an efficient code with the help of the method of Shannon.

Ans. (*a*) Refer to Chapter No. - 8.

Q8. (*a*) Consider (7,4) linear code whose generator matrix is:

$$G = \begin{bmatrix} 1 & 0 & 0 & 0 & \vdots & 1 & 0 & 1 \\ 0 & 1 & 0 & 0 & \vdots & 1 & 1 & 1 \\ 0 & 0 & 1 & 0 & \vdots & 1 & 1 & 0 \\ 0 & 0 & 0 & 1 & \vdots & 0 & 1 & 1 \end{bmatrix}$$

(*i*) Find all the code vectors of this code.

(*ii*) Find the parity check matrix for this code.

(*iii*) Find the minimum weight of this code.

(*iv*) Prove equation $CH^T = 0$.

Ans. Given (7, 4) linear code

$$G = \begin{bmatrix} 1 & 0 & 0 & 0 & \vdots & 1 & 0 & 1 \\ 0 & 1 & 0 & 0 & \vdots & 1 & 1 & 1 \\ 0 & 0 & 1 & 0 & \vdots & 1 & 1 & 0 \\ 0 & 0 & 0 & 1 & \vdots & 0 & 1 & 1 \end{bmatrix}$$

(*i*) Here $k = 4$ $n = 7$

K denotes no. of bits in a message. There are 16 possible combinations of 4 - bits ranging from 0000 - 1111. For each combination, the codeword is formed by

$$C = D\,G$$

$$C = [1111]\begin{bmatrix} 1 & 0 & 0 & 0 & \vdots & 1 & 0 & 1 \\ 0 & 1 & 0 & 0 & \vdots & 1 & 1 & 1 \\ 0 & 0 & 1 & 0 & \vdots & 1 & 1 & 0 \\ 0 & 0 & 0 & 1 & \vdots & 0 & 1 & 1 \end{bmatrix}$$

$$\therefore \quad C = 1\,1\,1\,1\,1\,1\,1$$

So on for other combinations codewords can be formed

(*ii*) Parity check matrix

$$H = [P^T / I_{n-k}]$$

$$H = \begin{bmatrix} 1 & 1 & 1 & 0 & \vdots & 1 & 0 & 0 \\ 0 & 1 & 1 & 1 & \vdots & 0 & 1 & 0 \\ 1 & 1 & 0 & 1 & \vdots & 0 & 0 & 1 \end{bmatrix}$$

(*iii*) Minimum Weight. First construct all codewords by C = D G

Weight (No. of nonzero)

0000000	0
0001011	3
0010110	3
0011101	4
0100111	4
0101100	3
0110001	3
0111010	4
1000101	3
1001110	4
1010011	4
1011000	3
1100010	3
1101001	4
1110100	4
1111111	7

Min weight = 3

(*iv*) Equation $CH^T = 0$. Choose any C = 0 0 1 0 1 1 0 from previous example.

$$H^T = \begin{bmatrix} 1 & 0 & 1 \\ 1 & 1 & 1 \\ 1 & 1 & 0 \\ 0 & 1 & 1 \\ 1 & 0 & 0 \\ 0 & 1 & 0 \\ 0 & 0 & 1 \end{bmatrix}$$

Now $\quad CH^T \Rightarrow [0\,0\,1\,0\,1\,1\,0] \Rightarrow [0010110] \begin{bmatrix} 1 & 0 & 1 \\ 1 & 1 & 1 \\ 1 & 1 & 0 \\ 0 & 1 & 1 \\ 1 & 0 & 0 \\ 0 & 1 & 0 \\ 0 & 0 & 1 \end{bmatrix}$

Q8. (*b*) A source is transmitting six message with probability 0.30, 0.25, 0.15, 0.12, 0.10 and 0.08 respectively.

(*i*) Find the binary Huffman code.

(*ii*) Determine its average word length, efficiency and redundancy.

Ans.

Message	*Probability*	*Code*	S_1	S_2	S_3	S_4
m_1	0.30	0 0	0.30	0.30	0.43 (1)	0.57 (0)
m_2	0.25	1 0	0.25	0.27 (01)	0.30 (00)	0.43 (1)
m_3	0.15	0 1 0	0.18 (11)	0.25 (10)	0.27 (01)	
m_4	0.12	0 1 1	0.15 (010)	0.18 (11)		
m_5	0.10 (110)	1 1 0	0.12 (011)			
m_6	0.08 (111)	1 1 1				

$$L = \sum_{i=1}^{6} P_i L_i = (0.3 \times 2) + (0.25 \times 2) + (0.15 \times 3) + (0.12 \times 3) + (0.1 \times 3) + (0.08 \times 3)$$

= 2.45 binary digits. The entropy $H(n)$ of the source is given by

$$(i)\ H(n) = \sum_{i=1}^{n} P_i \log \frac{1}{P_L} = 2.418 \text{ bits}$$

$$= 0.3 \log \frac{1}{0.3} + 0.25 \log_2 \frac{1}{0.25} + 0.15 \log \frac{1}{0.15} + 0.12 \log \frac{1}{0.12} + 0.70 \log \frac{1}{0.70}$$

$$+ 0.08 \log \frac{1}{0.08}$$

$$(ii)\quad L = \sum_{i=1}^{n} P_i L_i = (0.3 \times 2) + (0.25 \times 2) + (0.15 \times 3) + (0.12 \times 3) + (0.1 \times 3) + (0.05 \times 3) = 2.45$$

$$\eta = \frac{H(n)}{L} = 0.976 \text{ . redundancy } = 1 - \eta = 1 - 0.976 = \mathbf{0.024.}$$

INDEX